LIFE CYCLES
OF BRITISH & IRISH
BUTTERFLIES

by Peter Eeles

For my grandsons,
Edward and Charlie

Published 2019 by Pisces Publications

Copyright © Peter Eeles (2019)
Copyright © of the photographs remains with the photographers

All rights reserved. No part of this publication may be reproduced, stored in a retrieval system or transmitted, in any form or by any means electronic, mechanical, photocopying, recording or otherwise, without the prior permission of the publishers.

First published 2019.
Reprinted 2020 with minor amendments.

British-Library-in-Publication Data
A catalogue record for this book is available from the British Library.

ISBN 978-1-874357-88-9

Designed and published by Pisces Publications

Visit our bookshop
www.naturebureau.co.uk/bookshop/

Pisces Publications is the imprint of NatureBureau,
2C The Votec Centre, Hambridge Lane, Newbury, Berkshire RG14 5TN
www.naturebureau.co.uk

Printed and bound in the UK by Gomer Press Ltd

Contents

v • FOREWORD by Chris Packham CBE

vi • PREFACE

1 • INTRODUCTION
2 • Definitions
2 • Plant names of nectar sources and larval foodplants

4 • PAPILIONIDAE
6 • **Swallowtail** *Papilion machaon*

12 • HESPERIIDAE
14 • **Dingy Skipper** *Erynnis tages*
20 • **Grizzled Skipper** *Pyrgus malvae*
26 • **Chequered Skipper** *Carterocephalus palaemon*
36 • **Essex Skipper** *Thymelicus lineola*
42 • **Small Skipper** *Thymelicus sylvestris*
48 • **Lulworth Skipper** *Thymelicus acteon*
54 • **Silver-spotted Skipper** *Hesperia comma*
60 • **Large Skipper** *Ochlodes sylvanus*

66 • PIERIDAE
68 • **Wood White** *Leptidea sinapis*
74 • **Cryptic Wood White** *Leptidea juvernica*
80 • **Orange-tip** *Anthocharis cardamines*
86 • **Large White** *Pieris brassicae*
92 • **Small White** *Pieris rapae*
98 • **Green-veined White** *Pieris napi*
104 • **Clouded Yellow** *Colias croceus*
110 • **Brimstone** *Gonepteryx rhamni*

116 • NYMPHALIDAE
118 • **Wall** *Lasiommata megera*
124 • **Speckled Wood** *Pararge aegeria*

130 • **Large Heath** *Coenonympha tullia*
136 • **Small Heath** *Coenonympha pamphilus*
142 • **Mountain Ringlet** *Erebia epiphron*
148 • **Scotch Argus** *Erebia aethiops*
154 • **Ringlet** *Aphantopus hyperantus*
160 • **Meadow Brown** *Maniola jurtina*
166 • **Gatekeeper** *Pyronia tithonus*
172 • **Marbled White** *Melanargia galathea*
178 • **Grayling** *Hipparchia semele*
184 • **Pearl-bordered Fritillary** *Boloria euphrosyne*
190 • **Small Pearl-bordered Fritillary** *Boloria selene*
196 • **Silver-washed Fritillary** *Argynnis paphia*
202 • **Dark Green Fritillary** *Argynnis aglaja*
208 • **High Brown Fritillary** *Argynnis adippe*
214 • **White Admiral** *Limenitis camilla*
224 • **Purple Emperor** *Apatura iris*
232 • **Red Admiral** *Vanessa atalanta*
240 • **Painted Lady** *Vanessa cardui*
246 • **Peacock** *Aglais io*
252 • **Small Tortoiseshell** *Aglais urticae*
258 • **Comma** *Polygonia c-album*
264 • **Marsh Fritillary** *Euphydryas aurinia*
272 • **Glanville Fritillary** *Melitaea cinxia*
280 • **Heath Fritillary** *Melitaea athalia*

288 • RIODINIDAE
290 • **Duke of Burgundy** *Hamearis lucina*

296 • LYCAENIDAE
298 • **Small Copper** *Lycaena phlaeas*
304 • **Brown Hairstreak** *Thecla betulae*
310 • **Purple Hairstreak** *Favonius quercus*
316 • **Green Hairstreak** *Callophrys rubi*
322 • **White-letter Hairstreak** *Satyrium w-album*
328 • **Black Hairstreak** *Satyrium pruni*
334 • **Small Blue** *Cupido minimus*
340 • **Holly Blue** *Celastrina argiolus*
346 • **Large Blue** *Phengaris arion*
356 • **Silver-studded Blue** *Plebejus argus*
364 • **Brown Argus** *Aricia agestis*
370 • **Northern Brown Argus** *Aricia artaxerxes*
376 • **Common Blue** *Polyommatus icarus*
382 • **Adonis Blue** *Polyommatus bellargus*
388 • **Chalk Hill Blue** *Polyommatus coridon*

394 • BIBLIOGRAPHY

Foreword

by Chris Packham CBE

The love excited by natural science grows from a broad range of engagements. Pure physical beauty is the likely ignition—I recall kneeling over a Small Copper on a patch of wasteland behind my home and the whole world shrinking to exist only on those shimmering wings. Next come childhood 'miracles', watching a Tortoiseshell caterpillar munch nettles, knit a web, shed its skin, shrivel up and then weeks later split open and sprout wings … in a jam-jar … on the dining room table—contemplate this when under ten and you're hooked. But then eager minds need to know how and why the miracle works and suddenly that raw beauty becomes more than skin-deep—it has fertilised a lifelong desire to comprehend more about the things you've fallen in love with. And it's addictive—a craving for wonder fuels a search that demands an eye for detail, the last exquisite component of a naturalist's mental tool-kit.

Not since that lepidopteran legend F.W. Frohawk studied and etched them nearly one hundred years ago have all the stages in the life cycles of all of Britain and Ireland's butterflies been illustrated in one volume. No mean feat! And what a joy, because the devil in all this delicious detail is a clear devotion to a love of a precious group of insects fluttering upon a precipice. Our butterflies are in trouble. Small Coppers have long gone from behind my childhood home—it is most of our countryside which is a wasteland now. Only those who see, who look and who love, who understand, who care and who now must act, can ensure that these most beautiful miracles endure, and that is us.

Preface

As I watched a home-reared Garden Tiger moth emerge from its chrysalis as a 7-year-old boy, a Lepidopterist emerged with a lifelong passion for the immature stages—the egg, caterpillar and chrysalis—of our Lepidoptera. Several decades later, that passion has resulted in the book that you are reading.

In completing this work, a significant amount of travel across Britain and Ireland has been inevitable, although my most interesting journey has been the transition from photography to ecology, where my desire to capture the breathtaking beauty of our butterfly fauna has been superseded by an unquenchable thirst for knowledge, especially an understanding of the life cycle of each species. Most importantly, I wanted to create a work that would, in some small way, support the evidence-based conservation of butterflies through an improved understanding of their immature stages. I hope this book inspires those looking to follow a similar path.

Acknowledgements

I owe a huge debt of thanks to the reviewers, many of whom are experts in a given species. They have helped ensure that this book is as accurate, complete and up-to-date as possible and, naturally, I am responsible for any remaining errors. They are: Tim Bernhard, Nigel Bourn, Andy Brazil, Tom Brereton, Caroline Bulman, Jamie Burston, Andy Butler, Susan Clarke, Iain Cowe, Robin Curtis, Peter Cuss, John Davis, Bob Eade, Sam Ellis, David Green, Jesmond Harding, Paul Harfield, Dan Hoare, Stuart Hodges, Crispin Holloway, Julia Huggins, Neil Hulme, Rachel Jones, Paul Kirkland, Sarah Meredith, Lorraine Munns, Brian Nelson, Matthew Oates, Simon Primrose, David Simcox, Richard Soulsby, Andy Suggitt, Mike Williams and Robert Wilson.

I would also like to acknowledge the additional support provided by: Caroline Bulman, Andy Butler, Iain Cowe, Robin Curtis, Dan Danahar, Graham Dennis, Sam Ellis, Steve Ewing, Barry Fox, Richard Fox, Jesmond Harding, James Hogan, Neil Hulme, Rachel Jones, Angela Leaman, Darren Mann, Sarah Meredith, Matthew Oates, Guy Padfield, Chris Raper, David Simcox and Martin Warren.

I would like to specifically thank Mark Colvin and Vince Massimo, who reviewed each species description as it was written. I would also like to thank Mark for his excellent company on many field trips, and Vince for providing inspiration through his own detailed studies of immature stages.

I would also like to thank all of those at NatureBureau who have been involved in the creation of this book, and especially Peter Creed, who is a rare combination of designer and naturalist, and I thank him for his design wizardry, enthusiasm, ideas, advice and guidance.

Last, but certainly not least, I would like to thank my loving and supportive family for putting up with my obsession as I went about creating my life's work.

Photographers

While most of the images used are my own, this book would be incomplete were it not for those kindly provided by the following individuals, whose initials can be found in brackets after each image caption: Andy Butler [AB], Andy Brown [ABr], Butterfly Conservation [BC], Bob Eade [BE], Caroline Bulman [CB], Colin Knight [CK], David James [DJ], Dave Miller [DM], David Simcox [DS], Gary Norman [GN], Guy Padfield [GP], Iain Cowe [IC], Iain Leach [IL], Jim Asher [JA], James Arnott [JAr], John Bogle [JB], John Bingham [JBi], Jamie Burston [JBu], Jeremy Thomas [JT], Ken Dolbear [KD], Mark Colvin [MC], Mark Hayes Fisher [MHF], Marcin Sielezniew [MS], Matthew Oates [MO], Mike Williams [MW], Neil Freeman [NF], Neil Hulme [NH], Nigel Kemp [NK], Nigel Kiteley [NKi], Nick Morgan [NM], Paul Atkin [PA], Paul Brock [PB], Peter Creed [PC], Paul Chapman [PCh], Paul Harfield [PH], Peter Willmott [PW], Richard Carter [RC], Sam Ellis [SE], Steve Ogden [SO], Tim Bernhard [TB], Tracy Piper [TP], Vince Massimo [VM], Will Langdon [WL], Wendy Tobitt [WT] and Wolfgang Wagner [WW].

Additional images used on the following life cycle pages: Grizzled Skipper habitat [PC], Large White [CK], Green-veined White [VM], Clouded Yellow [PCh], Dark Green Fritillary [IL], Purple Emperor habitat [MO] and Brown Hairstreak [JA].

Peter Eeles, August 2019

Introduction

"Happiness is like a butterfly which, when pursued, is always beyond our grasp, but, if you will sit down quietly, may alight upon you". Nathaniel Hawthorne's prose could apply equally to my own search for a theme that would characterise this book. With a desire to produce something different and yet valuable, I chose to take a path less travelled by describing, in detail, the fascinating life cycles of the 59 butterfly species that are considered resident or regular migrants to Britain and Ireland.

Despite the relative paucity of species found on our shores (18,000 or so species are found worldwide), I believe that getting to know each in terms of their structure, behaviour and general ecology will provide a solid foundation that can be applied anywhere in the world. My reasoning is simple—there are six butterfly families in total (Hesperiidae, Papilionidae, Pieridae, Nymphalidae, Riodinidae and Lycaenidae) and we have representatives of each. If we consider subfamilies, then approximately 50% are represented. The limited number of species that we do have therefore represents an incredibly diverse set and there is much enjoyment to be had in getting to know each of the families and subfamilies, as well as each species.

While 59 species may appear a modest quantity, this book takes things much further by considering the immature stages as well as the adult butterflies. Since each species has four stages to consider, then this gives us 236 subjects of interest (59 species × 4 stages). Going further, the larva of each species goes through several 'instars' (the period between moults) and each instar of every species is also described here. You'll have to take it from me that the number of subjects described in this book stands at 468, and I've always liked a challenge.

Since this book is focused on species that are resident or regular migrants, it does not include scarce migrants (American Painted Lady *Vanessa virginiensis*, Bath White *Pontia daplidice*, Berger's Clouded Yellow *Colias alfacariensis*, Camberwell Beauty *Nymphalis antiopa*, Large Tortoiseshell *Nymphalis polychloros*, Long-tailed Blue *Lampides boeticus*, Monarch *Danaus plexippus*, Pale Clouded Yellow *Colias hyale*, Queen of Spain Fritillary *Issoria lathonia*, Scarce Tortoiseshell *Nymphalis xanthomelas* and Short-tailed Blue *Cupido argiades*) or extinctions (Black-veined White *Aporia crataegi*, Large Copper *Lycaena dispar* and Mazarine Blue *Cyaniris semiargus*).

F.W. Frohawk

Embarking on such a project may seem complete madness. In my defence, this is not the first time that such a work has been published. That honour goes to one of my personal heroes, the legendary naturalist Frederick William Frohawk, whose two-volume masterpiece *Natural History of British Butterflies* is one of my most treasured possessions. This seminal work was ultimately published in 1924 (it was originally available in 1914, but its publication was delayed due to the onset of war) and, in homage to the great man, the codename for this book was 'Frohawk 2.0' before I settled on a title.

That being said, this book should not be seen as a replacement for Frohawk's work, whose extensive descriptions of each of the larval instars, in particular, have stood the test of time for the most part. Rather, this book is my attempt at complementing Frohawk's work in two ways. The first is to include images that were unavailable to Frohawk. Specifically, one of the drawbacks of Frohawk's work is that many illustrations are drawn life size, making it difficult (for example) to distinguish one early instar larva from another. The second is to introduce more recent findings—there have been many discoveries since 1924, with hundreds of books and thousands of papers written, the latter magnified by the increasing awareness of species ecology and the importance of conservation.

Why is the subject of the book relevant?

This may sound like a strange question, but is one that I needed to answer first and foremost before investing the significant amount of time needed to write this book—aside from simply being up to date, how was the book going to make a positive difference?

Recent years have seen an increasing number of studies that focus on climate change and the impact that it is having on the natural world. A relatively recent study epitomises the rationale for writing this book, since it is focused on a butterfly life cycle—specifically, that of the Wall—a species that has been in worrying decline for some time. The authors of that study, Professor Hans Van Dyck of Louvain University and Dr Dirk Maes of the Flemish Research Institute for Nature and Forest, postulate that the Wall has fallen victim to a 'developmental trap'. They suggest that climate change has resulted in the Wall producing an increasingly frequent and larger third brood, late in the year, when there is insufficient time for larvae to reach their third overwintering instar. These larvae subsequently perish and there is a corresponding decline in the number of adult butterflies seen the following year.

What intrigued me about this study, and others, is the need to identify a particular instar—it is sometimes hard enough to identify the species of a given larva, let alone the instar. It then struck me that any guidance that could be provided to help identify the larval instar of a given species (and the immature stages more generally) would potentially benefit conservation efforts, especially in terms of understanding the impact of climate change on our butterfly populations.

How is the book organised?

The classification used in this book, including the order of families and species, is taken from the publication *Checklist of the Lepidoptera of the British Isles* by D.J.L. Agassiz, S.D. Beavan and R.J. Heckford, which has undergone several revisions since it was first published in 2013.

DEFINITIONS

Anal points – two projections on the last segment of the larva.
Androconia – specialised scent scales on the male forewing.
Coloured-up – an egg or pupa that has reached its final colour.
Cremaster – the often-hooked tail end of the pupa.
Imago – the adult stage (plural imagines).
Instar – the period between moults in the larva.
Larva – the caterpillar stage (plural larvae).
Metapopulation – a network of interconnected colonies.
Ovum – the egg stage (plural ova).
Pupa – the chrysalis stage (plural pupae).

PLANT NAMES OF NECTAR SOURCES AND LARVAL FOODPLANTS

Plant names are taken from *New Flora of the British Isles*, 4th edition, by Clive Stace and published in 2019.

Agrimony *Agrimonia eupatoria*
Alder Buckthorn *Frangula alnus*
alders *Alnus* spp.
Almond *Prunus dulcis*
Apricot *Prunus armeniaca*
Ash *Fraxinus excelsior*
Autumn Gentian *Gentianella amarella*
Barren Strawberry *Potentilla sterilis*
Beech *Fagus sylvatica*
Bell Heather *Erica cinerea*
bents *Agrostis* spp.
Betony *Betonica officinalis*
Bilberry *Vaccinium myrtillus*
birches *Betula* spp.
bird's-foot-trefoils *Lotus* spp.
Bitter-vetch *Lathyrus linifolius*
Black Medick *Medicago lupulina*
Blackthorn *Prunus spinosa*
Blue Moor-grass *Sesleria caerulea*
Bluebell *Hyacinthoides non-scripta*
Bog-myrtle *Myrica gale*
Bracken *Pteridium aquilinum*
brambles *Rubus* spp.
Bristle Bent *Agrostis curtisii*
Bristly Oxtongue *Helminthotheca echioides*
Broad-leaved Dock *Rumex obtusifolius*
Broom *Cytisus scoparius*
Buck's-horn Plantain *Plantago coronopus*
Buckthorn *Rhamnus cathartica*
Bugle *Ajuga reptans*
Bullace, Damson *Prunus domestica* ssp. *insititia*
burdocks *Arctium* spp.
Bush Vetch *Vicia sepium*
buttercups *Ranunculus* spp.
cabbages *Brassica* spp.
Charlock *Sinapis arvensis*
Cherry Plum *Prunus cerasifera*

clovers *Trifolium* spp.
Cock's-foot *Dactylis glomerata*
Common Bent *Agrostis capillaris*
Common Bird's-foot-trefoil *Lotus corniculatus*
Common Cottongrass *Eriophorum angustifolium*
Common Couch *Elymus repens*
Common Cow-wheat *Melampyrum pratense*
Common Dog-violet *Viola riviniana*
Common Fleabane *Pulicaria dysenterica*
Common Nettle *Urtica dioica*
Common Reed *Phragmites australis*
Common Restharrow *Ononis repens*
Common Rock-rose *Helianthemum nummularium*
Common Sorrel *Rumex acetosa*
Common Stork's-bill *Erodium cicutarium*
Common Vetch *Vicia sativa*
Cow Parsley *Anthriscus sylvestris*
Cowslip *Primula veris*
Crack-willow *Salix* × *fragilis*
Creeping Cinquefoil *Potentilla reptans*
Creeping Soft-grass *Holcus mollis*
Creeping Thistle *Cirsium arvense*
Cross-leaved Heath *Erica tetralix*
Cuckooflower *Cardamine pratensis*
currants *Ribes* spp.
Daisy *Bellis perennis*
Dame's-violet *Hesperis matronalis*
dandelions *Taraxacum* spp.
Devil's-bit Scabious *Succisa pratensis*
Dog-rose *Rosa canina*
Dogwood *Cornus sanguinea*
Dove's-foot Crane's-bill *Geranium molle*
Downy Birch *Betula pubescens*
Downy Oat-grass *Avenula pubescens*
Dwarf Thistle *Cirsium acaule*
Dyer's Greenweed *Genista tinctoria*
Early Hair-grass *Aira praecox*

elms *Ulmus* spp.
English Elm *Ulmus procera*
Evergreen Oak *Quercus ilex*
False Brome *Brachypodium sylvaticum*
False Oxlip *Primula veris* × *vulgaris*
Fennel *Foeniculum vulgare*
fescues *Festuca* spp.
Field Maple *Acer campestre*
Field Scabious *Knautia arvensis*
forget-me-nots *Myosotis* spp.
Foxglove *Digitalis purpurea*
Garlic Mustard *Alliaria petiolata*
Germander Speedwell *Veronica chamaedrys*
Glaucous Sedge *Carex flacca*
Goat Willow *Salix caprea*
Gorse *Ulex europeaus*
Great Fen-sedge *Cladium mariscus*
Greater Bird's-foot-trefoil *Lotus pedunculatus*
Greater Knapweed *Centaurea scabiosa*
Greater Stitchwort *Stellaria holostea*
Greengage *Prunus domestica* ssp. × *italica*
Grey Willow *Salix cinerea*
Ground-ivy *Glechoma hederacea*
Hairy Rock-cress *Arabis hirsuta*
Hairy Violet *Viola hirta*
Hare's-tail Cottongrass *Eriophorum vaginatum*
hawkbits *Leontodon* spp.
hawkweeds *Hieracium* spp.
hawthorns *Crataegus* spp.
Hazel *Corylus avellana*
Heath Bedstraw *Galium saxatile*
Heath Dog-violet *Viola canina*
Heather *Calluna vulgaris*
heathers *Calluna* and *Erica* spp.
Hedge Mustard *Sisymbrium officinale*
Hemp-agrimony *Eupatorium cannabinum*
Herb-Robert *Geranium robertianum*
Hoary Cress *Lepidium draba*
Hoary Rock-rose *Helianthemum oelandicum*
Hoary Willowherb *Epilobium parviflorum*
Holly *Ilex* spp.
Honesty *Lunaria annua*
Honeysuckle *Lonicera periclymenum*
Hop *Humulus lupulus*
Hornbeam *Carpinus betulus*
Horse-chestnut *Aesculus hippocastanum*
Horseshoe Vetch *Hippocrepis comosa*
Ivy *Hedera* spp.
Jointed Rush *Juncus articulatus*
Kidney Vetch *Anthyllis vulneraria*
knapweeds *Centaurea* spp.
Large Bitter-cress *Cardamine amara*
lavenders *Lavandula* spp.
Lesser Trefoil *Trifolium dubium*
mallows *Malva* spp.
Marram *Ammophila arenaria*
Marsh Thistle *Cirsium palustre*
Marsh Violet *Viola palustris*
Mat-grass *Nardus stricta*
Meadow Buttercup *Ranunculus acris*
Meadow Foxtail *Alopecurus pratensis*
Meadow Vetchling *Lathyrus pratensis*
meadow-grasses *Poa* spp.
Michaelmas-daisies *Aster* spp.
Milk-parsley *Thysselium palustre*
Nasturtium *Tropaeolum majus*
oaks *Quercus* spp.
Oil-seed Rape *Brassica napus* ssp. *oleifera*
Pale Dog-violet *Viola lactea*
Pedunculate Oak *Quercus robur*

Pellitory-of-the-wall *Parietaria judaica*
Plum *Prunus domestica* ssp. *domestica*
Primrose *Primula vulgaris*
Purple Moor-grass *Molinia caerulea*
Ragged-Robin *Silene flos-cuculi*
ragworts *Jacobaea* spp.
Raspberry *Rubus idaeus*
Red Campion *Silene dioica*
Red Clover *Trifolium pratense*
Red Fescue *Festuca rubra*
Red Valerian *Centranthus ruber*
restharrows *Ononis* spp.
Ribwort Plantain *Plantago lanceolata*
rushes *Juncus* spp.
Sainfoin *Onobrychis viciifolia*
Salad Burnet *Poterium sanguisorba* ssp. *sanguisorba*
Sea Beet *Beta vulgaris* ssp. *maritima*
Sea-kale *Crambe maritima*
Selfheal *Prunella vulgaris*
Sessile Oak *Quercus petraea*
Sheep's Sorrel *Rumex acetosella*
Sheep's-fescue *Festuca ovina*
Silverweed *Potentilla anserina*
Small Nettle *Urtica urens*
Small Scabious *Scabiosa columbaria*
Small-leaved Elm *Ulmus minor* ssp. *minor*
snowberries *Symphoricarpos* spp.
Spindle *Euonymus europaeus*
Sweet Chestnut *Castanea sativa*
Sweet Vernal-grass *Anthoxanthum odoratum*
Sycamore *Acer pseudoplatanus*
thistles *Carduus* spp. and *Cirsium* spp.
Thrift *Armeria maritima*
Timothy *Phleum pratense*
Tor-grass *Brachypodium rupestre*
Tormentil *Potentilla erecta*
Tufted Hair-grass *Deschampsia cespitosa*
Tufted Vetch *Vicia cracca*
Turkey Oak *Quercus cerris*
Turnip *Brassica rapa* ssp. *rapa*
vetches *Vicia* spp.
violets *Viola* spp.
Viper's-bugloss *Echium vulgare*
Water Mint *Mentha aquatica*
Water-cress *Nasturtium officinale*
Wavy Hair-grass *Deschampsia flexuosa*
Wayfaring-tree *Viburnum lantana*
White Clover *Trifolium repens*
White Dead-nettle *Lamium album*
Wild Angelica *Angelica sylvestris*
Wild Basil *Clinopodium vulgare*
Wild Cabbage *Brassica oleracea* var. *oleracea*
Wild Carrot *Daucus carota* ssp. *carota*
Wild Marjoram *Origanum vulgare*
Wild Mignonette *Reseda lutea*
Wild Plum *Prunus domestica*
Wild Privet *Ligustrum vulgare*
Wild Radish *Raphanus raphanistrum* ssp. *raphanistrum*
Wild Strawberry *Fragaria vesca*
Wild Teasel *Dipsacus fullonum*
Wild Thyme *Thymus polytrichus*
willows & sallows (broad-leaved willows) *Salix* spp.
Winter-cress *Barbarea vulgaris*
Wood Avens *Geum urbanum*
Wood Small-reed *Calamagrostis epigejos*
Wood Spurge *Euphorbia amygdaloides*
Wych Elm *Ulmus glabra*
Yarrow *Achillea millefolium*
Yellow Iris *Iris pseudacorus*
Yorkshire-fog *Holcus lanatus*

Papilionidae

This family contains the largest butterflies in the world, including the largest of them all—the Queen Alexandra's Birdwing *Ornithoptera alexandrae*. The Swallowtail is our sole representative of this family and also happens to be our largest native butterfly, with a wingspan that can reach 93 mm in our endemic subspecies, *britannicus*. Many of the 580 or so species that belong to this family possess tails although some, such as the birdwings, apollos and festoons, do not.

Family	Subfamily	Genus	Species	Vernacular name
Papilionidae	Papilioninae	*Papilio*	*machaon*	Swallowtail

IMAGO
Given the size of the members of this family, adults have relatively short antennae, whose clubbed tips are often curved and with a blunt apex. Unlike the members of the Nymphalidae and Riodinidae families, all six legs are fully developed in both sexes, and each leg ends in a pair of claws.

Antennae are often curved with a blunt apex

The Swallowtail is our sole representative of the Papilionidae family

The claws at the end of a Swallowtail leg

OVUM
Eggs are spherical or hemispherical with a granular surface and, as we might expect given the size of the butterflies, are large when compared with those from other families.

The spherical egg of the Swallowtail

LARVA
The caterpillar is quite remarkable in that it is usually in possession of a curious bright-orange, fleshy and forked organ called an 'osmeterium', that is found in the first thoracic

Life Cycles of British & Irish Butterflies

A Swallowtail larva with its osmeterium exposed

segment. This organ, found in all instars, is used as a defence mechanism and is protruded when the caterpillar is threatened, and gives off a pungent smell.

PUPA

The pupa is typically held upright on whatever surface it is formed, attached by a silk girdle around the thorax and by the cremaster that hooks into a silk pad. Larvae from several families attach themselves to a silk pad at their tail end prior to pupation and I am always amazed at how, during the last moult, the pupa is somehow able to extract itself from its old larval skin and then re-attach itself to the silk pad by the hooks on the cremaster. It is surely a minor miracle that those species of butterfly that suspend themselves head-down as a pre-pupation larva, such as many of those in the Nymphalidae family, do not lose their grip and perish while undertaking this precarious manoeuvre.

The final moult of a Swallowtail larva, revealing the pupa

Life Cycles of British & Irish Butterflies 5

Swallowtail
Papilio machaon

❖ SWALLOWTAIL IS SUPPORTED BY GREENWINGS WILDLIFE HOLIDAYS (greenwings.co.uk), KEVIN PARKER ❖
& SWALLOWTAIL AND BIRDWING BUTTERFLY TRUST

The Swallowtail is our largest native butterfly and its wingspan, which may reach up to 93 mm, is only eclipsed by the extremely rare migrant, the Monarch *Danaus plexippus*, whose wingspan can reach 100 mm. It is not surprising that several authors have tried to promote such a majestic butterfly to the aristocracy—William Petiver named it 'The Royal William' in 1695 and James Rennie referred to it as 'The Queen' in 1832. Whether it is seeing the adult butterflies constantly flit their wings while taking nectar, or watching them fly powerfully over the fens, this species is always near the top of the 'list of species to see' for most butterfly enthusiasts.

Subsp. *britannicus* ♂ has a wingspan of between 76 and 83 mm

Subsp. *britannicus* ♀ has a wingspan of between 86 and 93 mm

Subsp. *britannicus* ♂ and ♀ undersides are similar [IL]

Subsp. *gorganus* is lighter in appearance than *britannicus* [MC]

DISTRIBUTION

Two subspecies are found on our shores. The resident subspecies (whose distribution is shown on the map) is *Papilio machaon britannicus* that is endemic to the fens of the Norfolk Broads in East Norfolk, mainly in the Bure, Ant, Thurne and Yare valleys, where its restricted distribution is largely due to the availability of its primary larval foodplant, Milk-parsley, that is as local as the butterfly itself. The butterfly's former distribution included all of the East Anglian fens south of the Wash, but there has been a dramatic decline during the last century due to the draining of the fens and marshes that the butterfly depends upon.

In some years, there are also reports of the *gorganus* subspecies arriving from the continent, which is lighter in appearance and slightly larger than subspecies *britannicus*. Subspecies *gorganus* is also less

Life Cycles of British & Irish Butterflies

fussy than subspecies *britannicus* and will lay on several plant species, especially umbellifers such as Fennel, Wild Angelica and Wild Carrot, resulting in a much broader distribution. 2013 was an exceptional year for this subspecies, with sightings from Kent, Sussex, Hampshire, Dorset and Somerset, together with a single sighting in Buckinghamshire. These sightings included evidence of egg-laying and the resulting larvae and pupae were followed through to the spring of 2014 by members of the Sussex branch of Butterfly Conservation when the home-grown adults emerged and took flight. This unusual event was so remarkable that it made the national press and there is an excellent summary of it in *The Butterflies of Sussex* by Michael Blencowe and Neil Hulme. There are earlier periods where subspecies *gorganus* was considered resident in southern England, especially when there were very warm summers, and, with climate change, the hope is that the continental Swallowtail will recolonise and become resident once again.

On a worldwide basis, the Swallowtail is found throughout the Holarctic, namely through Europe to temperate Asia and Japan, in North Africa, and also in North America where it is known as the 'Old World Swallowtail'.

HABITAT

Subspecies *britannicus* is the more colonial of the two subspecies and inhabits open fens and marshes that are usually dominated by Great Fen-sedge and Common Reed and where the most widely used larval foodplant, Milk-parsley, grows vigorously. The periodic cutting of sedges and reeds, to produce roofing materials used by thatchers, creates a transitory habitat that is most suitable for the foodplant and therefore the Swallowtail. Pupae that are carried away with such roofing material occasionally results in sightings of adults outside of Norfolk. The needs of Milk-parsley are also supported through the annual cutting of the 'Marsh Hay' crops that are used as fodder and bedding for livestock. Thankfully for the butterfly, there has been a resurgence in these traditional practices following a period of steady decline up until the 1990s.

Reeds growing next to a walkway at Wheatfen Nature Reserve in Norfolk and (inset) the white flowers of Milk-parsley [TB]

Sedges, reeds, hay grasses and Milk-parsley, as well as some nectar sources used by the Swallowtail, also rely on sufficient water levels and, in the absence of appropriate management, there is a natural succession where the accumulation of matter in open water eventually leads to drier conditions in which alders, willows and other shrubs thrive, ultimately leading to further drying out and the establishment of other trees such as oaks and, ultimately, woodland. For this reason, the majority of suitable sites are those that are actively managed for the Swallowtail (many of which are nature reserves), where there is a focus on scrub removal and maintenance of the water table.

Subspecies *gorganus* is a much more mobile migrant and can be found almost anywhere, but is most frequently encountered in open grassland sites near the south coast of England.

STATUS

Subspecies *britannicus* has suffered greatly due to the loss of its specialised habitat, reaching an all-time low in the 1970s, due mainly to the draining of fens and marshes, and this loss was exacerbated as the demand for sedges and reeds, and hay crops, dwindled. However, in recent decades, targeted conservation management at its remaining locations has not only halted this trend but has managed to increase the amount of available habitat and the Swallowtail has bounced back accordingly.

The long-term trend shows a significant decline in distribution and a significant increase in abundance, with the more recent trend over the last decade showing a moderate decline in distribution and a moderate increase in abundance. Using the IUCN criteria, the Swallowtail is categorised as Near Threatened (NT) in the UK.

An example of habitat loss is provided by Wicken Fen Nature Reserve in Cambridgeshire, one of the most well-known Swallowtail collecting grounds for entomologists in its heyday, before the 1950s. Most museum collections have specimens from this site, and some believe that the Swallowtails found here were of a different race, since they exhibit certain physical characteristics that differ from the norm. It is thought that the collapse of the Swallowtail at this iconic site was the result of a reduction in available habitat as well as the partial drying out of the fen. Ultimately, the surrounding land was drained in order to make way for agriculture after the second world war and this resulted in a lowering of the water table and a shrinking of the soil. Wicken Fen is now an island, standing higher than the arable land that surrounds it, making it extremely difficult to maintain a water level that produces Milk-parsley of the quality required by the Swallowtail. Several reintroduction attempts have been made at this site although, unfortunately, all have failed.

LIFE CYCLE

Subspecies *britannicus*, whose phenology is shown on page 6, is on the wing from late May until early July, producing eggs and larvae through June and July. A small percentage typically go on to produce a second brood that flies in August, resulting in eggs and caterpillars in late August and September, although it is not known what proportion of these late developers are successful. In some years the flight period is continuous, with late first brood individuals flying alongside early second brood individuals, but there is usually a gap in the last two weeks of July (Andy Brazil, pers. comm.). It is thought that the key factor that influences the progression to another brood is day length during the larval stage and, consequently, there tends to be a larger second brood in years when the emergence of the first brood is relatively early.

Subspecies *gorganus* always goes through two full broods and an alternative subspecies name of *bigeneratus*, that is used by some authors, is an acknowledgement of its double-brooded nature.

Imago

The Swallowtail is an unmistakable insect and is a sight to behold as it flies over the fens and marshes of the Norfolk Broads. It is easiest to watch while it is nectaring in the morning or late afternoon although, much to the annoyance of photographers, it will intermittently beat its wings while doing so, making photography of this moving subject difficult. Both sexes are avid nectar feeders and have a preference for pink or mauve flowers such as those of Bluebell, Devil's-bit Scabious, Ragged-Robin, Red Clover, thistles and Wild Teasel. One important exception is the butterfly's fondness for the flowers of the spectacular Yellow Iris. Adults will occasionally fly out of the fens and into neighbouring gardens to nectar.

On the continent, subspecies *gorganus* is known for 'hill-topping' where, as the name implies, males and females congregate at a high point in the landscape as a mate-locating strategy. Of course, in the fens, such obvious features are often lacking in this very flat part of the world and so males, which emerge earlier than females, can be found patrolling in the vicinity of a surrogate prominent feature, such as a tree growing out of the reed and sedge beds. When a male encounters a female, then they initially hover before flying high into the air, ultimately coming back down and landing on a shrub or in the reedbed where they pair. The female is normally mated on the day she emerges and usually in the morning, with the pair remaining together for several hours.

When egg-laying, the female will fly low over vegetation looking for suitable plants, which are typically large flowering plants of Milk-parsley that stand above any surrounding vegetation, or plants that are isolated, where the vegetation around them has been flattened. Once settled on the chosen plant, the female will hold her wings over her back before bending her abdomen down to deposit a single egg, usually on the underside of the plant. Two or more eggs are found on occasion on favoured plants and these are usually the eggs of different females, or eggs from the same female that were laid at different times.

Ovum

The smooth spherical egg has a flattened base and is just under 1 mm in diameter. It is pale yellow when first laid, but darkens as it matures, when it develops brown blotches and ultimately turns a dark plum just before the larva emerges.

The egg is a pale yellow when first laid

The egg develops brown blotches as it matures

Prior to the larva emerging, the egg turns a dark plum

Larva

After a week or so, depending on temperature, the **first instar** larva emerges from the egg, eating some of the eggshell as its first meal. The spiny 3-mm long larva, which always rests with the front half of its body raised, is grey-black with a white saddle on the 6th and 7th segments, and convincingly mimics a bird dropping. However, it has been shown that young larvae are particularly prone to predation from spiders and other arthropods, resulting in losses of up to 80%. When fully grown, the first instar is 5 mm long.

After nine days or so, the larva moults into the **second instar** and eats its cast skin, as it does after every moult. It is similar in appearance to the first instar but now has a series of orange spots along each side, and the white saddle on the 6th and 7th segments is more pronounced. When fully grown, the second instar is 8 mm long.

1st instar: the larva resembles a bird dropping

2nd instar: the larva has a series of orange spots along its sides

3rd instar: the larva has a variegated head

After two weeks, the larva moults into the **third instar**. It is similar in appearance to the previous instar but now has some white showing on the anal segment. The larva may also exhibit some white on the 9th and 10th segments, as well as more extensive orange spotting along and below its back, although these characteristics vary from larva to larva. Also, the head is no longer black and has a more variegated appearance of black and white streaks. Just prior to moulting the third instar is 13 mm long.

After another week or so, the larva moults into the **fourth instar** and is now quite a different beast. Rather than the larva having a grey-black ground colour with white and orange markings, the ground colour is now a very pale greenish white with black and orange markings. In other words, rather than having an overall appearance of white on black, it now has the appearance of black on white. This change in colouring is clearly to act as a defence mechanism, where the first three instars are camouflaged as a bird dropping, whereas the last two instars adopt warning colouration that indicates, presumably, that they are distasteful to birds. However, this strategy is only partially successful, since the larva is known to be predated by the Reed Bunting *Emberiza schoeniclus*, Sedge Warbler *Acrocephalus schoenobaenus* and Bearded Tit *Panurus biarmicus*.

4th instar: the overall appearance of the larva is black on white

With this change of colour comes a change in behaviour—the larva now moves higher up the plant where it will feed on the developing flowerheads as well as the leaves. Given the size of the larva and its voracious appetite, it may run out of food on its current plant and will wander off in search of another plant on which to feed.

An important feature of the fourth instar, and one that sets it apart from the similar-looking fifth instar, is that it has two rows of very small black warts running along its back that each give rise to a number of short bristles. Just prior to moulting, the larva is just under 23 mm long.

After another week or so, the larva moults for the last time into the **fifth instar**. The larva now has a bright green ground colour, but lacks the warts and accompanying bristles found on the fourth instar, and is surprisingly smooth. When fully grown, the larva is just over 40 mm long when at rest and around 50 mm long when crawling.

The larva has a curious orange organ called an 'osmeterium' that is situated behind the first segment which is protruded when the larva is threatened, and gives off a strong odour that some liken to rotting pineapple. In his *The Lepidoptera of the British Islands* of 1893, Charles Barrett describes the familiarity that fen workers had with this smell: "*The Fenmen always assert that they know by the scent when a large specimen of the larva has fallen among the mown herbage, and that this assists them to find it*".

5th instar: the larva has a bright green ground colour and is surprisingly smooth

5th instar: a larva with its osmeterium protruded

5th instar: a larva preparing to pupate

When ready to pupate, the larva moves low down on the plant to locate a suitable surface, such as a reed or woody stem. The larva then spins a silk girdle around its waist, and a silk pad to which the pupal cremaster will ultimately attach.

Pupa

Once the larva has fixed itself in position then, after two days, the larva moults for the last time, revealing a 20-mm long pupa. If the butterfly goes on to produce another brood, then it emerges in around three weeks, otherwise the pupa overwinters. There are two main colour forms of the pupa that each blend into their surroundings—one form is greenish-yellow and the other is light brown with dark markings. An interesting fact is that the pupa is able to survive submersion in water for long periods, which is just as well given the vagaries of life in a fenland habitat that is prone to flooding.

The green form of pupa

The brown form of pupa

Life Cycles of British & Irish Butterflies

Hesperiidae

On a worldwide basis, approximately 4,000 of the 18,000 species of butterfly belong to the Hesperiidae and this family is, therefore, a substantial component of our butterfly fauna, despite the diminutive size of most of its members. The species found in Britain and Ireland are grouped into three subfamilies—the Pyrginae (the Dingy and Grizzled Skippers), the Heteropterinae (the Chequered Skipper) and the Hesperiinae (the golden skippers).

Family	Subfamily	Genus	Species	Vernacular name
Hesperiidae	Pyrginae	*Erynnis*	tages	Dingy Skipper
		Pyrgus	malvae	Grizzled Skipper
	Heteropterinae	*Carterocephalus*	palaemon	Chequered Skipper
	Hesperiinae	*Thymelicus*	lineola	Essex Skipper
			sylvestris	Small Skipper
			acteon	Lulworth Skipper
		Hesperia	comma	Silver-spotted Skipper
		Ochlodes	sylvanus	Large Skipper

IMAGO

The delightful members of this family are all known for the rapid, darting and dancing flight that gives them their collective name of 'skippers'. They are characterised as having a head that is at least as wide as the thorax, antennae that have a relatively wide separation at their base and antennal clubs that are hooked with a pointed apex. Each eye also has a distinct eyelash (a tuft of hairs that projects over the eye) and all legs are fully developed in both sexes, a characteristic that sets the Hesperiidae apart from the Nymphalidae and Riodinidae.

The head of a Dingy Skipper showing hooked antennae and eyelashes

The upperside of the male forewing often contains androconial scales, either in a costal fold (in the Pyrginae subfamily) or in a sex brand (in the Hesperiinae subfamily). An interesting characteristic that is unique to the members of the Hesperiinae subfamily is that the adults bask with the forewings and hindwings held in different planes, something not found in other subfamilies of the Hesperiidae or any other butterfly family.

A fold on the costa of the ♂ Dingy Skipper, a member of the Pyrginae subfamily, contains androconial scales

The sex brand is quite visible on the forewing of a ♂ Large Skipper, whose wings are held in different planes

OVUM

Eggs are quite variable in terms of their shape and there is no single characteristic that epitomises the Hesperiidae—eggs may be round (as in the Chequered Skipper) or oval (as in the Small Skipper), and smooth (as in the Essex Skipper) or ribbed (as in the Dingy Skipper).

LARVA

A larva of the Hesperiidae typically has very few hairs on its body, a relatively wide head and is tapered at both ends. The first thoracic segment (known as the 'prothoracic segment') often exhibits a pigmented plate, especially in early instars, and this segment is often narrow, the larva seemingly in possession of a neck. The larva also feeds within a shelter that is formed by spinning leaves of the foodplant together with silk (various grasses are used in the Heteropterinae and Hesperiinae subfamilies, and various dicotyledonous plants are used in the Pyrginae subfamily).

The round and smooth egg of the Chequered Skipper

The round and ribbed egg of the Dingy Skipper

The oval and smooth eggs of the Small Skipper

A 1st instar Small Skipper larva with a pigmented plate on the first thoracic segment

A final instar Lulworth Skipper larva within a grass tube

A Grizzled Skipper larva within its protective shelter of Agrimony leaves

In his *Natural History of British Butterflies*, F.W. Frohawk gives a detailed description of the anal comb that is used by larvae to eject frass from their protective shelters, which is most developed in the Hesperiidae: "*Just previous to the ejectment of the excrement the larva crawls backwards along its abode until its extremity is either at or slightly protruded beyond the tube, it then raises its anal segment, elevating the flap or lobe, and evacuates the faeces, which remain adhering to the anus. The comb is then brought down to the rim of the orifice and remains so fixed for a moment or two, as if to obtain a firm pressure with the tips of the tines, then, apparently with considerable power, it is suddenly released; spring-like, the comb flies up with a violent jerk, casting the pellet with remarkable force in an upward direction when it falls to the ground at a distance of two feet six inches to three feet away*".

PUPA

A pupa of the Hesperiidae is long and tapering, formed within a cocoon made from leaves, typically those of the foodplant. The Pyrginae use one or more leaves of the dicotyledonous plants on which their larvae feed, whereas the grass feeders (the Heteropterinae and Hesperiinae) form a cocoon within one or more grass leaves spun together.

A Grizzled Skipper pupa formed within several Agrimony leaves spun together

A Large Skipper pupa formed within a few grass leaves spun together

Life Cycles of British & Irish Butterflies

Dingy Skipper
Erynnis tages

❖ DINGY SKIPPER IS SUPPORTED BY STEPHEN & LUCY LEWIS, TREVOR & YVONNE SAWYER & BILL STONE, SUFFOLK BUTTERFLY RECORDER ❖

The **Dingy Skipper** does indeed live up to its name since it is the drabbest of all the butterflies of Britain and Ireland. The first common name of this species was given by a London apothecary and renowned botanist and entomologist James Petiver in the early 1700s, who named this butterfly 'Handley's Brown Butterfly', although this is equally unflattering. Despite these vernacular names, a newly emerged adult possesses a subtle mix of contrasting greys and browns that is quite beautiful, although most observers would concede that the butterfly does take on a more uniform and drab appearance with the passage of time as scales are lost. Although there is a superficial resemblance to the Grizzled Skipper, the Dingy Skipper is more likely to be confused with the Burnet Companion *Euclidia glyphica* or Mother Shipton *Euclidia mi* day-flying moths with which it often flies.

The ♂ has a wingspan of between 27 and 34 mm

The ♀ has a similar wingspan to the ♂ [IL]

A ♂ roosting on a dead flowerhead

♂ and ♀ have similar undersides [JAr]

Burnet Companion [PC]

Mother Shipton

Life Cycles of British & Irish Butterflies

DISTRIBUTION
The Dingy Skipper is our most widely distributed skipper and can be found in discrete colonies, often of no more than 50 adults at the peak of the flight period, throughout Britain and Ireland. Its distribution includes the Burren in Ireland, that is home to the endemic subspecies, *baynesi*, where the larva feeds on Common Bird's-foot-trefoil that can be found growing between and nearby the limestone slabs that make up this unique landscape. This butterfly is also found further north in Britain and Ireland than any other skipper, including the Chequered Skipper that is only found in north-west Scotland. On a worldwide basis, the Dingy Skipper is found throughout much of Europe and western Asia.

HABITAT
Despite the broad distribution of the Dingy Skipper, its strongholds are undeniably the south-facing chalk and limestone grasslands in central and southern England in the counties of Wiltshire, Hampshire, Surrey and Sussex, where the butterfly can be found basking on patches of bare ground or on stones in full sun. Other habitats include the open parts of woods (such as clearings and rides), disused quarries, sand dunes, undercliffs, brownfield sites and disused railway lines. All habitats are sunny and sheltered where the larval foodplant grows in a sparse grass sward interspersed with patches of bare ground, along with taller plants that the adults use when roosting at night and during inclement weather.

Yellow flowers of Common Bird's-foot-trefoil, a larval foodplant, at Greenham Common, Berkshire

STATUS
It is a sad fact that most of our butterfly species are in decline and the Dingy Skipper is, unfortunately, no exception. The main threats to this species are habitat loss due to inappropriate management of grasslands, woodlands and brownfield sites, and agricultural intensification. These losses subsequently lead to a fragmented landscape that hampers the ability of this species to reach new areas and for recolonisations following local extinctions. Consequently, conservation efforts focus on maintaining a network of habitat patches over a large area.

The long-term trend shows a significant decline in distribution and a moderate decline in abundance, with the more recent trend over the last decade showing a moderate increase in distribution and a significant increase in abundance. It would therefore appear that the status of this delightful, albeit drab, butterfly can be considered stable. Using the IUCN criteria, the Dingy Skipper is categorised as Vulnerable (VU) in the UK and Near Threatened (NT) in Ireland.

LIFE CYCLE

The butterfly is on the wing in May and June and males may even emerge in the middle of April in favourable years, with females always emerging around a week later. Good summers may also result in a partial second brood in southern England that emerges in late July and August (a second brood is the norm in Southern Europe), and this may become a more frequent and widespread phenomenon in Britain and Ireland with a changing climate. The spring emergence at calcareous sites is often accompanied by sightings of both Grizzled Skipper and Green Hairstreak that have similar habitat requirements, the latter sharing a key larval foodplant of Common Bird's-foot-trefoil.

Imago

Like all skippers, the adult butterfly is relatively small and, when flying rapidly and low over the ground, is extremely difficult to track. Even when a male conspicuously darts up from the ground to examine a passing insect, in the hope that it is a virgin female, he is still difficult to follow. In terms of appearance, male and female differ very little, although the male does possess a difficult-to-locate fold on the costa (the leading edge of the forewing) that contains specialised scent scales known as 'androconia' that exude pheromones and are used during courtship. When feeding, both sexes seem to prefer yellow flowers such as those of bird's-foot-trefoils, buttercups, hawkweeds and vetches, although they are also known to nectar on Bugle and Ragged-Robin.

In the late afternoon and early evening, and during periods of dull or inclement weather, the butterfly will roost in a moth-like pose with its wings wrapped around a dead flower or grass head. Searching dead flowerheads at suitable sites can be a very effective way of locating the adult butterfly and this can come as a surprise to observers that have not seen this behaviour before. I once came across a dead flowerhead of Wild Teasel at Greenham Common in Berkshire that had five Dingy Skippers roosting at its base.

When searching for sites to lay, the female will fly slowly over stretches of foodplant, dipping down every now and again to inspect a plant for suitability. The most commonly used foodplant is Common Bird's-foot-trefoil, but Greater Bird's-foot-trefoil is also used on damp sites, as is Horseshoe Vetch on chalk grassland sites. Even when a carpet of foodplant is available, the female will choose a surprisingly small subset of plants on which to lay and there have been many occasions where I have found three or more eggs on the same plant, despite there being sufficient foodplant nearby. Favoured spots are typically those that are sheltered by surrounding vegetation and with relatively long stems of the foodplant growing over bare ground, such as a rabbit scrape or other hollow, although I have watched females lay on plants in an area that is devoid of bare ground on occasion. When egg-laying, the female will crawl over the plant before depositing a single egg on the upperside of a leaflet, close to its base.

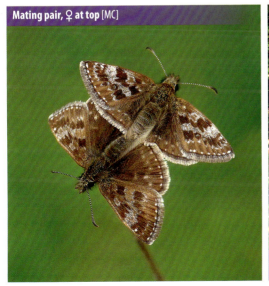

Mating pair, ♀ at top [MC]

Three eggs were found on this Common Bird's-foot-trefoil growing over a hollow with bare ground

Ovum

The egg is 0.5 mm high and almost spherical, with 12 or 13 ribs that are quite conspicuous when viewed with a hand lens. When first laid, the egg is a very pale yellow and gives the impression of being light green when viewed against the surface of the leaflet. After several days, however, the egg turns a very conspicuous orange, much like the colour change in the egg of the Orange-tip that many observers will be familiar with. It is not clear why this colour change occurs in the egg of the Dingy Skipper, but it could be that it alerts other females to the presence of an egg so that they do not lay another, which may make all the difference to the survival of the resulting larva if the foodplant is in short supply. As the egg develops it becomes darker in colour, with the black head of the larva becoming increasingly conspicuous at the top of the egg.

The egg is pale yellow when first laid

The fully coloured-up egg is bright orange

Larva

After nine or 10 days, or longer if the weather is not particularly favourable, the **first instar** larva emerges from the egg by eating a hole in the crown. When newly emerged, it measures 1.2 mm long and is pale yellow with a distinctive black head, a small number of hairs and a dark brown band on the first segment, just behind the head, that is a characteristic of skipper larvae in general. Without feeding on the eggshell, the larva immediately starts to construct a protective shelter from two or three leaflets of the foodplant, drawn together with silk.

The early instars feed by eating the epidermis on the undersides of the leaflets, but leave their shelter structurally intact. Later instars require a much more substantial meal and will eat the edges of the leaflets that make up their shelter itself. As soon as the shelter is considered insufficient, which can be triggered by the larva being exposed to direct sunlight, then a new abode is created under the cover of darkness. As the larva grows it proceeds to create larger shelters, ultimately requiring the silking together of more than one leaf (and associated leaflets).

The leaflets that comprise the shelter are not always bound tightly together, and the silk strands can catch the eye as they glisten in the sun. One possible reason for this loose binding of leaflets is to allow the larva to eject its frass from the shelter. The force with which frass is ejected cannot be underestimated—the legendary Lepidopterist J.W. Tutt, who was presumably discussing one of the later instars in his *A Natural History of the British Butterflies*, says the following when describing his experiences of rearing this species in captivity: "*They keep the tent beautifully clean, shooting their excrement far away. One larva, noticed doing this, shot the pellet on to a window-pane thirteen inches away with such force that the pellet rebounded some inches*".

When fully grown, the first instar is 2 mm long. In all instars, the larva lays down a carpet of silk within the shelter that is not just used to maintain a hold, but is also essential when the larva undergoes a skin change, after which the old skin is eaten.

After about three weeks, the young larva moults into the **second instar**. It is now pale green and similar in overall appearance to the first instar, with a black head and a slightly subdued brown band on the first segment. The fully grown second instar is 4 mm long.

After a couple of weeks, the larva moults in to the **third instar**. The body is now covered in a large number of tiny black dots that result in a pale olive hue, despite the body being largely green in colour. The key distinguishing feature from earlier instars is that there is no longer a brown band on the first segment. When fully grown, the third instar is 7 mm long.

After two weeks or so, the larva moults into the **fourth instar**. The general colour is now a rather dull green and the lobes of the head have subtle rust-red markings, but the larva is similar to the previous instar in all other respects. The fully grown fourth instar is 12 mm long.

After 10 days or so, the larva undergoes its final moult into the **fifth instar**. The head now appears large in relation to the first segment and is of a distinctive and blotchy rust-brown colour. After two weeks, the larva is fully grown, measuring 19 mm long when at rest and 22 mm long when crawling. The larva will wander for a while before spinning together several leaflets of the

1st instar: a larva making its first shelter

2nd instar: the larva has a subdued brown band on the first segment

2nd instar: a larval shelter

3rd instar: the larva no longer has a brown band on the first segment

4th instar: the larva has subtle rust-red markings on the lobes of the head

5th instar: a larva with its distinctive and blotchy rust-brown head

foodplant or other vegetation, such as dead leaves, close to the ground. The larva then spins a strong oval-shaped cocoon within this shelter and overwinters in the resulting hibernaculum until the following spring.

At the end of March and early April, the larva awakens from hibernation and typically remains in its overwintering cocoon, although it may leave this abode and wander a while before constructing a new shelter and cocoon. After a couple of weeks, the larva pupates within its cocoon without feeding further. This butterfly therefore spends an incredible eleven months as a larva, with nine of those months spent in hibernation.

Pupa

The pupa is 14 mm long and is attached to the cocoon by the hooks on the cremaster. It is, however, most remarkable for its two-tone colouring that is especially noticeable shortly after the pupa has formed. The head, thorax and wings are olive green, while the abdomen is a deep chestnut-red. Although the pupa develops less contrast as it matures, it is always rewarding to find. Like most species, the pupa darkens prior to emergence and the adult butterfly emerges after four or five weeks.

5th instar: hibernaculum formed from leaflets of Common Bird's-foot-trefoil

A pupa formed in a cocoon within a dead leaf

Life Cycles of British & Irish Butterflies

Grizzled Skipper
Pyrgus malvae

❖ GRIZZLED SKIPPER IS SUPPORTED BY MARK BUNCH, CERIN POLAND & KATRINA WATSON ❖

	January	February	March	April	May	June	July	August	September	October	November	December
IMAGO												
OVUM												
LARVA												
PUPA												

Like most skippers, the Grizzled Skipper is extremely difficult to follow when in flight although, thankfully, it frequently stops to feed from various flowers, when the unmistakable black and white pattern on its wings is revealed. Although we only have one species of the *Pyrgus* genus on our shores, this does make identification of this species relatively simple—on the continent this genus and related genera cause a significant identification challenge even for experienced observers, with many similar species to contend with.

Early entomologists mistakenly thought that the Grizzled Skipper, like many of its continental cousins, fed on mallows and not species of the Rosaceae family such as Agrimony and Wild Strawberry, and Linnaeus gave this species the specific name of *malvae* accordingly in the 10th edition of his *Systema Naturae* of 1758 that also introduced the binomial naming system. Given the rules of precedence regarding the naming of species, this name has stuck.

The ♂ has a longer and narrower abdomen than the female

Like the ♂, the ♀ has a wingspan of between 23 and 29 mm [MC]

The underside is a beautiful mix of browns and greens on a white background [MC]

DISTRIBUTION

This butterfly is found locally throughout southern and central England, with scattered colonies further north and in Wales. This species is absent from Scotland, Ireland, the Isle of Man and the Channel Islands. On a worldwide basis, the butterfly is found throughout most of Europe and eastward into Asia as far as China and Korea.

While some colonies can be measured by the hundred at the best sites and at the peak of the flight period, most are relatively small, containing less than 100 individuals. Much of what we know regarding the Grizzled Skipper's ecology is thanks to Tom Brereton who studied this species for his 1997 PhD thesis entitled 'Ecology and conservation of the butterfly *Pyrgus malvae* (Grizzled Skipper) in south-east England'. Brereton suggested that large and small colonies may form a metapopulation structure, although he also determined through a mark-release-recapture (MRR) study that the butterfly has limited mobility when faced with substantial barriers, such as dense woodland or a desert of improved grassland. Of course, there are always exceptions, and individuals have been known to cross a 100-m stretch of woodland—an example close to home is of individuals from a colony on the north side of Greenham Common in Berkshire that colonised a woodland clearing approximately 150 m further north in the Wildlife Trust reserve at Bowdown Woods. In the absence of such barriers, Brereton found that the butterfly is seemingly more mobile, with several individuals from his study flying over 1 km, and the record holder travelling just over 1.5 km.

HABITAT

This species occurs in a variety of habitats that are all characterised by warmth, shelter, and sparse vegetation, and includes calcareous grassland; woodland clearings and wide rides; and industrial habitats such as disused quarries, brownfield sites and disused railway lines. The butterfly is found less frequently on acid grassland, heathland, shingle and sand dunes. Brereton considered suitable habitat to contain an abundance of spring nectar sources, an abundance of larval foodplant growing in short turf and over bare ground, and the presence of patches of taller vegetation, together with scrub or a woodland edge. Larval foodplants growing in short vegetation over bare ground is preferred by females when egg-laying, and Brereton considered this to be the limiting factor in terms of the butterfly's distribution and abundance.

Stockbridge Down in Hampshire has patches of Wild Strawberry growing in short turf and over bare ground

STATUS
The long-term trend shows a significant decline in distribution and a moderate decline in abundance, with the more recent trend over the last decade showing both a stable distribution and abundance, suggesting that the butterfly is now holding its own. Using the IUCN criteria, the Grizzled Skipper is categorised as Vulnerable (VU) in the UK.

At woodland sites, the main threat to this species is inappropriate management, such as a lack of coppicing that would otherwise create the open areas and the accompanying early successional vegetation that this butterfly needs. Both undergrazing and overgrazing can be detrimental at grassland sites and, at industrial sites, the lack of any management regime results in the sites becoming ultimately unsuitable as they become scrubbed over. Another threat to this species is the increasing isolation of colonies due to barriers between them, including land that has undergone agricultural improvement, preventing any recolonisation after local extinctions.

LIFE CYCLE
The butterfly emerges throughout April and flies until the end of June, although the butterfly may be found from mid-March in early years and until mid-July in late years. As with most species, those colonies found further north emerge a little later than their more southern counterparts. There is usually a single brood each year, although there may be a small second brood with sightings in August in some years (which is the norm in southern Europe) when weather conditions are favourable. Brereton suggested that sightings that are reported in late June and July may actually be the emergence from an early second brood rather than a delayed emergence from the first brood.

Imago
Both sexes are active throughout the day and this warmth-loving butterfly will even fly into the evening if the temperature is high enough. Both sexes bask in the sun for long periods, typically on a stone, dead leaf or bare earth, as well as on tall grass blades and dead flowerheads in late afternoon. This butterfly is also an early riser—Brereton once recorded an adult basking as early as 6.36 am in early summer.

The butterfly can be found roosting on dead flowerheads and grasses during cool weather and at night, between one and two feet off the ground. They are also remarkably well camouflaged, with the forewings tucked below the hindwings and the speckled underside closely resembling a dead flowerhead. When looking for roosting adults at a known site for this species, it soon becomes clear that it is best to look for the dead flowerheads and then, and only then, look to see if a Grizzled Skipper is present.

A ♂ roosting on a dead flowerhead during a period of cold weather

A ♂ basking on a dead leaf

A mating pair, with ♀ at top [VM]

Nectar is an important resource for both sexes, and they can be found feeding extensively throughout the day, even between bouts of egg-laying or when defending a territory. Favourite nectar sources include buttercups, Bugle, Common Bird's-foot-trefoil, Daisy, dandelions, Ground-ivy, Ragged-Robin, Wild Thyme and Wild Strawberry.

Brereton found that the territorial males will employ either perching or patrolling behaviour, depending on the availability of suitable larval foodplants. When habitat quality is poor and larval foodplants are few and far between, then males can be found perching in bays above scrub edges that offer good visibility, often at the base of a hill or along a track, where they will await a passing female. However, when habitat quality is good and there is an abundance of larval foodplants then the males intermingle freely with females.

Males will see off other insects, but an encounter with another male, often that from an adjoining territory, results in a spiralling flight that may last for over half a minute before each male returns to a favourite perch in their respective territories. A female entering the territory, however, is courted for a short period, with the male showering the female with pheromones that are emitted from the specialised scent scales ('androconia') that are found in a fold in the basal half of the leading edge of the forewings. If the female is receptive, then mating follows, the pair remaining coupled for between 30 and 60 minutes.

As in many species, a female looking to lay her eggs seems to take on a different persona, and the female Grizzled Skipper is no exception. Rather than the rapid buzzing flight of most adult skippers, the female appears much more sedentary as she investigates the ground flora for suitable sites to lay. The female is quite fussy when egg-laying and pays a lot of attention to plants growing in direct sunlight, especially those growing over bare ground. According to Brereton, these plants are situated in a favourable warm microclimate, but may also be chosen for their elevated nutritional (nitrogen) content. As if to prove the point, he found 22 eggs on one such plant and estimated that nearly half of all eggs that he found during his studies were laid on less than 5% of the available plants. These observations provide anyone who is looking to find Grizzled Skipper eggs with some excellent clues as to where to search.

A variety of larval foodplants from the Rosaceae family are used, and favourites include Agrimony, Creeping Cinquefoil and Wild Strawberry. Barren Strawberry, brambles, Dog-rose, Raspberry, Salad Burnet, Silverweed, Tormentil and Wood Avens are also used on occasion. Eggs are laid singly near the edge of the underside of a leaf and a female may use more than one of the available foodplants if they are available at a given site.

A ♀ egg-laying on Wild Strawberry

Of all the species I have studied, the Grizzled Skipper is one whose development is difficult to predict in terms of timing and the durations given here can only be considered rough estimates. This was confirmed by Brereton who made a similar observation, even at the egg stage: *"There was considerable variation in egg development time, even among eggs laid on the same oviposition flight: for example five eggs laid within a few minutes of each other at Rewel Wood took between 10 and 18 days to hatch"*.

Ovum

The dome-shaped egg is 0.6 mm high and 0.5 mm wide, with between 18 and 20 ribs that run from the crown to the base. The egg is light green when first laid, but gradually becomes paler and then greyer as the larva develops within, with the head of the larva showing at the top of the egg just prior to the larva emerging.

The newly laid egg is light green

The egg becomes grey as the larva develops within

Larva

After between 10 and 18 days, the 1.2-mm long **first instar** larva leaves the egg by eating a hole in the crown, and immediately moves to the upperside of the leaf without eating the remainder of the eggshell, where it spins silk across the midrib (that may draw the two sides of the leaf together) to create a shelter within which it feeds and rests. The larva feeds only on the leaf surface at first but, as it matures, will eat through the leaf surface while leaving the structural veins intact, resulting in tell-tale blotches and holes in the leaf. The larva is covered in a number of white hairs, has a distinctive black collar just behind the head, and has a largely uniform colour that distinguishes it from the granular appearance of later instars—it is a very pale grey-brown when it emerges from the egg but soon turns a light green-brown after it starts feeding. Larvae feed primarily during the day, with a peak in the middle of the day. Brereton observed that: *"Larvae kept in captivity at room temperature were regularly active throughout the night, indicating that temperature rather than daylight limits larval activity"*. The fully grown first instar is 2 mm long.

1st instar: the larva turns light green-brown after feeding

1st instar: a newly emerged larva is a pale grey-brown

1st instar: a rolled-up leaf of Agrimony with the vacated eggshell

After between one and two weeks, while still in its shelter, the larva moults into the **second instar**. The larva retains the black collar found in the first instar, but it now has many more white hairs covering its surface and a granular appearance that is the result of white markings against a largely olive ground colour. The larva continues to feed from under a silk web, although this is more substantial than in the first instar and may be spun between two adjacent leaves of the foodplant. The fully grown second instar is around 3.5 mm long.

After between two and three weeks, the larva moults into the **third instar**. The larva now creates a larger shelter than earlier instars by either spinning the edges of a leaf together, or by spinning a number of leaves together, even incorporating those from non-foodplants on occasion. Whatever the construction, the larva always leaves the ends of the shelter open, from where it ejects frass and

2nd instar: the larva now has a granular appearance

3rd instar: apart from its size, the larva is similar in appearance to the 2nd instar

4th instar: the larva has wavy lines along the top of its sides

from where it feeds on the foodplant, including the edges of the shelter itself, creating new shelters as needed. Late-instar larvae may also leave their shelter completely for a short time to feed, before returning to digest their meal. The larva spends a great deal of time resting, rather than feeding, and development is correspondingly slow. Apart from its size, the third instar is similar in appearance to the second instar, although it does have longer white hairs on its body. The fully grown third instar is around 6.5 mm long.

After two to three weeks, the larva moults into the **fourth instar**. The larva retains its rough, granular appearance, although it now has prominent wavy lines along the top of its sides, that run the length of the body. This appearance comes from the largest of the greenish-white spots from which the hairs emanate. Another distinguishing feature is that the collar just behind the head is formed from two separate halves, with a clear division between them, and is not the continuous band found in earlier instars. When fully grown, the fourth instar is 13 mm long.

Brereton found that more mature larvae will not only switch plant, but will also select another species of foodplant so long as it is high in nutrients, even selecting plants growing in rank vegetation that is below 50 cm in height. What is more surprising is that Brereton found that larvae used a wider range of foodplants when feeding (seven were recorded) than the females used when egg-laying (four were recorded). Larvae will also select much coarser plants, especially bramble, which, according to Brereton, may be the principal foodplant of more mature larvae at some sites and it may even act as the sole foodplant on occasion.

5th instar: the larva lacks a collar behind the head and has a prominent dark line along its back

5th instar: the larval tent is constructed from several leaves of the foodplant

After a further two to three weeks, the larva moults for the last time into the **fifth instar**. The ground colour is now a pale green and the larva has a distinctive dark stripe running the length of its back. Another distinguishing characteristic is that the larva no longer has a collar behind its head.

The final instar is the most mobile and Brereton recorded larvae moving a linear distance of up to 4 m between shelters. The fully grown larva is around 20 mm long and, when ready to pupate, turns a pale pink and will wander away from the foodplant before constructing a loose cocoon within vegetation (not necessarily among stems of the foodplant) and close to the ground. In total, the larval stage lasts between two and three months.

The pupal cocoon is formed low down among vegetation

Pupa
After around two weeks, the larva sheds its skin for the last time, revealing a 13-mm long pupa that is secured within the cocoon by protrusions on the body and the cremaster, and the pupa overwinters for around nine months. Brereton found that adults emerged earlier from pupae found in short vegetation than those found in longer vegetation, due to the difference in temperature and therefore the rate of development of the adult within the pupa. Whatever the timing, finding a newly emerged Grizzled Skipper is always rewarding.

An opened cocoon showing the pupa

Chequered Skipper
Carterocephalus palaemon

❖ CHEQUERED SKIPPER IS SUPPORTED BY MARK COLVIN, LEE SLAUGHTER & BARRIE, ANITA & ELLIOTT STALEY ❖

The Chequered Skipper is the most colourful of our skippers, with an upperside that exhibits a mixture of light browns and creams on a dark brown ground colour, resulting in a beautiful chequered appearance that gives this butterfly its name. The underside is equally colourful. Male and female differ little in appearance, although the spots on the male forewing are a brighter orange than the female, and the abdomen is longer and narrower than that of the egg-bearing female, a difference that is noticeable in most butterfly species.

The ♂ with a wingspan of 29 mm

♂ underside

The ♀ with a wingspan of 31 mm

♀ underside

Of all the butterflies found in Britain and Ireland, the complete life cycle of the Chequered Skipper is one of the most rarely observed, since the distribution of the butterfly is restricted to north-west Scotland where the level of recording is relatively low. Also, while many enthusiasts have made pilgrimages to see the adult butterfly, very few have put in the significant effort needed to locate all the immature stages. In 2014, I made a concerted effort to record all stages, including all larval instars, an adventure that took me three years to complete, and that took me to many sites in this beautiful part of the world, including the Butterfly Conservation reserve at Allt Mhuic, Ariundle National Nature Reserve, Glasdrum National Nature Reserve, Glen Loy, Glen Nevis, the head of Loch Etive and the area around Spean Bridge. The key inspiration for my study was Neil Ravenscroft's 1992 PhD thesis, 'The Ecology and Conservation of the Chequered Skipper Butterfly *Carterocephalus palaemon* (Pallas)', which describes in some detail his observations made at Ariundle NNR and Glasdrum NNR, and which went on to inform several publications, including the Chequered Skipper Species Action Plan in 1996 and a modest Butterfly Conservation booklet in the same year.

DISTRIBUTION

The history of the Chequered Skipper in Britain and Ireland is one of the most fascinating of all our species. Up until the mid-1900s, it was thought that this species was only found in England where it inhabited the larger oak woods of central and eastern England. However, much to the surprise of lepidopterists, Lt-Col Mackworth-Praed announced, in 1942, that he had found the butterfly near Fort William in Scotland. It subsequently emerged that the butterfly had, in fact, been found a few years earlier in June 1939 at Loch Lochy in West Inverness-shire by a Miss C. Ethel Evans, but this

was not made public at the time. The butterfly is now known to exist over a wide area that is centred on Fort William. Looking further afield, the butterfly is found across the northern hemisphere, covering Europe, Asia and North America.

The butterfly was declared extinct in England in 1976, with its last sites in north-east Northamptonshire, although (at the time of writing) a reintroduction programme is currently underway in Rockingham Forest, also in Northamptonshire. There were several differences between the English and Scottish colonies. In Scotland, the foodplant of the larva is Purple Moor-grass, whereas English colonies used False Brome, and the adult butterflies in Scotland are also considered to be somewhat distinct in both appearance and behaviour from those in England.

HABITAT

Ravenscroft suggests an ideal site for the adult Chequered Skipper: *"open woodland, usually dominated by* [Sessile] *Oak,* Quercus petraea, *or* [Downy] *Birch,* Betula pubescens, *on gently sloping hillsides, often by the side of lochs, in sheltered clearings that catch the sun"*. Many sites, such as Glasdrum NNR, fit this profile perfectly, but the Chequered Skipper can also be found at the edge of woodland in areas of scrub, such as the area south of the River Strontian at Ariundle NNR. The butterfly can also be found in wet meadows and is not, therefore, entirely a woodland butterfly in Scotland. Whatever the site, there are some obvious and not-so-obvious habitat requirements of adults and immature stages if the site is to play host to the Chequered Skipper.

The first requirement is that the site provides areas that are sheltered, allowing adults to raise their body temperature to maintain the level of activity that they exhibit, such as males defending their territories. The second requirement is that the site is rich in appropriate nectar sources, given that both sexes are very active and need to constantly top up their energy reserves. Both sexes feed avidly and at length on the flowers of various species, especially those with blue or purple flowers—Bluebell, Bugle and Marsh Thistle are particular favourites, when available. This food supply is almost certainly an important factor in lifespan, the number of eggs laid and the development of eggs in the female, and may be the limiting factor in terms of population distribution given that the area available for larvae is usually considerably larger than that available for adults.

Suitable habitat must also provide places where the male butterfly can set up a territory that allows him to perch and intercept passing females, and such territories should exhibit certain characteristics. The first is that the territory should offer shelter, such as a bay at a woodland edge or an area surrounded by scrub. The second is that the territory provides good visibility—territories

Prime Chequered Skipper habitat in the wayleave at Glasdrum NNR

Woodland edge habitat at Ariundle NNR

Scrub habitat at Ariundle NNR, south of the River Strontian

usually have well-spaced perches that are above the level of underlying vegetation and provide good visibility of passing insects. The third is that the territory promotes high temperatures and is therefore usually south facing and has relatively dry and sparse vegetation. In addition to providing shelter, good visibility and high temperatures, territories are usually close to nectar sources, but not so close that they interfere with the purpose of a territory by inadvertently attracting other males, other insects and females looking only to nectar.

So why has the Chequered Skipper adopted a system where males defend territories in areas that have no resources attractive to females, such as nectar sources, despite concentrations of females occurring elsewhere? Firstly, from the perspective of the female, she is not looking to encounter a male unless looking for a mate. For example, while there is no conflict when the two sexes nectar together, since they are seemingly focused on feeding and oblivious to one another's presence, it has been shown that interference from other individuals will drive a female away from an area when laying eggs. Secondly, from the perspective of the male, while he could wait for a female at a nectar source, she may go unnoticed when he is also aware of all the other insect life that may be present. He could also wait near an area where eggs are laid but, since eggs are deposited at low density, this would not be an effective mate-locating strategy. Ravenscroft notes similarities between a Chequered Skipper territory and a lek, used by species such as Black Grouse *Tetrao tetrix*, in that there is no competition for resources used by females in their territories and that if a female enters such areas then she is probably looking for a mate.

♂ perching in his territory

Given that a proportion of eggs are laid within the site itself, then it goes without saying that any suitable site should also contain the larval foodplant which, in Scotland, is almost exclusively Purple Moor-grass although there are a few records of the Chequered Skipper using False Brome, the foodplant used by the former English population. The larva feeds for over four months before entering hibernation and therefore needs foodplant that remains in good condition until it has finished feeding. The vast majority of Purple Moor-grass, a widespread and abundant plant in north-west Scotland, does not meet this criterion, becoming unsuitable before the larvae have finished feeding, resulting in their slow development and ultimate demise.

Several factors determine the longevity and quality of suitable Purple Moor-grass. These include a good water supply and a soil that is well aerated, high in nutrients and neither too acidic nor too alkaline. In such preferred conditions, leaf blades are longer and wider. The best larval survival rates are on plants that have few or no flowers, since the period after flowering results in a decline in nutrients in the leaves. These conditions give rise to several other plants, such as Bog-myrtle, birches, alders and sallows, which provide an indication that the nearby Purple Moor-grass is suitable for the development of the larva. Consequently, suitable sites for egg-laying are in damp, scrubby and overgrown areas and are often in partial shade at the edge of a clearing or under light woodland. Although Purple Moor-grass may grow profusely in other areas, alongside plants such as Cross-leaved Heath, Heather or rushes, it is avoided by egg-laying females.

Ravenscroft says that larval development at Ariundle NNR was most successful in areas that had Bog-myrtle, Downy Birch or Bracken present, all of which enrich the soil, with the usual combinations being all three, Bog-myrtle and Downy Birch, or Downy Birch and Bracken. Knowing that these combinations play host to Chequered Skipper larvae provides one of the best visual clues when searching out the immature stages of this species.

Ideal larval habitat of Purple Moor-grass surrounded by Downy Birch (left), Bog-myrtle (foreground) and Bracken (right)

STATUS

There was much concern about this butterfly following the extinction of the English colonies and it was given full protection in the Wildlife and Countryside Act of 1981. However, further surveys of the Scottish colonies revealed a wider distribution than previously thought and, in 1986, the Chequered Skipper was downgraded from full protection to protection 'for sale only', thereby requiring a license for its sale. In Scotland, the main threats are the development of dense scrub that shades out areas for the adults as well as overgrazing of the larval habitat.

The long-term trend shows a moderate decline in distribution, with the more recent trend over the last decade showing a moderate increase in distribution. No figures are available for relative abundance. The status of this most charismatic of our skippers can therefore be considered stable although, using the IUCN criteria, the Chequered Skipper is categorised as Endangered (EN).

LIFE CYCLE

The Chequered Skipper is single brooded and is typically on the wing for a relatively short period from the third week of May until the third week of June, although this varies somewhat from year to year and site to site. For example, in 1984, the flight season started incredibly early on 4 May and finished on 5 June, and so anyone looking to see the adult butterfly would do well to monitor relevant sources of sightings and be flexible in their plans. There is also a level of variability between sites and the emergence at one of this butterfly's most popular and most southerly sites, Glasdrum NNR, is always a week to 10 days ahead of Ariundle NNR, one of the most westerly sites. At all sites, males emerge several days before females.

Imago

Most species that have precise habitat requirements, such as Silver-studded Blue and Heath Fritillary, form close-knit colonies. The Chequered Skipper, however, is not considered colonial, occurring in low densities in an open population structure and will freely disperse to satisfy its needs, such as finding nectar. The precise boundaries of a given population are therefore difficult to define. The implication is that certain sites may be reused year after year not because of any colonial properties, but because the conditions are right and adults migrate into the area each year.

Based on a mark-release-recapture exercise, Ravenscroft showed that male Chequered Skippers remain relatively immobile once they have moved from their emergence site to a favoured area, whereas females become more mobile and disperse once they have mated, and presumably lay wherever conditions are deemed favourable. When describing the distances travelled by the more-mobile female, Ravenscroft says *"Just flying from the eclosion site to nectar supplies, then to a male territory and back to egg laying areas may take a female > 1km. in Ariundle"*. The conclusion is that populations are dynamic and that this behaviour makes the butterfly a good colonist of new areas.

Conversely, and based on my own observations at both Ariundle NNR and Glasdrum NNR, conditions that meet the majority of the needs of the butterfly may also exist in the core flight areas. For example, there is one area at Glasdrum NNR that is roughly 2 m × 2 m in which I have found both males and females nectaring, a mating pair, two eggs (one of which was followed through to pupation) and three final instar larvae. Clearly, this particular spot is rather special since it would seem to satisfy most of the needs of the butterfly within a very small area indeed.

Both sexes rest with their wings tightly shut during dull weather and at night, although they may be active from 7 am on warm and sunny days in sheltered areas that catch the early morning sun, until 7 pm, again on the warmest days. Males are active throughout the day, with peaks late morning and early afternoon, and females are usually seen earlier in the day and later in the afternoon. There is therefore a drop in butterfly numbers in the middle of the day.

Mating pair, ♀ on left | Mating pair, ♀ on right

Male Chequered Skippers are very easy to find at suitable sites since they are conspicuous when protecting their territories. They will dart up at any passing object, including other species, an intrusion by another insect lasting only a few seconds, but an intrusion by another male leads to a spiralling flight that can be quite prolonged and which can result in the pair flying 20 to 30 m outside of the occupant's territory before one of the males, usually the resident, returns. If the male encounters a virgin female, then she will lead him away in a short flight before landing, when the two mate, staying coupled for around 45 minutes. A female that has already mated, however, will simply drop to the ground, ignoring the advances from her suitor by remaining motionless, whereas the male will constantly flutter around her before eventually losing interest and flying off. Female Chequered Skippers are by far the more difficult to find of the two sexes. This is almost certainly because females are more mobile than males, often moving out of the core areas once they have mated.

When egg-laying, a sight I have been privileged to witness only a few times, the female alights on a grass blade, shimmies down it a short way, curls her abdomen under the leaf, then deposits a single white egg before flying off almost immediately, the entire event lasting no more than 15 seconds. Eggs are laid approximately halfway up the leaf and never at the leaf tip or base. Most Purple Moor-grass plants will be home to a single larva, although there are exceptions of course, usually when two different females have laid on the same plant.

Eggs are always laid on the upperside of a grass blade (the side without the midrib), but many authors state that the underside is used. My own observation is that eggs are always laid on the surface that is facing the ground and I always assumed this to be the underside of the leaf. However, Ravenscroft shows that this is not so, with 42% of eggs laid on leaf uppersides facing up, and 58% laid on the uppersides facing down. My conclusion is that there is some confusion over the meaning of 'upperside' between authors.

Ovum

Eggs are dome-shaped with a sunken micropyle, and a shell surface that exhibits a very subtle network pattern primarily made up of hexagons. They are approximately 0.5 mm high and 0.6 mm wide and are pure white when first laid. A few days after being laid, the egg starts to colour up, becoming a very pale lemon-orange. A couple of days before hatching, the black head of the larva becomes progressively more visible through the eggshell near the micropyle.

A newly laid egg in context | A newly laid egg | A coloured-up egg | An egg just prior to hatching

Larva

Between 10 to 15 days after the egg was laid, typically towards the end of June, a 2-mm long **first instar** larva emerges by eating away the crown before going on to eat the remainder of the eggshell, although a detectable 'halo' can subsequently be seen at the egg site. The newly emerged larva has a shiny black head with a black collar that is characteristic of many skipper larvae, and the body is a light lemon colour, with a small number of short fine hairs. Once the eggshell has been eaten, the larva proceeds along the leaf (up or down), stopping when it finds the right width from which to form a protective tube, and it may move to a nearby grass blade before finding a suitable site. The tube is formed by using silk to bring the edges of the leaf together and the young larva will spend a considerable amount of time in this process, moving its head in successive side-to-side movements as it places silk on each leaf edge. As the silk dries it contracts, resulting in the edges of the leaf drawing together, thereby forming a tube in which the larva resides and from which it emerges to feed.

The construction of a protective tube is a common theme of most instars. Only when fully grown, in the fifth instar, will the larva rest openly on a blade of grass. Up until that point, the

1st instar: a larva eating its eggshell | 1st instar: a larva laying down silk to form a tube | 1st instar: a larva extending its tube | 1st instar: larval feeding pattern

Life Cycles of British & Irish Butterflies

2nd instar: a protective tube and characteristic feeding damage

2nd instar: a larva feeding below its protective tube

3rd instar: a larva at the end of a leaf with characteristic feeding damage

3rd instar: a larva with an hour-glass mark on the last segment

larva seeks shelter in a tube that is constructed from one (early instars) or several (later instars) blades of the foodplant. The role of the tube is almost certainly to avoid predation from spiders, harvestmen and sucking bugs, although the tube may also provide a local microclimate (whose higher temperature is beneficial to the development of the larva), as well as control water loss.

The larva will feed both above and below the tube, resulting in characteristic feeding damage that allows the observer to locate larvae more easily in the wild, especially those of the later instars when the tubes and associated feeding damage are more extensive. The feeding can sometimes be so extensive that only the midrib of the grass blade remains intact and the tube is long enough to just about conceal the larva. Once this point has been reached, then the larva will typically move to a new location and create a new tube, possibly on the same grass blade. One curious behaviour of the larva is that a gentle tap of its tube may cause it to quickly move forwards or backwards out of the tube, remaining motionless for some time before returning once the apparent danger has gone. The larva also reverses out of this tube to eject frass.

After 10 to 14 days, in early July, the fully grown 3.5-mm long first instar moults into the **second instar**. The larva is similar in appearance to the previous instar in that it has a black, shiny head, a black collar, a body that is covered in a small number of short hairs and has no distinctive markings. When fully grown, the second instar is 6 mm long.

After a further nine to 16 days, in mid-July, the larva moults into the **third instar**. The larva now exhibits several subtle longitudinal stripes, but the most noticeable change is the addition of a black mark on the last segment, in the shape of an hour glass. The fully grown third instar is 12 mm long.

After another 10 to 13 days, at the end of July, the larva moults into the **fourth instar**. It is similar in appearance to the third instar, with the same markings, although the black mark on the final segment is particularly conspicuous. The larva is now of a size where a tube may be constructed from multiple leaves, rather than a single leaf, and the fully grown fourth instar is 17 mm long.

After 18 to 20 days, in the second half of August, the larva moults for the last time into the **fifth instar**. The larva is initially a very pale green, but turns a deeper green as it grows. However, the most significant change is that the head is no longer black, but a very pale green, with a faint line separating the two lobes of the head. Also, the black collar and black mark on the anal segment are absent. When fully grown, around the beginning of October, the larva is approximately 24 mm long. As the larva grows it starts to live a more open existence and can be found sitting out in the open on a silk pad that it clings to when inactive. It also wanders more than younger larvae, and will more frequently be found on a

4th instar: a larva head-down in a tube formed from multiple leaves

4th instar: a larva with an extensive hour-glass mark on the last segment

5th instar: protective tube

5th instar: a larva with a pale green head above typical feeding damage

5th instar: a fully grown larva out in the open

5th instar: a hibernaculum

neighbouring plant. The feeding damage is also characteristic at this late stage, when the larva no longer creates tubes in which it resides, but simply eats a notch at the base of a blade before feeding above it. It is believed that this results in higher nutrients being maintained in the grass above the notch for a few days.

In late October, the larva prepares itself for hibernation by fixing the edges of two or more leaves together with silk, before sealing both ends. Given the Scottish weather, and the fact that the Purple Moor-grass dies back in the winter (it is a deciduous perennial), then the hibernacula do not always stay upright, and it is not unusual to discover that they have been blown horizontal with the leaves on which they were formed.

Around the end of March and early April, the larva emerges from hibernation, revealing an amazing evolutionary trick—what was a largely green larva has transformed and is now a pale straw colour, perfectly matching the colour of the dead Purple Moor-grass leaves.

The author attempting to relocate post-hibernation 5th instar larvae [MC]

5th instar: a post-hibernation larva starting to show itself

5th instar: a post-hibernation larva

5th instar: a larva preparing its pupal tent

5th instar: a pre-pupation larva

Unfortunately, the post-hibernation larvae are not only extremely well camouflaged—they will also wander to adjacent plants, making relocation very difficult. The overwintering larva first shows itself while still in its hibernaculum, and it may remain in this state for several days before finally leaving its winter home. Once it does leave, the larva can sometimes be found resting on or under a grass blade, although it does not feed post-hibernation.

A week or so after emerging from hibernation, the larva constructs a flimsy 'tent' from several dead grass blades. The larva then creates a silk pad at its tail end (which the cremaster of the pupa will eventually attach to) before creating a silk girdle to hold itself, and the subsequent pupa, in position. The larva remains in this position for approximately a week before pupating, the fifth instar lasting approximately 230 days, with the larval stage lasting around 290 days in total.

Pupa

The pupa is formed in early April and is approximately 16 mm long. A newly formed pupa has a series of brown longitudinal stripes along its back and sides, which darken over the next few days, and a noticeable 'beak' on its head. The resulting pupa is extremely well camouflaged against the dead and withered grass, turning almost black as the time for emergence draws near. The pupa may also be found lying horizontal since it is at the mercy of the plant to which it is attached, and the dead grass blades are often blown over.

Although I have never been fortunate enough to see an adult emerge from the pupa, despite many hours waiting for the big event, I do believe that it is possible to sex the pupa based on the width of the abdomen, which is noticeably wider in the female. The butterfly remains in the pupal stage for slightly over 40 days before this most colourful skipper emerges.

Pupae may be found lying horizontal on flattened blades of Purple Moor-grass

A newly formed ♂ pupa

♂ pupa | ♂ pupa prior to emergence | ♀ pupa | ♀ pupa prior to emergence

Life Cycles of British & Irish Butterflies

Essex Skipper
Thymelicus lineola

❖ ESSEX SKIPPER IS SUPPORTED BY NICK BALLARD, LARAINE DEAR & JIM HUME ❖

	January	February	March	April	May	June	July	August	September	October	November	December
IMAGO												
OVUM												
LARVA												
PUPA												

Although the Essex Skipper was identified as a new species in 1808, based on specimens from Germany, it was not recognised as a native of our shores until 1890 when a Mr F.W. Hawes published the following in *The Entomologist*: "The specimens, – three in number, all males, – were taken by me during the month of July, 1888, in one of the eastern counties, and remained in my cabinet, merely as curious varieties of Hesperia thaumas [Small Skipper]. *Happening to turn over those plates in Dr. Lang's 'Butterflies of Europe,' on which the genus* Hesperia *is figured, I was struck with the great resemblance of my specimens to* H. lineola [Essex Skipper]". The location was eventually given as Hartley Wood, St Osyth in Essex and subsequent searches of collections turned up specimens that had been caught in the early 1860s. This butterfly is the most recent addition to our resident butterfly fauna with one exception—the Cryptic Wood White is a more recent discovery in the Wood White 'complex'.

The reason for such a late discovery is the butterfly's similarity to the Small Skipper, which is recognised in two early names for this species, namely 'New Small Skipper' and 'Scarce Small Skipper'. It was not until 1906 before the butterfly was given the name we use today by Richard South in his classic work *The Butterflies of the British Isles*. An earlier vernacular name, and the specific name, both use '*lineola*' which stems from the Latin '*linea*', meaning 'line' or 'streak', and refers to the sex brands found on the forewing uppersides of the male. When compared with the Small Skipper, the sex brands of the male Essex Skipper are shorter and straighter, and run parallel with the leading edge of the forewings rather than at an angle.

The shape of the sex brand is of no help, of course, when attempting to determine if a female butterfly is an Essex or Small Skipper. While there are other clues (for example, the Essex Skipper flies slightly later than the Small Skipper and individuals have a fresher appearance as a result), it is usually necessary to resort to looking at the underside of the tips of the antennae, which have an orange-brown tint in the Small Skipper, but are jet black in the Essex Skipper. Even experienced observers make the mistake of claiming a butterfly to be an Essex Skipper when it is not, simply because they are looking at a black area of the antennae, but not the underside of the antennal tips. In early evening, the adults bask communally, often in good numbers, and this is often the best time to separate Small and Essex Skippers.

♂ with its distinctive sex brand on the forewing

Both ♂ and ♀ (shown) have a wingspan of 26 to 30 mm

The underside is plain with no distinctive markings

Essex Skipper: underside of the tips of the antennae are jet black

Small Skipper: underside of the tips of the antennae are orange [MC]

Life Cycles of British & Irish Butterflies

DISTRIBUTION

While the butterfly may originally have had a distribution centred on Essex, it now has a distribution that covers most of the English counties south-east of Cheshire and Yorkshire, with more isolated colonies in Cornwall, Devon, and the Welsh border counties, and it was first recorded in Wales as recently as 2000. Curiously, it has also a small foothold in County Wexford in south-east Ireland, where it was first recorded in 2006. The butterfly is not found in Scotland or the Isle of Man, but is found in the Channel Islands.

On a worldwide basis, the butterfly has a Holarctic distribution—it is not only found across Europe and Asia and in parts of North Africa, but was also accidentally introduced as overwintering eggs to Ontario in Canada in 1910 in a shipment of Timothy hay that was used as commercial packing material, some of which was disposed of in a local dump. This was undoubtedly just one of many instances of overwintering eggs being inadvertently transported and the butterfly quickly established itself throughout North America as hay bales were moved across the continent. There it is known as the 'European Skipper' and has become a pest of hay crops. In their 1977 paper 'Transport of hay and its importance in the passive dispersal of the European Skipper, *Thymelicus lineola*', published in *The Canadian Entomologist*, Jeremy McNeil and Raymond-Marie Duchesne estimated that Timothy hay bales from areas of high skipper densities contained an incredible 5,000 eggs per bale.

The butterfly's marked increase in distribution in Britain and Ireland is undoubtedly influenced by under-recording in previous decades due to its similarity with the Small Skipper, although most authorities agree that this butterfly has genuinely expanded its range. The cause of this expansion is unknown, although several suggestions have been put forward, which may act in combination. The first is that this may be a relatively new species to our shores and it is simply making use of the habitat available. The second is that climate change is allowing the butterfly to move into areas that were previously too cool. The third is that, as in the Canadian example, overwintering eggs are transported in hay. The fourth is that the butterfly's expansion has been assisted by the grass-covered embankments often found next to motorways, major trunk roads and railway lines that have acted as corridors, allowing this species to reach new locations more easily.

The Essex Skipper can be very common where it does occur and, at many sites, is the dominant species, even when it flies in the company of the equally ubiquitous Small Skipper. Although the butterfly forms discrete colonies, these vary in size from a handful of individuals to several thousand at the best sites.

HABITAT

This species is found in open and sunny areas of rough and well-drained grassland where the sward grows fairly tall, such as woodland clearings, rides and glades; mixed scrub; calcareous and acid grasslands; field margins and set-aside; road verges and embankments; and disused railway lines.

An area dominated by Cock's-foot at Farley Mount Country Park in Hampshire

STATUS

The Essex Skipper is one of the few species whose distribution is expanding rapidly, particularly in the north of its range, and it is not, therefore, a species of conservation concern. The official status of this species is slightly confusing, however, with a long-term trend showing a significant increase in distribution and a significant decline in abundance, with the more recent trend over the last decade showing a stable distribution and a significant decline in abundance. The apparent decline in abundance may simply be a misinterpretation of available data, given the potential for mis-recording Essex Skipper as Small Skipper. Using the IUCN criteria, the Essex Skipper is categorised as Least Concern (LC) in the UK.

LIFE CYCLE

The butterfly has a single generation each year, with adults emerging in the second half of June and flying throughout July and into the first half of August. The butterfly overwinters as a fully formed larva within the egg, remaining in this stage for between eight and nine months.

Imago

Both sexes are nectar-loving, and are often found visiting the flowers of thistles, although many other nectar sources are used, including clovers, Common Fleabane, heathers, ragworts and Wild Marjoram. Much of what we know about this butterfly is thanks to the work of Kenneth Pivnick and Jeremy McNeil, who studied this butterfly extensively in Canada in the late 1980s and early 1990s, and, in one of their articles, they demonstrated that the number of eggs laid by a female was directly related to the concentration of sucrose obtained from flowers.

A mating pair [PB]

Males will also feed on damp earth and animal droppings, where they take on minerals and salts, with a particular need to top up their reserves of sodium, since the male transfers a third of his sodium to the female during mating, which is subsequently transferred to any eggs laid. Sodium is beneficial for several reasons: it increases the number of fertile eggs that are laid, it improves the drought resistance of eggs, and it increases the number of pairings that can be achieved by a male.

As in most butterfly species, the male emerges first, and is the more active of the two sexes, spending half of his time patrolling in search of a virgin female, at least in Canada according to Pivnick and McNeil, although a strategy of perching within a territory may predominate in Britain and Ireland. Males will approach other males, females and even mating pairs, suggesting that mate locating is based primarily on visual signals. It has also been shown that the male is able to fly in cooler conditions than the female since he has less 'wing load' (he is not weighed down by eggs) and that sexual selection may favour males that are better-able to take to the wing in suboptimal weather conditions.

When courting, the male hovers around the female for a second or so, showering her with pheromones as he goes, and then lands behind her, and slightly to one side, before moving forward in an attempt to mate. If initially unsuccessful, the male may fly off and return soon after, which he repeats two or three times, before he ultimately succeeds, the pair remaining coupled for around an hour. An unreceptive female, on the other hand, will vibrate her wings rapidly, with the male immediately getting the message and flying away. Pivnick and McNeil showed that females choose males based on the quantity and quality of the pheromone released from their sex brands since this provides an indication of the male's age and mating history, with the female preferring younger males that have not already mated.

When egg-laying, the female will fly slowly between long and dead grass stems of Cock's-foot or another foodplant, before alighting on a stem and then 'shimmying' down it, probing the sheath as she goes. When a suitable opening in the furled sheath is found, she will lay several eggs between the stem and the sheath, attaching them to the sheath. The female Essex Skipper selects tighter leaf sheaths than the Small Skipper, which may explain the difference in primary foodplant (the Small Skipper uses Yorkshire-fog rather than Cock's-foot). In addition to Cock's-foot, the female Essex

Skipper will lay in the sheaths of Common Couch, Creeping Soft-grass, False Brome, Meadow Foxtail, Timothy and Tor-grass. The female also lays lower down the stem than the Small Skipper where they are often hidden from view, and this may explain why there a fewer sightings of egg-laying Essex Skippers.

Ovum

The oval-shaped eggs are 0.8 mm long and 0.3 mm wide and are pale yellow when first laid, gradually deepening to yellow-cream after a few days. After three weeks the egg turns paler, when the head of the larva becomes clearly visible as a dark spot through the transparent shell, while its body is coiled head-to-tail around the edge of the shell. The fully grown 'pharate' larva overwinters inside the egg until the following April.

Newly laid eggs in a sheath of Cock's-foot

The black heads of the larvae can be seen through the transparent shell

Larva

The 1.6-mm long **first instar** larva eats its way out of the egg in the spring, but does not eat the remainder of the shell, which is typical of those species that overwinter in the egg stage and whose eggshells are particularly tough and not particularly nutritious. The newly emerged larva initially positions itself on the midrib of the leaf upperside where it rests, only moving away to feed at the edge of the leaf. After a few days the larva forms a protective tube by spinning the edges of the leaf together with silk, emerging to feed both above and below its new abode. The straw-coloured first instar has a large black head and a cylindrical body that tapers at its rear end, and which is covered in a small number of short white hairs. The distinguishing feature of this instar, however, is the significant dark brown band found on the first segment. When fully grown, the first instar is just under 3 mm long, when it has developed a greenish hue.

After two weeks, the larva moults into the **second instar** and continues to inhabit a protective tube. The body is pale yellow, has a series of dull olive stripes running along its length and is sprinkled with numerous dark warts. The larva retains its black head and has a barely visible and subdued band on the first segment. When fully grown, the second instar is around 5 mm long. On several occasions, I have found a larva that has completely eaten each side of the grass blade below its protective tube, leaving only the midrib intact, and this could be a deliberate strategy to ensure that nutrients are retained in the grass blade above, and is a behaviour seen in other skippers, such as the Chequered Skipper.

After another 12 days or so, the larva moults into the **third instar**. The body is similar in appearance to that of the second instar, although the ground colour is greener, the longitudinal stripes

1st instar: the larva has a distinctive brown band on the first segment

1st instar: after a few days the larva constructs a protective tube

2nd instar: the brown band on the first segment is barely visible

2nd instar: typical feeding damage both above and below the protective tube have more contrast, and there is no brown band on the first segment. The main distinguishing feature, however, is the head, which is now a mix of browns (with black mouth parts), rather than the black of earlier instars, with a thin and dark brown line that separates the two lobes of the head. The third instar continues to feed within a protective tube and is 8 mm long when fully grown.

After another 10 days, the larva moults into the **fourth instar**. While the larva has slightly more contrast in its markings, especially those found on the head, the key distinguishing feature is size—the fully grown fourth instar is 14 mm long, which is almost twice the size of a fully grown third instar.

After just over a week, the larva moults for the last time into the **fifth instar**. Immediately after moulting, the head is pale green, which distinguishes this instar from earlier instars, with a white streak that is bordered with light brown on its outer edge on each lobe. The resulting markings on the head have much more contrast than those found in the third and fourth instars. In all other respects, the fifth instar is similar in appearance to a fourth instar apart from the difference in size—the fully grown fifth instar larva is 20 to 24 mm long. In his *Natural History of British Butterflies* of 1924, F.W. Frohawk writes: "*The general colouring of the larva with its longitudinal stripes harmonizes well with the grass blades, and the brown and white banded head resembles a withered eaten end of a blade*". This camouflage is so convincing that fully grown larvae are able to bask in the open on a grass blade without attracting the attention of avian predators.

3rd instar: a fully grown larva

4th instar: apart from size, the larva is identical to a third instar

5th instar: a fully grown larva

3rd instar: the head is a mix of browns
4th instar: the head now shows more contrast
5th instar: the head is now green with a white and brown streak on each lobe

When ready to pupate, after 11 days or so, the larva binds a few grass blades together with silk at the base of the foodplant before spinning a silk pad at its tail end and a silk girdle around its waist.

Pupa

After three days, the larva sheds its skin and the 16-mm long green pupa is held in place by its cremaster and the silk girdle. The pupa is similar in colour to the final instar larva, with several pale stripes running the length of its back, and it is also in possession of an obvious 'beak' on its head. After two to three weeks, depending on temperature, this delightful butterfly emerges.

A cocoon prised apart to reveal the green pupa
A fully coloured-up ♀ pupa one hour before emergence

Small Skipper
Thymelicus sylvestris

❖ SMALL SKIPPER IS SUPPORTED BY PAUL BRIGGS, MAIRI McINTOSH & PIETER VANTIEGHEM ❖

The **Small Skipper** is one of our commonest 'golden' skippers and can be seen in high summer wherever tall grasses grow in flowering clumps, when it can be found basking with wings half open or making short buzzing flights amongst the grass heads. Despite its name, the Small Skipper is larger than the Grizzled, Essex and Lulworth Skippers, and is the same size as the Dingy Skipper. Like several closely related species, the male can be distinguished from the female thanks to a prominent black sex brand on each forewing, which is visible as a slightly curved line of specialised scent scales. This butterfly is almost identical to the Essex Skipper and several identification aids are discussed in the description of that species.

Early entomologists referred to skippers as 'hogs' due to their plump bodies, and this is reflected in the first names given to this species by James Petiver in his *Gazophylacii naturae et artis* of 1704, where he named the male 'The Streakt Golden Hogg' and the female 'The Spotless Hogg'. Moses Harris brought the two sexes together under the name we use today in *The Aurelian* of 1766, although he confused male and female, with the latter incorrectly described as *"having a black stroke on the middle part of each superior wing, which is wanting in the male"*.

The ♂ possesses a sex brand on each forewing

Both ♂ and ♀ (shown) have a wingspan of between 27 and 34 mm

Both sexes have plain undersides

DISTRIBUTION

This widespread butterfly is found throughout England and Wales, and has expanded its range in recent years—it was first found in County Kildare in Ireland in 2006, and it also reached Scotland in 2007, although it remains absent from the Isle of Man and the Channel Islands. Given that the butterfly's larval foodplants are widespread, then a logical conclusion is that the butterfly's northern range is limited by climate, and that warmer conditions are allowing it to expand further north. The butterfly has a Palearctic distribution and is found throughout Europe, western Asia and North Africa.

The Small Skipper lives in discrete colonies which vary enormously in size, usually in proportion to the amount of suitable habitat. In the study documented in 'Some aspects of the population ecology and dispersal of the Small Skipper butterfly *Thymelicus sylvestris* (Poda) in a series of linked grasslands', published in *The Entomologist* in 1995, Stuart Warrington and John Brayford showed that adults moved only 17 m per day on average, demonstrating that the butterfly is not particularly mobile and, while distances of 300 m have been recorded on occasion, these numbers do not explain the extent of the butterfly's expansion in recent decades.

HABITAT

Given the butterfly's simple requirement for tall flowering grasses (especially its primary larval foodplant of Yorkshire-fog that is found commonly throughout Britain and Ireland), it can be found in open areas of a wide variety of habitats, including calcareous and acid grasslands; woodland clearings, rides and glades; mixed scrub; hedgerows; meadows and pasture; disused

Tall flowering grasses at Harewood Forest in Hampshire [TB]

quarries; undercliffs; arable field margins and set-aside; brownfield sites; road verges; and disused railway lines.

STATUS

The status of this widespread butterfly is considered stable, with the long-term trend showing a stable distribution and a significant decline in abundance, with the more recent trend over the last decade also showing a stable distribution but with a moderate increase in abundance. Using the IUCN criteria, the Small Skipper is categorised as Least Concern (LC) in the UK.

The greatest threat to this butterfly is habitat loss due to development, agricultural intensification and scrub invasion. Overgrazing and regular cutting of the long grasses that the Small Skipper requires for its larvae also have some negative impact, with the elimination of small patches of rough grassland resulting in a corresponding loss of the butterfly.

LIFE CYCLE

Adults are on the wing from the middle of June, through July and into August. This emergence is usually a week or two earlier than the Essex Skipper with which this species is often confused although, when the Essex Skipper is also on the wing, the relative condition of an adult can provide a clue as to which species it is, with Small Skippers looking slightly faded in comparison with the bright colours of newly emerged Essex Skippers.

Imago

Adults feed from a variety of nectar sources, including Betony, brambles, Common Bird's-foot-Trefoil, Common Fleabane, dandelions, Devil's-bit Scabious, hawkbits, knapweeds, Red Clover, restharrows, Sainfoin, Selfheal, thistles, vetches and Wild Marjoram. The butterfly is known for its tendency to feed from the same plant species when that particular nectar source is available, only switching to a different plant species if its original choice becomes less productive.

Many observers will also notice just how manoeuvrable this butterfly is as it flies among grass heads, seemingly changing direction on a whim, making it extremely difficult to track when in flight. One of the best times to see the butterfly is with the onset of evening, when adults can be found roosting in groups high up on grass stems, making them fairly easy to find.

The male Small Skipper is territorial and will wait on a favourite perch, often near a patch of flowers, from

A ♂ feeding on knapweed

A mating pair

A ♀ egg-laying in a sheath of Yorkshire-fog

which it will intercept any passing butterfly in the hope of encountering a virgin female. The male will also investigate any butterfly that resembles a female even when nectaring. As in closely related species, the male will fly around the female before landing behind her and, if she is receptive, moves forward to mate.

The female is more sedentary than the male and exhibits unusual behaviour when egg-laying. After buzzing around tall grasses, she will alight on a stem and then slowly revolve backwards down it, probing the sheath with her abdomen as she goes. When a suitable opening has been found, she closes her wings over her back, points her antennae forward, and spends a couple of minutes laying between three and eight eggs in a row inside the sheath. The eggs are not attached to any surface and simply rely on the pressure of the closed sheath to hold them in place and eggs will drop to the ground if the sheath is opened. Once she has finished laying, the female rests for a few seconds before flying off, leaving the eggs completely concealed within the grass sheath. The Small Skipper lays most frequently on Yorkshire-fog, although several other grasses have been recorded as foodplants, including Cock's-foot, Creeping Soft-grass, False Brome, Meadow Foxtail and Timothy.

Ovum

As many as eight flattened oval eggs may be laid, and these are more rounded than the eggs of both the Essex or Lulworth Skippers, being 0.85 mm long and 0.65 mm wide. They are white when first laid, but gradually turn pale yellow and, just prior to hatching, the dark head of the larva can be seen through the shell. On one occasion, I watched a female laying in a sheath of Yorkshire-fog and, after she had flown off, was amazed to find 17 eggs within the sheath that, based on their colour and

A batch of seven eggs laid in a sheath of Yorkshire-fog

Eggs are flattened ovals in shape

17 eggs laid in three batches by different ♀s

Life Cycles of British & Irish Butterflies

position, were in three separate batches that had been laid at different times, and almost certainly by different females. As is often the case with our butterfly fauna, females seem to select a surprisingly small subset of available plants when laying, despite there being hundreds of seemingly suitable grass stems in the vicinity in this particular case.

Larva

Eggs hatch in around three weeks and each 1.8-mm long **first instar** larva will eat most of its eggshell. It spins a dense oval cocoon around itself on the site of the egg, attached to the inside of the grass sheath. The larva hibernates within its cocoon, alongside those of its siblings, and its dark brown head remains somewhat visible. The larvae emerge from their cocoons in April and disperse to separate grass blades, where each spends a period of time eating notches in the edge of the leaf and resting along the middle of the leaf when not feeding. The larva eventually silks the edges of the leaf together, which forms a protective tube as the silk dries. The larva rests within this tube, emerging only to feed, or when moving to create a larger tube. The first instar has a pale greenish-yellow ground colour with a dark green line running down the middle of its back. It also has a dark brown head along with a dark brown band on the first segment, and these distinguish this instar from all others. When fully grown, the first instar is 3 mm long.

1st instar: a larva within its grass tube

1st instar: larval cocoons inside a grass sheath

1st instar: the larva has a black band on the first segment

A few weeks after emerging from its cocoon and feeding, the larva moults into the **second instar**. The larva has a pale greenish-brown ground colour when newly moulted, developing a green hue as it feeds, and retains the dark green line along its back, along with a whitish line about half way up each side of the body. Its head, however, is now a light brown and markedly different from that of the first instar, and it also lacks the black band on the first segment. The fully grown second instar is around 5 mm long.

2nd instar: a newly moulted larva outside of its protective tube

2nd instar: the larva develops a green hue as it matures

3rd instar: the head is now a light greenish-brown

After two weeks, the larva moults into the **third instar**. The body now has a light green ground colour, and the dark line running along the back and the whitish lines along each side are still present, although these are much more prominent than in the previous instar. The head is now a greenish-brown, rather than the light brown of the previous instar, and the fully grown third instar is 7 mm long.

After around eight days, the larva moults into the **fourth instar** and has a more whitish ground colour than the previous instar, along with a more developed pale line along each side just below the spiracles, but is similar in all other respects. The main distinguishing feature is size—the fully grown fourth instar is 14 mm long, which is twice the length of the third instar.

4th instar: the larva is similar to the previous instar

5th instar: the larva now has a globular and green head

5th instar: a cocoon among grass blades and other vegetation

After 10 days or so, the larva moults for the final time into the **fifth instar** and now starts to live more openly, when it can occasionally be found resting on a grass blade. The ground colour is a light bluish-green, with the dark green line along the back and the white lines along each side still present, although the white line along each side, just below the spiracles, is now especially prominent. Also, the tapering of the body towards the rear and an anal flap are most pronounced in this instar. A key distinguishing feature, however, is the head, which is now particularly 'globular' and a uniform green. After two weeks and when preparing to pupate, the fully grown 21- to 25-mm long larva will move to the base of the grass tussock and construct a flimsy cocoon between several grass blades and other vegetation. The larva then attaches itself to a blade within the cocoon using a silk girdle around its waist and a silk pad at its tail end.

Pupa

After around 11 days in the fifth instar, the larva sheds its skin for the last time and the resulting 18-mm long pupa is held in place by the silk girdle and the cremaster, whose hooks attach to the silk pad. The pupa is yellowish-green when it is first formed but darkens as the adult butterfly develops, with the wings of the butterfly becoming quite visible through the pupal case just prior to the butterfly emerging. As in several closely related species, the pupa has a small conical 'beak' at the top of its head, and the casing that houses the proboscis is extremely long, extending beyond the apex of the wings. After two or three weeks, depending on temperature, the adult butterfly emerges, and can occasionally be found hanging low down within a grass tussock as it dries its wings before taking to the air.

An opened cocoon showing a newly formed pupa

The pupa colours up as the adult butterfly develops

Life Cycles of British & Irish Butterflies **47**

Lulworth Skipper
Thymelicus acteon

❖ LULWORTH SKIPPER IS SUPPORTED BY RICHARD CARTER, ANGELA SHARP & STUART WOODLEY ❖

In the grand scheme of things, the Lulworth Skipper is a relatively late addition to the butterfly fauna of Britain and Ireland, no doubt a result of its very local nature in the south of England and its similarity to both the Small and Essex Skippers. The butterfly was originally discovered in England by the legendary entomologist James Charles Dale in 1832 in the area around Lulworth Cove in Dorset, but this was not made public until John Curtis made mention of the discovery in volume 10 of his *British Entomology* of 1833, when he named the butterfly the 'Lulworth Skipper'. What is most astonishing about the discovery is that Dale kept detailed diaries (which are now housed in the Oxford University Museum of Natural History) and, in the relevant entry, simply wrote "Acteon", the specific name given to this butterfly by S.A. von Rottemburg in 1775 based on a German type specimen. It can only be assumed that Dale knew of the butterfly's existence on the continent but was, perhaps, unaware of the significance of his find on British soil.

The butterfly is the smallest of our 'golden' skippers and is most easily distinguished by the faint pale orange crescents that are found on the forewings and that are especially prominent in the female. The butterfly is also slightly darker than other skippers, especially the male which develops an olive-brown colour as scales are lost. In common with the Small and Essex Skippers, the male has sex brands on the upperside of its forewings that show as very fine lines that are barely discernible in many cases.

The ♀ has prominent crescents on the forewings and a wingspan of 25 to 28 mm [MC]

The ♂ is generally darker than the ♀ and has a wingspan of 24 to 27 mm [NF]

The undersides of both sexes are a plain orange-brown

DISTRIBUTION

As the butterfly's name suggests, the distribution of this species is centred around Lulworth on the south coast of Dorset, on the strip of calcareous grassland between Weymouth to the west and Swanage on the Isle of Purbeck to the east, with smaller colonies some 20 km further west. There are historical records from Devon where it now appears to be extinct, with the butterfly last recorded there in the 1930s. There are also a few records from Cornwall, although the butterfly's status in that county is disputed.

This species is at the northern limit of its range in Britain and is rarely found more than five miles from the coast and, although it is not a maritime species, its distribution must surely lend itself to some of the most enjoyable butterfly watching, where it can be seen from the Dorset coastal paths with glorious views over the sea. On a worldwide basis, the butterfly can be found throughout central and southern Europe, extending into Asia Minor, and in several countries in North Africa, where the butterfly is not confined to coastal sites as it is in Britain. Perhaps surprisingly, the butterfly is not found in the Channel Islands.

The butterfly forms discrete colonies with limited dispersal and the distribution of the butterfly has changed little since Dale's first record in 1832. Where the butterfly does occur, colonies can be very large and contain many thousands of individuals, especially where the primary larval foodplant, Tor-grass, grows tall as a result of little or no grazing. Egg-laying females search out these lush plants and it is thought that their availability is the limiting factor in terms of the butterfly's distribution. The species was at its peak in the late 1970s when over 40 new sites were discovered at the time—in *The Butterflies of Britain and Ireland*, Jeremy Thomas writes *"During the 1970s, nearly 400,000 adults emerged annually on Bindon Hill, with perhaps a million individuals on the ranges as a whole"*.

Such numbers were attributed to an increase in the amount of suitable Tor-grass, thanks to a reduction in the tight grazing caused by sheep as well as the decline of the rabbit population when myxomatosis was introduced in the 1950s, a disease that was welcomed by many who considered rabbits to be a serious agricultural pest. This change in habitat management has, however, come at a price for those species that like short turf, such as the Adonis Blue. I must admit, I was once rather surprised to find a fully grown Adonis Blue larva crawling on the ground on the fringes of lush growth of Tor-grass at Bindon Hill, above Lulworth Cove, although the butterfly does seem to be holding its own there despite the tall sward.

HABITAT

Most colonies are found on south-facing, sheltered calcareous grassland slopes, cliff tops and undercliffs, where tall patches of Tor-grass grow in swards that reach a height of between 30 and 50 cm. Such plants are favoured by egg-laying females who generally never select plants that are under 10 cm in height.

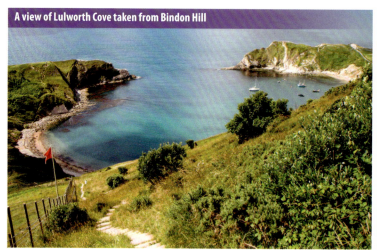

A view of Lulworth Cove taken from Bindon Hill

The shiny leaves of Tor-grass stand out from the surrounding vegetation

STATUS

The long-term trend shows a significant decline in both distribution and abundance, with the more recent trend over the last decade showing a moderate increase in both distribution and abundance. The butterfly's status can therefore be considered relatively stable in Britain, although it is considered a priority species due to its decline in northern Europe, where it is now extinct in the Netherlands. Using the IUCN criteria, the Lulworth Skipper is categorised as Near Threatened (NT) in the UK.

Like many of the skippers, eggs and larvae are often situated relatively high on the plant, making them susceptible to heavy grazing by horses and cattle. Another threat to this species is grazing by rabbits, whose numbers have been steadily increasing following their decline due to myxomatosis.

LIFE CYCLE

The butterfly has one generation each year, with a flight period that has become extremely protracted in recent decades. In *The Lepidoptera of the British Islands* of 1893, Charles Barrett writes: *"On the wing from the end of June till August or even September"*. Subsequent works compress the flight period even further, suggesting that adults were never seen before mid-July. During my childhood (I won't give my age away, other than to say that The Beatles were still in the charts), I always considered this species a butterfly of late summer, where holidays spent at Swanage in Dorset inevitably meant a trip to the 'Great Globe' at Durlston Country Park where I would wander off in search of butterflies found in the area with my trusty *The Observer's Book of Butterflies* to hand. Due almost certainly to climate change, there is now a good chance of finding this butterfly at early sites, such as Bindon Hill, at the end of May—a full month earlier than Barrett had stated—and at later sites until mid-September. The butterfly may even turn up at the end of April in good years.

Imago

This is a sun-loving butterfly and, while it remains inactive in dull weather, makes up for any lethargy when the sun comes out by doing what skippers do—flying so rapidly over vegetation that they are extremely difficult to track when darting between flowers. Favourite nectar sources include thistles, Greater Knapweed and Wild Marjoram on chalk grassland, and Common Bird's-foot-trefoil on coastal undercliffs. In the hot summer of 2018, adults were also seen taking moisture from damp patches of ground 150 m away from their core habitat (Rachel Jones, pers. comm.).

When courting, the male lands behind the female and gradually moves towards her. An unresponsive female will flutter her wings until the male loses interest and flies off, whereas an encounter with a receptive female quickly leads to the pair mating. The detailed triggers for one outcome or the other are not well understood, although any interaction almost certainly involves pheromones emitted from the androconial scales on the sex brands of the male.

When egg-laying, the female will alight on a stem of mature flowering Tor-grass growing in a sunny and sheltered spot, before moving backwards down the flower stem, probing the sheath as she goes. When a suitable opening is found, she will point her antennae forward and lay up to 15 eggs inside the sheath, although five or six eggs is more typical. The sheaths used are much looser than those required by the Small and Essex Skippers, so much so that I remember watching, through close-focusing binoculars, a female lay her eggs and seeing each as it was laid.

A mating pair [RC]

A ♀ laying eggs in a sheath of Tor-grass

Ovum

The smooth, oval, pearly white eggs are around 1.5 mm long and similar to those of the Small and Essex Skippers. As the larva develops, the egg takes on a glossy appearance and the shell becomes quite transparent when the larva is fully formed, whose black head becomes quite visible and, under high magnification, even the collar that is found on the first segment of the larva can be made out. I monitored some fully formed larvae in their eggs for a couple of days to determine if there was any pattern to which way they faced when in this state and all theories were blown away when, on my second visit, I found that most of them had turned around in the egg 180 degrees and others had turned 90 degrees so that their heads were now in the middle of the egg.

Eggs are laid in loose grass sheaths

Eggs are oval in shape

As the larva develops, its head becomes quite visible through the eggshell

Life Cycles of British & Irish Butterflies

Larva

After around three weeks, the 2.1-mm long and pale yellow **first instar** larvae eat their way out of their eggs and immediately spin individual cocoons where they remain for around eight months until the following spring.

1st instar: larvae in the process of hatching from their eggs

1st instar: a grass sheath held open to reveal the larval cocoons

1st instar: a protective tube

1st instar: a close-up of the larva showing the distinctive collar on the first segment

The larvae typically bore their way out of their cocoons in April, but the precise timing depends on whether they inhabit an 'early' or 'late' site. Each larva, which is now paler and straw-yellow in colour, begins its solitary existence by crawling away from its siblings to a grass blade where, after an initial meal taken from the edge of the leaf, it silks together the edges which then come together as the silk dries and contracts, forming a protective tube. The larva will also put down a layer of silk on which to rest and this becomes particularly visible in later instars.

The larva feeds by night, eating the leaf both above and below the tube, and will ultimately move to a new leaf blade and form a new tube as it outgrows its current abode. After each bout of feeding, the larva retreats to its protective tube where it digests its meal. The larvae leave characteristic notches in the grass blade as they feed, making their detection much easier. When feeding outside of its tube, the first instar can be distinguished from a second instar by the presence of a black collar on the first segment and, when fully grown, the first instar is 4 mm long.

After around 10 days, the larva moults into the **second instar**. It retains its black head, which distinguishes the second instar from later instars, but, as already mentioned above, is now lacking the black collar on the first segment. Its habits are similar to those of the first instar since it continues to construct a protective tube from a single blade of grass, moving to a new blade once it has outgrown its old tube. The fully grown second instar is 6 mm long.

After a couple of weeks, the larva moults into the **third instar**. The larva now has a very different appearance, primarily because the head capsule is no longer black but a very light brown. The body has also changed appearance and is now a mix of greens with very faint lines that run longitudinally.

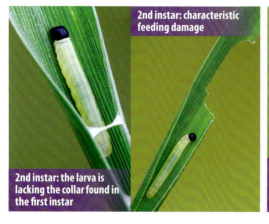

2nd instar: the larva is lacking the collar found in the first instar

2nd instar: characteristic feeding damage

3rd instar: the larva now has a light brown head and continues to construct protective tubes

3rd instar: characteristic feeding damage below the tube

Life Cycles of British & Irish Butterflies

Later instars create characteristic notches on both sides of the blade below the protective tube, and this may ensure that nutrients are retained higher up the leaf where the larva subsequently feeds. When fully grown, the third instar is around 8 mm long.

After another week or so, the larva moults into the **fourth instar** and, when fully grown in this instar, is just under 13 mm long. Apart from the obvious difference in size, it is similar in appearance to the previous instar although the various longitudinal lines on the body are more distinct.

After around two weeks, the larva moults into the final and **fifth instar**. The larva is similar in appearance to the previous instar and, apart from the difference in size, now has a light green, rather than a light brown, head. Tubes are now formed from two or more leaves and, as the final instar matures, it becomes more free-living and can be found resting openly on leaf blades. When fully grown, the larva is around 24 mm long.

4th instar: the lines running the length of the body are more distinct

When preparing to pupate, the larva forms a loose tent within the base of the Tor-grass tussock by drawing several grass blades together with silk, where it fixes itself in position with a silk girdle and a silk pad at its tail end.

5th instar: a final instar larva preparing to pupate

5th instar: the larva now makes a protective shelter from two or more leaves

5th instar: the larva now has a predominantly green head

5th instar: the characteristic notches are particularly visible in this instar

Pupa

After around four days, the larva moults for the last time, revealing an 18-mm long pupa that is attached by the silk girdle and the cremaster. The pupa initially exhibits the same pattern of longitudinal stripes seen in the larva, with the thorax, head and wings a translucent green, and the abdomen a light yellow-green. A key characteristic of the Lulworth Skipper pupa, however, is the presence of a prominent 'beak' on the head. The pupa darkens as the adult develops and, after between two and three weeks, is almost black just before the butterfly emerges.

The pupa is initially green in colour and has a prominent 'beak'

The pupa gradually turns brown as the butterfly develops

The pupa is almost black just before the adult emerges

Life Cycles of British & Irish Butterflies

Silver-spotted Skipper
Hesperia comma

❖ SILVER-SPOTTED SKIPPER IS SUPPORTED BY SARAH TUTTON, JOHN V & KATRINA WATSON ❖

	January	February	March	April	May	June	July	August	September	October	November	December
IMAGO												
OVUM												
LARVA												
PUPA												

The Silver-spotted Skipper, with an unmistakable spotted underside that distinguishes it from all of our other skippers, is one of the few butterflies that is steadily increasing in both distribution and abundance, and this warmth-loving butterfly has undoubtedly benefited from climate change. Like all skippers, it is hard to follow as it flies rapidly around its habitat and, even when it has landed among vegetation or on bare ground, can be hard to relocate as it blends in with its surroundings. Like our other 'golden' skippers, the male is distinguished from the female by a sex brand on each forewing that contains specialised scent scales.

The butterfly was severely declining in the mid-20th century, and a landmark study of this species was undertaken in the late 1970s and early 1980s by Jeremy Thomas, Chris Thomas, David Simcox and Ralph Clarke. In their article 'Ecology and declining status of the Silver-spotted Skipper butterfly (*Hesperia comma*) in Britain', published in *Journal of Applied Ecology* in 1986, they highlighted two important factors—the butterfly has limited dispersal ability and egg-laying females require foodplants growing in particularly warm spots. This study went on to inform conservation measures that have since contributed to the species' recovery.

In *The English Lepidoptera* of 1775, Moses Harris gave this butterfly the name 'The Pearl Skipper' that recognises the white spots found on the underside of the hindwings and William Lewin, in his *The Papilios of Great Britain* in 1795, gave the butterfly the name 'August Skipper' to recognise the butterfly's relatively late flight period. It was not until 1803 that the butterfly was given the name we use today, when Adrian Haworth included it in his *Lepidoptera Britannica*.

The ♀ has a darker ground colour than the ♂ and a wingspan of 32 to 37 mm

The ♂ has a prominent sex brand on each forewing and a wingspan of 29 to 34 mm [NF]

♂ and ♀ have similar undersides [MC]

DISTRIBUTION

This is a warmth-loving butterfly and is restricted to closely grazed chalk downland sites in southern England where its larval foodplant, Sheep's-fescue, grows. Core populations are found in the borders of Dorset, Wiltshire and west Hampshire; the southern Chilterns; the South Downs of Hampshire and Sussex; and the North Downs of Surrey and Kent. The butterfly is not found in Wales, Scotland, Ireland, the Isle of Man or the Channel Islands. On a worldwide basis, the butterfly has a Holarctic distribution and is found from Europe through temperate Asia to China and Japan, in North Africa and in North America. Thanks to higher temperatures, in many of these regions the butterfly is able to inhabit areas that are more overgrown than those used in Britain, and it also uses a wider range of grasses as its larval foodplant.

Most interchange is between sites that are less than 1 km apart and, in their article 'Partial recovery of a skipper butterfly (*Hesperia comma*) from population refuges: lessons for conservation in a fragmented landscape', published in *Journal of Animal Ecology* in 1993, Chris Thomas and Theresa Jones found that colonisation increased when suitable habitat covered a large area, and when sites were in close proximity, with no evidence of interchange between sites that were more than 9 km apart. An example of the butterfly's limited dispersal ability was given by Jeremy Thomas in

The Butterflies of Britain and Ireland, where he writes: *"After suitable habitat had been restored to Old Winchester Hill, in Hampshire, it took 17 years for the butterfly to cross the Meon Valley – a distance of just 3 km – which separated it from Beacon Hill, where more than a thousand adults flew every year"*. However, there are several examples of the butterfly flying much further. In *The Butterflies of Sussex*, Neil Hulme cites an example from 2007 of the butterfly reaching Chantry Hill from Newtimber Hill, a distance of 18 km, although such 'leaps' are the exception.

These studies also suggest that colonies are not self-sustaining but reside within a metapopulation structure—a loose network of both permanent and temporary sites that are spread over a large area and with interchange between them.

HABITAT

The butterfly is on the north-western edge of its European range in Britain and requires chalk grassland that is sufficiently warm for its needs, especially the successful development of the immature stages. Established thinking suggests that typical sites have a relatively short and sparse sward of the larval foodplant, Sheep's-fescue, growing among bare ground, such as rabbit scrapes, and often on south-facing hillsides. Habitat management is focused on maintaining these conditions and, at some sites, this is achieved naturally through grazing by rabbits, although grazing by cattle or sheep can also be beneficial, with cattle grazing having the additional benefit of breaking up the turf to create bare ground.

However, a warming climate has resulted in some remarkable changes for this species over the last few decades, that to some extent refines this thinking. In a nutshell, because ambient temperatures are now warmer, more of the available habitat meets the butterfly's stringent requirements, with a reduced need for bare ground, the potential to use plants with a taller sward and the potential to inhabit several aspects of a given site, not only those that are south-facing.

Broughton Down in Hampshire [TB]

STATUS

The butterfly steadily declined over the 20th century with some sites lost to agriculture and development, while others became unsuitably overgrown and relatively cool as grazing was abandoned, and as the rabbit population declined as a result of myxomatosis. The butterfly had also been found on limestone sites as well as chalk. In their 1986 article, Jeremy Thomas and colleagues determined that the butterfly had reached its low point in the 1970s when it was reduced to about 46 colonies.

Thankfully, the recovery of the rabbit population, the reintroduction of grazing at former sites (often the result of conservation efforts) and a warming climate have allowed the butterfly to make a spectacular comeback, and this is one of the few species that is now increasing its range, although it has yet to reach its former distribution that extended as far north as south Cumbria and Yorkshire.

Thorough studies of the species over the last decade by Callum Lawson, Chris Thomas, Jenny Hodgson, Jonathan Bennie, Robert Wilson, Zoe Davies and many others has confirmed the butterfly's positive outlook. Between 1982 and 2000 there was a 10-fold increase in the area of occupied habitat, and a four-fold increase in the number of populations, although more recent studies show that the rate of expansion has slowed, particularly when conditions in August have been cool or wet, and in landscapes in which suitable areas of grassland have become overgrown or are isolated and several kilometres from the nearest colony.

The long-term trend shows a moderate increase in distribution and a significant increase in abundance, with the more recent trend over the last decade showing a stable distribution and a moderate increase in abundance. Using the IUCN criteria, the Silver-spotted Skipper is categorised as Near Threatened (NT) in the UK.

LIFE CYCLE

This is one of our latest butterflies to emerge, first appearing in the second week of July at some sites, and flying until mid-September, with males emerging slightly before females. On 5 August 2011, at Malling Down in Sussex, Crispin Holloway (the Silver-spotted Skipper species champion for the Sussex branch of Butterfly Conservation) found adults, an egg, a late-developing larva (on the north-facing slope) and a pupa, all on the same day, showing just how dependent on warmth this species is for its development, and explaining the gradual emergence of adults over the flight period.

The butterfly has also adapted to its environment in different parts of the world and, in Alaska, for example, some sources suggest that the butterfly takes two years to complete its development and may also overwinter as a larva or pupa, as well as an egg, which is the exclusive overwintering stage in Britain.

Imago

Both sexes spend the majority of their time either basking or feeding, and a wide variety of nectar sources is used, including Autumn Gentian, Common Bird's-foot-trefoil, dandelions, Field Scabious, knapweeds, Red Clover and Wild Basil, although thistles are a particular favourite, especially Dwarf Thistle. The butterfly requires warmth and, unlike many of our butterfly species that will fly if the temperature is above 13°C, remains inactive unless the temperature has reached around 20°C. In order to raise its body temperature, the butterfly can be found basking on the warmest patches of ground, such as paths, rabbit scrapes, hoof prints and other bare ground that has been baked by the sun.

When seeking out a mate, the male rests on a suitable sunlit perch, and investigates any passing butterfly. An encounter with another male results in the pair spiralling around one another until one of the males breaks off and flies away, while the other returns to the original spot. An encounter with a female results in a similar spiralling flight, although the male is more persistent and she soon lands, with the male landing slightly behind her. The pair simply sit together for a minute or so and, if the female flies off, then the male will follow her and repeat the process. Eventually, the male moves forward with wings twitching and, if the female is receptive, mates with her, the pair remaining coupled for around two hours.

An egg-laying female is quite conspicuous as she flies low over the ground in search of a suitable patch, which is often next to bare ground and in a hollow, such as a rabbit scrape. After tasting the plants with her feet to confirm suitability, she lays a single egg, either on Sheep's-fescue or on nearby vegetation. Females are very choosy and will only select the small subset of available plants that meet their particular needs, and it is not uncommon to find several eggs laid on the same plant. The studies in the 1980s showed that the female preferred plants that are 1 to 4 cm tall, around 2.5 cm in diameter,

A mating pair with ♀ on the left [BE]

A ♀ egg-laying on Sheep's-fescue next to bare ground

and surrounded by bare ground although, as mentioned earlier, taller and larger plants are being increasingly used with a warming climate. Eggs overwinter and are easily displaced and fall to the ground, and it is thought that this might protect them from incidental grazing.

In their paper 'Changing habitat associations of a thermally constrained species, the Silver-spotted Skipper butterfly, in response to climate warming', published in *Journal of Animal Ecology* in 2006, Zoe Davies, Robert Wilson, Sophie Coles and Chris Thomas show that females will not lay if the ground temperature is below around 25°C, with eggs laid more frequently in higher temperatures. The same study also showed that, between 1982 and 2002, the percentage of bare ground providing the optimum conditions required for egg-laying had shifted from 41% to 21%, with the butterfly using areas that are now suitably warm thanks to a warming climate.

Ovum

The dome-shaped egg is 0.9 mm in diameter at its base and 0.7 mm high, has a sunken crown, and is quite conspicuous against the green of the foodplant. The egg is white when first laid but turns slightly darker with a yellowish tinge after a couple of weeks, with the larva fully formed within the egg before the onset of winter.

Eggs laid on a favoured Sheep's-fescue plant [DM]

The larva overwinters in its dome-shaped egg

Larva

The 2-mm long **first instar** larva emerges from the egg in March by eating an untidy hole in the crown, but it does not eat the remainder of the eggshell. It immediately goes about constructing a loose tube from a cluster of fine stems of Sheep's-fescue, leaving an opening at the top. If more than one egg has been laid on the same plant, then this may result in several larvae sharing the same tube. These tubes are extremely difficult to find in the wild since they are not only formed within a tuft of the foodplant, but the larva spends much of its time hidden away at the base of its new abode.

1st instar: a larva emerging from its egg

The larvae are fascinating to watch, even in their early instars, when they seem to spend all of their time either resting, feeding or tube building. They also seem to be very wary and will wriggle backwards into their abode at the slightest disturbance. It is here where they clearly feel the most secure, and I have watched larvae reach out from their tube, fell a blade of Sheep's-fescue, and draw it back into their tube where they continue to feed upon it. Watching the larvae get rid of their frass is also an interesting event, which cannot be claimed of most species. The larva, which is normally facing 'head up', moves to the base of its tube, reverses out and, when its back end sits proud of its tube, ejects its frass with quite some force.

The first instar has a black head, a black collar on the first segment and a straw-yellow ground colour, growing paler as it matures. It has several rows of black spots that run the length of the body and each bears a very short hair. There are also two long and pale hairs that project backwards from the anal segment. When fully grown, the first instar is just over 4 mm long.

The larva develops extremely slowly, and it is almost a month before it moults into the **second instar**. The ground colour is pale straw-yellow, as in the previous instar, although this now has a greenish tinge. The larva now has a more extensive covering of hair-bearing black spots and, while the black head and collar are as in the first instar, the front half of the first segment is particularly elastic and lilac, a colour that is revealed when the larva extends its neck as it moves and feeds. The larva resides within a tube as in the previous instar and, when fully grown, the second instar is 7 mm long.

After three or four weeks, the slow-growing larva moults into the **third instar** and is now a pale greyish-green and, apart from size, is similar to the previous instar in all other respects. When fully grown, the third instar is 9.5 mm long.

After between two and three weeks, the larva moults into the **fourth instar** and now has an olive-green ground colour and, while there are very faint lines that run along the face in the previous instar, these are much more prominent in this instar. When fully grown, the fourth instar is 16 mm long.

After around two weeks, the larva moults into the final and **fifth instar** and is similar in all respects to the previous instar, apart from size. After two weeks or so, and around 100 days in this stage, the larva is fully grown, when it is 25 mm long at rest, and 28 mm long when crawling. Curiously, a white, granular and waxy substance develops under the 10th and 11th segments although its purpose is unknown. The larva may wander a little before it builds a robust cocoon from fine grass blades, that is lined loosely with silk.

Pupa

The 19-mm long pupa has a two-tone colouring, with dark brown wing cases and a light brown abdomen. Surprisingly, the pupa has prominent hairs on its head and down its back to the cremaster.

Some of the hairs on the head are hooked and, along with the hooks on the cremaster, hold the pupa in a more-or-less upright position when they are attached to the silk of the cocoon. The male pupa is easily identified since it has a bulging ridge on each wing case that covers the sex brand of the forewing. The casing of the wings, legs, antennae and proboscis are also covered in a waxy lilac film that is easily rubbed off and its purpose, like the waxy substance observed in the larva, is unknown. After two weeks this characteristic butterfly of chalk downland emerges.

Large Skipper
Ochlodes sylvanus

❖ LARGE SKIPPER IS SUPPORTED BY ANDREW COOPER, PETER MARREN & ANDY & DI SAWYER ❖

The **Large Skipper** is the largest of our 'golden' skippers and is found in sheltered areas where coarse grasses grow tall. Like most skippers, the adults have a swift and whirring flight, especially the male, although they are arguably easier to follow than closely related species, thanks to their colour and size. As in related species, the male can be distinguished from the female by the presence of conspicuous sex brands found on their forewings. A subtler distinction that has been noted by members of the UK Butterflies website is that the knobs on the antennae are black in the female but are lined with orange on the inside in the male, and this is also true of the Silver-spotted Skipper.

The ♂ has distinctive sex brands and a wingspan of 29 to 34 mm

The ♀ has a wingspan of 31 to 36 mm

In his *Gazophylacium naturae et artis* of 1704, James Petiver gave male and female different names, presumably considering them different species, calling them 'The Checker-like Hogg' and 'The Checkered Hogg' respectively. In his *Papilionum Britanniae Icones* of 1717, Petiver changed the names to 'Streakt cloudy Hog' and 'Cloudy Hog', the former acknowledging the male's sex brands. It was not until 1766 that the two sexes were brought together, when Moses Harris named this butterfly the 'Large Skipper' in his *The Aurelian* where he writes: "*They delight to fly in woods, and lanes near woods*", recognising a key habitat of this species. This habitat preference was also recognised by George Samouelle in his *The Entomologist's Useful Compendium* of 1819, where he names this butterfly the 'Wood Skipper'. A similar connection is made in the specific name of this species, *sylvanus*, which is a reference to the Roman god of woods and fields.

♂ and ♀ have similar undersides

DISTRIBUTION

The butterfly is found throughout England and Wales but, in Scotland, is confined to the south-west and the south-east. It is absent from the highest ground throughout its range. It is not found in Ireland (despite some erroneous records claiming otherwise), the Isle of Man and the Isles of Scilly, and is restricted to Jersey in the Channel Islands. On a worldwide basis, the butterfly is found throughout most of Europe, through Asia as far as China and Japan, and in parts of North Africa.

This species forms discrete colonies which vary in size and may occupy a site that is as small as 0.5 ha (70 m × 70 m) and that contains only a few dozen adults. Large numbers can build when conditions are suitable, such as the first five years after a woodland has been cleared and left to regenerate, when native coarse grasses grow tall. The butterfly is somewhat mobile and has extended its range in recent decades, reaching Ayrshire in south-west Scotland, and the Scottish Borders in south-east Scotland.

Life Cycles of British & Irish Butterflies

HABITAT

This species is found in sheltered and often damp areas of grassland, including woodland clearings, rides and glades; mixed scrub; hedgerows; calcareous and acid grassland; grassy heathland; river banks; arable field margins and set-aside; brownfield sites; road verges; railway embankments; parks and churchyards. Suitable sites contain vegetation that the butterfly is able to use for basking and for the territorial male to perch on, in addition to nectar sources and larval foodplants.

A woodland edge at Pamber Forest in Hampshire

STATUS

The status of this butterfly is considered stable, despite the loss of sites as a result of the agricultural intensification of unimproved grassland and the often unnecessary tidying of roadside verges and field margins that not only removes breeding habitat, but also destroys eggs, larvae and pupae that might be present on tall grasses. Rotational cutting is therefore desirable, when a proportion of grasses remains uncut. The long-term trend shows a moderate decline in both distribution and abundance, with the more recent trend over the last decade showing a moderate increase in both distribution and abundance. Using the IUCN criteria, the Large Skipper is categorised as Least Concern (LC) in the UK.

LIFE CYCLE

This single-brooded butterfly is on the wing from mid-May, peaks at the end of June and early July, and flies until mid-August, with a few stragglers hanging on until the end of the month. On rare occasions, fresh individuals may be found in the second half of September, suggesting that the butterfly does produce the occasional second brood individual.

Imago

The butterfly's exceptionally long proboscis allows it to feed on a wide variety of flowers, although it has a particular fondness for brambles and thistles. Other nectar sources include Bugle, clovers, Common Bird's-foot-Trefoil, dandelions, Devil's-bit Scabious, Field Scabious, Kidney Vetch, knapweeds, Ragged-Robin, vetches, Wild Privet and Yarrow. Like all 'golden' skippers, the butterfly basks on flowers, vegetation or tall grasses with forewings and hindwings held in different planes, but closes its wings when resting and roosting.

In their article 'Mate location behavior of the Large Skipper butterfly *Ochlodes venata* [*sylvanus*]: flexible strategies and spatial components', published in *Journal of the Lepidopterists' Society* in 1987, R.L.H. Dennis and W.R. Williams describe, in detail, the strategies applied by the male when searching for a mate, based on observations made at Lindow Common near Wilmslow in Cheshire.

In early morning, males will initially adopt a strategy of perching, where they await a passing female. On clear and sunny days, with the onset of warmer temperatures and with energy levels high, males will switch to a strategy of patrolling from around 10 am when they seek out newly emerged females, flying low over the ground and among grass stems where the females will have pupated, locating their target by both sight and smell. Dennis and Williams provided a sense of the power of the scent from the female: *"One male was observed searching a 1 m tall birch seedling systematically for 15 min, weaving in and out of the twigs and leaves, before it found the female"*. Males revert to a strategy of perching from mid-day to around 4 pm, and it is thought that this change of strategy may be initiated due to a paucity of unpaired females. Males are also opportunistic and will approach females while feeding. The switch from perching to patrolling behaviour is delayed in cooler conditions, and patrolling may not commence until after mid-day, possibly to coincide with a delayed emergence of females that day.

A typical perch is low down on vegetation, and always less than a metre from the ground, such as a leaf of young growth of oak or birch, or a bramble leaf. Each perch is normally close to nectar sources, in a sunlit spot and at a suitable vantage point such as edge habitat or the junction of two paths, where females are likely to fly. Favoured perches are reused by different males in the same year, and by different males in different years. Many observers will know of such spots and I know of one area at Stockbridge Down in Hampshire that has been used by perching males every year for over a decade.

The male will chase off any other insect when perching, and will vigorously defend his patch against another male, when the pair fly around one another as they rise upwards before ultimately separating, with the occupant returning to his territory and often to the same perch. If a female is encountered, then the male will pursue her in a rapid flight until she settles, landing behind her. An unresponsive female will flutter her wings rapidly and may even raise her abdomen, with the male losing interest and flying off. However, if the female is responsive, then the pair will sit together for a short while before the male moves forward to mate.

A courting ♂ behind a ♀

A mating pair, with ♂ on the left

The female tends to lay her eggs early in the afternoon and is easy to spot as she flies gently over tall and lush grasses, frequently landing as she investigates each tussock. Favoured plants are those that are sheltered, in full sun, and abutting an open area, such as those next to a path or at a scrub edge. When a suitable spot has been found, then she will alight on a blade that is around 30 cm above the ground and move around it before curling her abdomen underneath the blade to lay a single egg on its underside. She may then fly to another tussock nearby and repeat the process before flying off. While a number of coarse grasses are used, Cock's-foot is the sole foodplant at many sites, with Purple Moor-grass used on more acid soils. The butterfly has also been observed using False Brome, Torgrass and Wood Small-reed on occasion.

The head of the larva is clearly visible just prior to hatching

The egg is a pearl-white when first laid

Ovum

The dome-shaped egg is 0.9 mm in diameter at its base and 0.5 mm high, and the top of the egg is slightly flattened. While the egg appears quite smooth to the naked eye, it is actually covered in a very fine network pattern all over its surface. The egg is a pearl-white when first laid but develops a faint yellowish-orange tinge as the larva develops, fading back to white just prior to the egg hatching, when the black head of the larva is clearly visible through the eggshell.

Life Cycles of British & Irish Butterflies

Larva

The Large Skipper has seven larval instars and, in Britain and Ireland, only the Glanville and Heath Fritillaries have as many. Apart from size, the different instars can be distinguished based on the colour and shape of the head capsule, although other features may also provide a clue.

The seven instars can be distinguished by the colour and shape of the head capsule

1st instar

2nd instar

3rd instar

4th instar / 5th instar / 6th instar / 7th instar

After around 18 days, the 2.5-mm long **first instar** larva emerges from the egg and proceeds to consume its eggshell, before it immediately goes about constructing a protective tube by drawing the edges of a leaf blade together with silk. The larva emerges to feed on the edge of the leaf above the tube, and also crawls backwards out of the tube to get rid of its frass, which it flicks some way from its abode. The larva also lays down silk within the tube, on which it rests. The larva is initially pale yellow but develops a greenish tinge as it starts to feed on the foodplant. It has a black, shiny and almost hairless head, with a black collar on the first segment and, when fully grown, the first instar is 5 mm long.

1st instar: a newly emerged larva eating its eggshell

1st instar: a larva in its protective tube

1st instar: the fully grown larva has a greenish tinge

After around a week, the larva moults into the **second instar** and is now a greenish-yellow. It still has a black, shiny and relatively hairless head, but now has a brownish patch on the anal segment that differentiates it from the previous instar. The larva continues to reside within a protective tube and, when fully grown, the second instar is 7.5 mm long.

After eight days or so, the larva moults into the **third instar** and is similar to the previous instar, but the head is hairier and the ground colour is a deeper green, with a fine dark green line running along the back. The larva continues to live in a protective tube, although this may now be formed from two blades of grass, with the larva now feeding on the uppermost part of the tube itself. When fully grown, the third instar is 9 mm long.

2nd instar: the larva is now greenish-yellow

3rd instar: the larva is now a darker green

After another eight days, the larva moults into the **fourth instar** and now has a very pale yellowish-green ground colour, a very fine black collar on the first segment and the faintest of pale lines along each side. In this instar the larva starts to develop a chestnut colour on the two lobes of the head and these pale areas become more extensive in later instars. The larva also has a conspicuous anal flap on its last segment that is raised as the larva ejects its frass. When fully grown, the fourth instar is 11 mm long.

After just over a week, the larva moults into the **fifth instar** and now has a deep green ground colour, and the collar on the first segment is barely discernible. The head now has a granulated appearance and light reddish-brown patches in the centre of each lobe, with dark brown surrounds. The larva prepares for hibernation in September by silking together several leaf blades, forming a substantial protective tube that is closed off at each end.

4th instar: the black collar is now very fine

The larva does not feed during the winter, but remains in its shelter until March, when it emerges from its hibernaculum and wanders for a short while before creating a new protective tube. It feeds for a couple of weeks and the fully grown fifth instar is just under 13 mm long

In the second half of March, or in early April, the larva moults into the **sixth instar** and is similar to the previous instar, although the faint lines on each side of the body are less prominent and the pale patches on the head are more extensive. It continues to seek safety within a tube made from several grass blades, and feeds above and below its abode, usually at night. The fully grown sixth instar is 22 mm long.

After around five weeks, the larva moults for the final time into the **seventh instar** and is now in possession of a head capsule that is more conical than round. The pale colouring on each lobe is somewhat variable, ranging from dull grey to light brown. After around three weeks, the fully grown 28-mm long larva prepares for pupation by silking together several grass blades to create a tube, 10 to 30 cm above the ground, within which it creates a loose cocoon.

5th instar: the larva has two pale lines on each side

6th instar: the fully grown larva has a plump appearance

7th instar: fully grown larva with light brown markings on the head capsule

Pupa

The 19-mm long pupa is predominantly black, with a dark grey-green abdomen, and has an extraordinarily long casing that houses the proboscis, that is detached beyond the ends of the wing cases, and that extends as far as the anal segment. While the pupa is darker than that of the Silver-spotted Skipper, it is similar in several other respects—it has prominent hairs on its head and down its back that, along with the hooks on the cremaster, hold it in place within its cocoon, and it is also covered in a thin film. It is also easy to sex the pupa since the male pupa has a bulging ridge on each wing case over the sex brand of the forewing, which is absent in the female pupa. After three weeks, this denizen of the woods emerges.

A ♀ pupa

Life Cycles of British & Irish Butterflies

Pieridae

The **Pieridae** are found in all zoogeographical regions of the world and comprise approximately 2,000 species, most of which are double-brooded. The species found in Britain and Ireland are grouped into three subfamilies—the Dismorphiinae (the wood whites), the Pierinae (all other whites) and the Coliadinae (the yellows).

Family	Subfamily	Genus	Species	Vernacular name
Pieridae	Dismorphiinae	*Leptidea*	sinapis	Wood White
			juvernica	Cryptic Wood White
	Pierinae	*Anthocharis*	cardamines	Orange-tip
		Pieris	brassicae	Large White
			rapae	Small White
			napi	Green-veined White
	Coliadinae	*Colias*	croceus	Clouded Yellow
		Gonepteryx	rhamni	Brimstone

IMAGO

Adult butterflies are typically of a moderate size and usually exhibit sexual dimorphism in that the male and female differ somewhat in appearance, with androconial scent scales found on the upperside forewings of the male that are used during courtship. All butterflies are mostly white or yellow in colour and newly emerged adults may also give off a faint perfume. Unlike the Nymphalidae and Riodinidae families, all legs are fully functional in both sexes, and each leg ends in a pair of claws. Many females of this family are known for their posture when rejecting the advances of an amorous male, when they raise their abdomen at an acute angle that makes pairing physically impossible.

OVUM

Eggs are characteristically tall, bottle-shaped and strongly ribbed. All of our species lay their eggs singly except for the Large White, which lays its eggs in batches with the resulting larvae often decimating their foodplant.

A ♀ Orange-tip raising her abdomen to reject the advances of the ♂ [NF]

A Large White egg batch laid on the underside of a leaf

A bottle-shaped Green-veined White egg

LARVA

Larvae of the Pieridae are usually green, devoid of spines and have short and insignificant hairs. They are therefore relatively smooth when compared with the larvae of other families. Unsurprisingly, given that eggs are usually laid singly, larvae typically lead a solitary life (which is just as well in the Orange-tip, whose larvae are cannibalistic), although larvae of the Large White, whose eggs are laid in batches, live gregariously throughout their early instars.

The Orange-tip larva is quite smooth

Large White larvae are gregarious in early instars

Wood White larvae lead a solitary existence

PUPA

The pupa typically has a pointed head and is often secured in an upright position, supported by a silk girdle around the 'waist' (a segment of the thorax) and by a silk pad into which hooks found on the cremaster attach.

A Clouded Yellow pupa

A coloured-up ♂ Orange-tip pupa

Life Cycles of British & Irish Butterflies

Wood White
Leptidea sinapis

❖ WOOD WHITE IS SUPPORTED BY HARRY E. CLARKE, JONATHAN PHILPOT & KIRSTY PHILPOT ❖

	January	February	March	April	May	June	July	August	September	October	November	December
IMAGO												
OVUM												
LARVA												
PUPA												

With one of the slowest and most delicate flights of all the butterflies of Britain and Ireland, the Wood White is always a delight to behold as it dances along woodland rides or scrub edges. Despite this apparent fragility, it is surprising just how pristine Wood Whites remain throughout their life—it is very rare to come across a butterfly with tattered wings or more serious damage. Furthermore, this 'dainty' butterfly, and especially the male, is able to undertake prolonged flights that test the endurance of even the most ardent enthusiast, since they never, ever, seem to stop flying as they continuously patrol no more than a metre off the ground in search of females. This butterfly is, therefore, one that sees observers up at the crack of dawn so that they can find adults when they are relatively docile in early morning, and before the heat of the day recharges their batteries.

With the exception of the closely related Cryptic Wood White, the Wood White is easily distinguished from most other 'whites' when at rest—the wings are always held closed and the forewings have rounded tips. When at rest, and even when in flight, the male Wood White can be distinguished from the female by the presence of a black spot found at the apex of each forewing, which is much less prominent in the female. The male also has a white patch on the inside of the tip of each antenna that is brown in the female, although male and female have a similar wingspan of around 42 mm. Males of the second brood have darker wing spots than those of the spring brood, meaning that this butterfly is not only sexually dimorphic, but seasonally dimorphic too.

Much of what we know about the Wood White comes from the studies of Christer Wiklund, Martin Warren (who studied this species for his PhD thesis in 1981), Stephen Jeffcoate, and a team comprising Susan Clarke, David Green, Jenny Joy, Kate Wollen and Ian Butler.

The ♂ has dark spots at the apex of each forewing [DM]

A ♂ resting on the unopened flowers of Meadow Vetchling A ♀ feeding on Greater Stitchwort

DISTRIBUTION

This butterfly is found in discrete colonies in central and southern England and also in western Ireland in the Burren of County Clare and adjacent areas of limestone pavement in County Galway. This species is not found in Scotland, the Isle of Man or the Channel Islands and, in Britain and Ireland, is at the northern limit of its European range. On a worldwide basis, the butterfly is found throughout Europe and across Asia to China, Korea and Japan.

As documented in the paper 'Leptidea sinapis (Wood White butterfly) egg-laying habitat and adult dispersal studies in Herefordshire', published in *Journal of Insect Conservation* in 2010, Clarke and colleagues found that the butterfly is able to overcome habitat barriers (such as flying over the top of mature woodland or across a road)

and can also fly some distance—one first-brood male flew 2,087 m and one first-brood female flew 828 m, and the mean for the population was just under 400 m for both broods. The study concluded that, while most adults spend most of their time in a single habitat patch, a proportion of both sexes will disperse and that they will quickly find newly available habitat within 500 m of the present patch, but there is less chance of rapid colonisation of areas more than 1 km away without any interim 'stepping stones'. The 'stepping stones' theory may explain records from Surrey which imply a dispersal up to 10 km on occasion.

HABITAT

Suitable habitat is characterised as being warm, sheltered and damp, and where both larval foodplants and nectar sources can be found in abundance. While a few colonies persist on coastal undercliffs and disused railway lines, the butterfly has two primary habitats.

In England, the butterfly breeds primarily in the margins of wide and sunny rides in woodland (many of these in conifer plantations), where tall vegetation and an amount of scrub grow, and where there is light shading from surrounding trees. Based on a study at Yardley Chase in Northamptonshire and Monks Wood in Cambridgeshire, Warren determined that the vast majority of Wood Whites preferred habitat with a shade level of 20–50%, although Clarke and colleagues found that the butterfly preferred areas with less than 20% shade, including no shade at all. This discrepancy would suggest that shelter, rather than shade, is the key requirement for this species.

In Ireland, the butterfly is found on the limestone pavement of the Burren that provides a mosaic of bare rock, herb-rich vegetation and Hazel scrub. This habitat type exhibits similar qualities to suitable woodland in being open but sheltered, and where ample larval foodplants and nectar sources are to be found.

An area in Chiddingfold Forest in Surrey that has a high concentration of larval foodplant

The Wood White can be found near Hazel scrub in the Burren in Ireland

STATUS

Worryingly, the long-term trend shows a significant decline in both distribution and abundance, with the more recent trend over the last decade showing a moderate decline in both distribution and abundance. Using the IUCN criteria, the Wood White is categorised as endangered (EN) in the UK and near threatened (NT) in Ireland. Consequently, the Wood White is one of our most threatened species and a priority for conservation efforts.

In Britain, this butterfly has suffered due to changes in woodland management and, in particular, the reduction in coppicing that would otherwise allow new woodland clearings to develop that provide the 'edge habitat' and conditions required by this species. Unfortunately, improvements in habitat management will not guarantee that the species will reappear from areas where it has been lost, if the nearest colony is some distance away. Today, the butterfly typically survives in larger woods where there is a cycle of clear-felling and replanting, or where there is deliberate habitat management for this species in place that ensures that sufficiently wide rides are maintained. The population is possibly less threatened in Ireland, given the stability of its preferred habitat there.

LIFE CYCLE

The English colonies emerge in early May and fly until the end of June, whereas the emergence in Ireland starts in late May, with adults flying until the middle of July. Depending on location, the butterfly may produce a second brood that typically emerges during the second half of July and throughout August, and which may be more substantial than the first. A second brood is the norm at more southern sites and, based on their work at Forestry Commission woodland in Shropshire during 2007, Susan Clarke and David Green confirmed that a second brood now occurs even in the most northern part of the butterfly's range. Interestingly, the butterfly has three broods in the most southern parts of its range in Europe.

♂s taking mineral salts from damp earth in hot weather [NH]

Imago

As in most species, males emerge slightly earlier than females, and both sexes are avid nectar feeders. In 'Seasonal variation in the use of vegetation resources by *Leptidea sinapis*, a multivoltine species in southern Britain: implications for its conservation at the edge of its range and in the context of climate change', published in the *Entomologist's Gazette* in 2006, Stephen Jeffcoate provides a detailed analysis of this butterfly's behaviours at Chiddingfold Forest in Surrey. He found that the first brood nectared primarily on bird's-foot-trefoils, Bitter-vetch, Bugle, Greater Stitchwort and Herb-Robert, whereas those of the second brood nectared primarily on bird's-foot-trefoils, knapweeds, Meadow Vetchling, Selfheal and Tufted Vetch. In hot weather, males can also be found taking mineral salts from puddles and damp earth.

Many observers that have encountered this butterfly will also have witnessed its elaborate courtship that was unravelled by Christer Wiklund, who documented his findings in the classic paper 'Courtship Behaviour in Relation to Female Monogamy in *Leptidea sinapis*', published in *Oikos* in 1977. While it subsequently turned out that Wiklund was observing the Cryptic Wood White (that was unknown at the time), the key findings from his work hold true for both species, with the description below adjusted for the Wood White accordingly.

The male lands in front of the female and faces her. He then starts to wave his fully extended proboscis and antennae either side of the female's head, while emitting pheromones and flashing open his wings at irregular intervals. The female, on the other hand, quietly sits there with her antennae

The ♂ (on the left) waving his proboscis and antennae either side of the ♀ [IL]

reclined, while also intermittently flashing open her wings. If the female is receptive, then she bends her abdomen toward the male and the pair mate, staying coupled for between 30 minutes and an hour, and sometimes longer. If she is not receptive then she simply stays put, ignoring the advances of the male, with the whole charade often lasting for several minutes before the male flies off. Unlike some other members of the Pieridae, the female Wood White exhibits no rejection signal, such as raising her abdomen to indicate to the male that she is uninterested.

Those observers that have been lucky enough to witness a pairing report that it is a rapid affair of only a couple of seconds—so rapid that the male may not even have started the courtship behaviour before the female has shown a willingness to mate.

Females fly slowly amongst low vegetation when egg-laying and land on a variety of plants, testing each for suitability using chemoreceptors on their feet. A variety of legumes is used

A mating pair (♂ at top) [MC]

Life Cycles of British & Irish Butterflies

as larval foodplants, including Bitter-vetch, Common Bird's-foot-trefoil, Greater Bird's-foot-trefoil, Meadow Vetchling and Tufted Vetch, with any preference determined not by the species of plant, but by their suitability in terms of growth and location. Jeffcoate found that, in spring, the first brood females laid on plants growing in full sun in the warmest parts of the site but that, in the heat of summer, the second brood females laid eggs on plants growing in partial shade. In 2018, Susan Clarke and David Green also found females egg-laying along a totally shaded ride (95% to 100%) at a Devon site (Susan Clarke, pers. comm.).

When a suitable plant is found then the female bends her abdomen under a leaf and lays a single egg. She then immediately flies off to either find a place to rest, to feed, or to inspect other plants in the vicinity. Observations in the West Midlands during 2016 showed that a female may lay at least 50 eggs in a single day, with between one and nine eggs laid between stops to rest or nectar (David Green, pers. comm.). Clarke and colleagues found that eggs were laid at various heights on the foodplant, from those growing on bare ground to those that stood over 1 m tall in thick vegetation and scrub. However, in all cases, eggs were laid towards the top of the foodplant, irrespective of the height of the plant, which confirms the findings from earlier studies on this species, including those undertaken by Martin Warren.

A ♀ laying a single egg near the top of the foodplant [MC]

The egg is pale yellow

Eggs may have a slight curve to one side

Ovum

The 1.3-mm high eggs are skittle-shaped, like those of other members of the Pieridae family, and each egg has 11 ribs that run from top to bottom. Unusually, eggs are often asymmetrical and curve slightly to one side. The egg remains a relatively constant pale yellow until just before hatching, when it darkens slightly.

1st instar: the ground colour is pale yellow

2nd instar: the larva now has a pale line at the base of each side

Larva

The 1.5-mm long **first instar** larva emerges from the egg after between 10 and 20 days, depending on temperature, and eats part of the eggshell as its first meal. The young larva has a pale brown head and a pale yellow ground colour, but soon develops a greenish hue as it consumes the foodplant, which is often the leaflet on which the egg was laid. The larva also has six rows of bifurcated white hairs that each resemble a letter 'T', with longer hairs on the first and last segments that project forwards and backwards respectively. The first instar always rests and feeds while outstretched along the edge of a leaf. Prior to moulting, the larva is 3 mm long and the palest of white markings appears along the base of each side that provide a hint of the white stripe found along each side of the second instar.

After five days, the larva moults into the **second instar** and now has a yellowish-green ground colour, with a darker green line along its back and a pale white line running along the base of each side of the body. The head is also a yellowish-green, matching the colour of the body, and the larva continues to feed on the edge of the leaf, but may position itself against a stem prior to moulting into the third instar. The 'T'-shaped hairs are still present but are greatly reduced in size. When fully grown, the second instar is 6 mm long.

After around a week, the larva moults into the **third instar** and, apart from size, is similar to the second instar in most respects, although the white line found along the base of each side is slightly bolder. Behaviourally, the larva will now move to the top of the plant to feed on the young growth, gradually working its way down the plant as it grows, and it will also position itself along the underside of the midrib of a leaflet, or on a stem, when resting or feeding. The fully grown third instar is 13 mm long.

3rd instar: the larva now has bolder markings

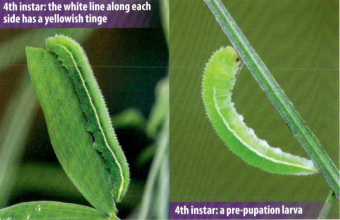

4th instar: the white line along each side has a yellowish tinge

4th instar: a pre-pupation larva

In around a week, depending on temperature, the larva moults into the **fourth instar** and is identical in all respects to the third instar, although the white line along the base of each side now has a yellowish tinge. When fully grown, the fourth instar is 19 mm long. As the time for pupation draws near, the larva moves off the foodplant and will wander for some time, over a day in some cases, travelling several metres in the process. When it finds a suitable site, it attaches itself to its chosen surface with a silk girdle and a silk pad at its rear end.

The descriptions given above are of a larva that went through four instars in total, although literature is peppered with examples of larvae going through five instars, as well as four. It is not clear which factors influence the number of instars, but obvious candidates are temperature, rate of development, brood, the propensity to diapause, or some combination of these and other factors.

Pupa

After a day or so, the larva pupates, revealing a 16-mm long pupa that is attached by the silk girdle and by the cremaster that hooks into the silk pad spun by the larva. The pupa is initially green, and the most beautiful pink markings develop on its edges and on the veins of the wing cases as it matures. If the pupa does not over winter, then the adult will emerge in around eight days. Pupae are extremely difficult to find in the wild, although Warren found 20 during the course of his studies, with most found 10 to 70 cm above ground and never on a larval foodplant (pupae were found attached to grasses, wild roses and rushes). According to Warren, the pupa also has two colour forms, green and brown, with green being the commonest by far.

The pupa is adorned with beautiful pink markings

A pupa one hour before the adult emerged

Life Cycles of British & Irish Butterflies

Cryptic Wood White
Leptidea juvernica

❖ CRYPTIC WOOD WHITE IS SUPPORTED BY JOHN MATHERS, BRIAN NELSON & MAURICE HUGHES & JAMES O'NEILL ❖

	January	February	March	April	May	June	July	August	September	October	November	December
IMAGO												
OVUM												
LARVA												
PUPA												

Early writings on the Wood White are littered with questions surrounding the differences between the British and Irish populations—the butterfly in Ireland seemed to have different habitat preferences and seemed much more stable. The Irish specimens were even given subspecific status and named *juvernica* (a variation of '*Hibernia*' that is Latin for the island of Ireland). In 2001, this conundrum was finally solved, although a short history lesson is in order.

In 1988, based on an examination of the genitalia of specimens from the French Pyrenees, the Wood White was 'split', with a new species emerging that was ultimately named Réal's Wood White *Leptidea reali*. In the article '*Leptidea reali* Reissinger 1989 (Lep.: Pieridae): a butterfly new to Britain and Ireland', published in *The Entomologist's Record and Journal of Variation* in 2001, Brian Nelson, Maurice Hughes, Robert Nash and Martin Warren showed that all Irish specimens outside of the Burren were of this new species, including specimens taken in the 19th century. As if these events were not incredible enough, the heritage of the two species was subsequently examined further and revealed a third species. In the paper 'Unexpected layers of cryptic diversity in wood white *Leptidea* butterflies', published in *Nature Communications* in 2011, Vlad Dincă, Vladimir Lukhtanov, Gerard Talavera and Roger Vila showed that, based on a DNA analysis of specimens, the three species shared a common ancestor, but split into two lineages—one that went on to divide to produce the Wood White and Réal's Wood White, and the other that represented the new species, which was the butterfly found in Ireland outside of the Burren (with Réal's Wood White not found at all in Britain and Ireland). Conforming to naming conventions, the authors used the oldest name that had ever been applied to this new species, resulting in *Leptidea juvernica*, and proposed the vernacular name of Cryptic Wood White.

Unsurprisingly, the Cryptic Wood White shares many of the characteristics of the Wood White—the wings are always held closed when at rest, the forewings have rounded tips, the male has a prominent black spot at the apex of each of forewing and a white patch on the inside of the tip of each antenna, and both sexes have a wingspan of around 42 mm. There are, however, some differences between the two species, none of which concern the butterfly's appearance. The first is that, in Ireland, the distributions of Cryptic Wood White and Wood White do not overlap. The second is that the courtship between the two species differs since, unlike the Wood White male, the Cryptic Wood White male does not flick his wings open while waving his proboscis and antennae either side of the female's head. The third is that, based on my own observations that are discussed below, the Cryptic Wood White may also be wholly single-brooded throughout its range in Ireland, whereas the Wood White is often double-brooded.

The ♂ has white on the inside of the tips of the antennae

A ♀ feeding on Bush Vetch

DISTRIBUTION

This species is only found in Ireland on our shores, where it is widespread, although it is not found in the Burren of County Clare and adjacent areas of limestone pavement in the west of Ireland, where the Wood White is found. The precise distribution of the Cryptic Wood White is an evolving picture as specimens across Europe and elsewhere are examined. Bernard Watts provides a very useful summary of current research into the *Leptidea* group in his privately published *European Butterflies: a Portrait in*

Photographs, where he describes the Cryptic Wood White distribution as "*found across Europe north of the Pyrenees and Central Alps*".

In the paper 'Demographics and spatial ecology in a population of Cryptic Wood White butterfly *Leptidea juvernica* in Northern Ireland', published in *Journal of Insect Conservation* in 2018, James O'Neill and Ian Montgomery shared their findings from a study undertaken at a site in Edenderry, Portadown, County Armagh, Northern Ireland. Based on their observations of the butterfly overcoming physical barriers, such as tall scrub and trees, as well as their analysis of the butterfly's mobility, they conclude that the butterfly exists within a metapopulation structure with interchange between colonies, and has the potential to colonise new areas.

HABITAT

This butterfly is found in more open habitats with lusher vegetation than the Wood White, and includes mixed scrub, hedgerows, flower-rich grassland, disused quarries, sand dunes, bog edges, road verges, brownfield sites and disused railway lines. Like the Wood White, the butterfly seems to prefer sites that are sheltered, and where there is an abundance of nectar sources and larval foodplants.

Scrubby grassland at Craigavon Lakes in County Armagh, Northern Ireland

STATUS

The recent separation of this species from the Wood White has led to difficulties in interpreting its true status. A study in the 1960s by Henry Heal showed that the butterfly was using the disused railway lines of Counties Armagh and Down in Northern Ireland to fuel a northward expansion, although a more recent analysis shows that the butterfly has suffered declines in some areas. The recent trend over the last decade shows a moderate decline in distribution, although there are no figures for long-term trends or relative abundance. Using the IUCN criteria, the Cryptic Wood White is categorised as Least Concern (LC) in Ireland.

LIFE CYCLE

Adults fly throughout May and June, with individuals often seen at the end of April and in early July. The voltinism (number of broods) of the Cryptic Wood White in Ireland has been the subject of much debate. While it would be logical to think that the butterfly would produce a second brood on occasion in Ireland, especially since it is double-brooded in parts of the continent, no concrete evidence has been forthcoming. As it happens, the Wood White population found in the Burren in Ireland regularly produces a second brood.

In 2018 I had the fortune of rearing, from eggs, both the Cryptic Wood White (from County Kildare) and Wood White (from Surrey). All larvae were fed on Meadow Vetchling and thrived, with no losses. What is most surprising about this experiment is that, in the hot summer of 2018, none of the Cryptic Wood White went into a second brood, whereas all of the Wood White did. While this study cannot be considered conclusive, it would suggest that the voltinism of the Cryptic Wood White is genetically controlled and that the population found in Ireland is single brooded.

Imago

The first thing that strikes enthusiasts that are familiar with the Wood White in Britain, is that the Cryptic Wood White flies in much more open areas, and away from woodland. This conundrum puzzled entomologists for decades until, of course, the Cryptic Wood White was identified as a different species, albeit one that was identical to the Wood White in appearance. The situation in Ireland is a little more complex, however, since the Wood White is not found in woodland, but flies in Hazel scrub growing among the limestone pavement of the Burren and, at the edge of its range, can occasionally be found flying in the lusher habitat that is more typical of that used by the Cryptic Wood White.

O'Neill and Montgomery found that males had an affinity for the linear features of sheltered habitat, such as hedgerows and woodland edges, whereas females were more frequently sighted in more open areas that contain the short sward preferred by females looking for egg-laying sites, allowing males to locate them more easily and for females to spend less time looking for somewhere to lay their eggs once mated. O'Neill and Montgomery also confirmed my own observations that long vegetation in sheltered areas are important for roosting—whenever I visit Craigavon Lakes in Northern Ireland, the first sightings of the day are almost always of adults emerging from tall and lush grasses where they will have spent the night.

Unsurprisingly, the Cryptic Wood White has much in common with the Wood White—males emerge earlier than females and both sexes feed avidly on nectar. In his MSc thesis 'The ecology and conservation of *Leptidea reali* (Real's Wood White) in Northern Ireland', published in 2008, Neal Warnock (who was actually studying the Cryptic Wood White) showed that the butterfly has a particular fondness for the purple flowers of Bush Vetch, an observation confirmed by O'Neill and Montgomery. Other nectar sources observed include Common Bird's-foot-trefoil, Common Vetch, Germander Speedwell, Hoary Willowherb, Meadow Buttercup, Meadow Vetchling, Ragged-Robin, Red Clover and Tufted Vetch. In the privately published *Discovering Irish Butterflies and their Habitats*, Jesmond Harding also mentions that, in hot weather, males will feed from damp ground to absorb dissolved minerals—an activity known as 'puddling'.

Even the courtship of the Cryptic Wood White is remarkably similar to that of the Wood White, with some subtle differences. Like the Wood White, the male faces the female and waves his fully extended proboscis and antennae either side of the female's head. However, unlike the male Wood White, the male Cryptic Wood White does not flash open his wings at all while undergoing this ritual, which is a much more protracted affair in the Cryptic Wood White. In 'Courtship Behaviour in Relation to Female Monogamy in *Leptidea sinapis*', published in *Oikos* in 1977, Christer Wiklund (who was studying the Cryptic Wood White, despite the title of his paper) documents an example of a

A courting couple with ♂ on the right

A mating pair with ♂ on the left

male that courted for an incredible 35 minutes before giving up on an unreceptive female! As in the Wood White, a successful pairing is a rapid affair involving minimal courtship, and an event that few observers have witnessed, with the pair remaining coupled for up to an hour.

So why such an elaborate courtship? In Europe there are five species of *Leptidea*—Wood White *Leptidea sinapis*, Cryptic Wood White *Leptidea juvernica*, Eastern Wood White *Leptidea duponchelli*, Fenton's Wood White *Leptidea morsei* and Réal's Wood White *Leptidea reali*, with further *Leptidea* species found in western Asia. These species are not only similar in appearance, but also have overlapping distributions to some degree. It is thought that the elaborate courtship exhibited by these species is necessary to ensure that pairings only occur between males and females of the same species, despite there being differences in their genitalia that would make it physically impossible for certain species to interbreed.

Studies show that it is the female that determines whether or not the pair mate. In the paper 'Female mate choice determines reproductive isolation between sympatric butterflies', published in *Behavioral Ecology and Sociobiology* in 2008, Magne Friberg, Namphung Vongvanich, Anna-Karin Borg-Karlson, Darrell Kemp, Sami Merilaita and Christer Wiklund analysed all aspects of male-female interaction during courtship, including the structure and colour of the wings, the structure and colour of the antennae and the behaviour of both male and female. The one characteristic that stood out as being unique between the two species studied—the Wood White and Cryptic Wood White—was the combination of chemicals (pheromones) given off by the male during courtship, and it is this that distinguishes a male Wood White from a male Cryptic Wood White. However, their conclusion was that the situation is more complex than this and that a combination of signals may come into play. Interestingly, in *The Butterflies of Britain and Ireland*, Jeremy Thomas writes: *"In Sweden, where our two species overlap on some sites, the male of neither wood white seems able to distinguish between his own female and that of the sibling species. Each will court the other's females with the same persistence as his own"*. Ongoing research into the *Leptidea* will hopefully unlock the remaining secrets of this mysterious group.

A ♀ egg-laying on Meadow Vetchling

Females fly slowly over vegetation when looking to lay and will land on a plant growing in a sheltered area and use chemoreceptors on their feet to test it for suitability. Various legumes are used as larval foodplants and include Common Bird's-foot-trefoil, Greater Bird's-foot-trefoil, Meadow Vetchling and Tufted Vetch. All eggs I have ever seen laid were on plants growing in full sun and with a good amount of surrounding shelter, such as scrub or a hedgerow. Eggs are laid near the top of the foodplant and on the underside of a leaf, with the female typically moving off to nectar or rest before she repeats the process. Eggs are almost always laid singly, although Warnock observed an instance of a female laying three eggs together on the same leaf.

Ovum

All of the stages are similar in appearance to those of the Wood White, and the egg is no exception. Each egg is of the skittle shape that is typical of the Pieridae family, 1.3 mm high, and has 11 ribs that run along its sides. The egg is pale yellow, but darkens slightly just before the larva emerges.

Larva

The egg is laid on the underside of a leaf of the foodplant

Each larval instar is identical in appearance to the corresponding instar in the Wood White and my experience of rearing the butterfly in captivity is that the larva goes through four instars in total. However, the larva may go through five instars on occasion (Jesmond Harding, pers. comm.), although the triggers that determine the number of instars are unknown.

The **first instar** larva emerges after between one and three weeks, is 1.5 mm long when it first emerges from the egg and eats the eggshell as its first meal. It has a pale brown head and initially has a pale yellow ground colour, although the larva develops a greenish hue as it feeds. There are six rows of T-shaped white hairs running along the length of the body, with longer hairs on the first and last segments that project forwards and backwards respectively. The first instar rests and feeds while outstretched along the edge of a leaf and, when fully grown, is 3 mm long.

The rate of development of the larva is very much driven by temperature and, in as little as just under a week, the larva moults into the **second instar** and now has a yellowish-green ground colour, with a dark green line running along the middle of its back. The most obvious

1st instar: the larva has a pale yellow ground colour

2nd instar: the larva has a dark line along its back

difference with the first instar is the presence of a pale white line that runs along the base of each side of the body. The head is also yellowish-green that distinguishes this instar from the first instar. The larva feeds on the edge of the leaf but may move onto a stem to moult and, when fully grown, this instar is 6 mm long.

After another week or so, the larva moults into the **third instar** and the white line found along the base of each side is now slightly bolder. Apart from its size, it is similar in all other respects to the second instar and, when fully grown, the third instar is 13 mm long.

3rd instar: the larva has bolder white lines along each side

4th instar: the white line along each side now has a yellowish tinge

4th instar: a pre-pupation larva

The larva moults into the **fourth instar** after a week or so. The white line found along each side now has a yellowish tinge, although the larva is similar in all other respects to the third instar and after a week or two, when fully grown in this instar, is 19 mm long. The larva will wander off the plant in order to find a suitable location to pupate, and positions itself upright on its chosen surface with a silk girdle and a silk pad at its tail end.

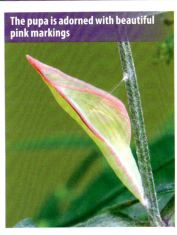

The pupa is adorned with beautiful pink markings

Pupa

The pupa is attached by the silk girdle and by the cremaster, which hooks into the silk pad created by the larva. The pupa is primarily green, with pink markings that vary in intensity along its edges and the veins of the wing cases. The resulting colouring is quite beautiful, and the butterfly remains as a pupa over the winter, with this dainty butterfly emerging the following spring.

Orange-tip
Anthocharis cardamines

❖ ORANGE-TIP IS SUPPORTED BY RICHARD LEWINGTON, NICK MORGAN & MARION WILLIAMS ❖

The Orange-tip is a true sign of spring since it is one of the first species to emerge from its overwintering pupa, signalling a definitive end to the winter as temperatures steadily increase. This is also one of our most popular butterflies, thanks to the beautiful contrast of the orange tips of the male against its otherwise white upperside. The female, on the other hand, is much less flamboyant since she lacks these orange tips and is often misidentified as a Green-veined White or Small White which often fly in the same habitats. On extremely rare occasions, nature may throw out a 'bilateral gynandromorph' where an individual has one side male and one side female, and this is never more striking than in the Orange-tip.

This difference between the two sexes is so striking that early authors gave each sex its own name. In his *Musei Petiveriani* of 1699, James Petiver named the male 'The white marbled male Butterfly' and the female 'The white marbled female Butterfly', which both acknowledge the delicate marbling of the hindwing uppersides. The butterfly's name that is used today is an obvious reference to the male, but an earlier name of 'The Lady of the Woods', given to this butterfly by Benjamin Wilkes in his *Twelve New Designs of English Butterflies* of 1742, is an equally appropriate name given that this butterfly is often found flying in woodland rides and glades.

It is thought that the orange tips of the male are an example of warning colouration, since the butterfly is not particularly palatable to predators thanks to the mustard oils in its body which accumulated as it fed on its foodplant as a larva. The female is not as active as the male, and this may explain why she does not exhibit the same colouring, preferring to remain relatively inconspicuous, despite being equally distasteful to birds.

The ♂ has a wingspan of between 40 and 53 mm

The ♀ has a similar wingspan to the ♂

A ♂, superbly camouflaged on a Cuckooflower flowerhead

The green found on the underside is an illusion created by yellow and black scales

Both sexes have an amazing underside pattern of green blotches formed by a combination of yellow and black scales and, when at rest on a flowerhead of Garlic Mustard, Cuckooflower or Cow Parsley, this butterfly is so well camouflaged that an adult resting just a few feet away is easily overlooked, even by an experienced observer. However, there is a plus side to knowing that the butterfly often settles with wings closed on these plants during overcast weather and when roosting—simply look carefully at each flowerhead and you'll be surprised at how often an Orange-tip is found.

There are two named subspecies of Orange-tip in Britain and Ireland. The subspecies *britannica* is found throughout England, Scotland and Wales and the subspecies *hibernica* is found throughout Ireland. The *britannica* subspecies was assigned by the Italian entomologist and physician Roger Verity in 1908 in his *Rhopalocera Palaearctica*, although its validity is questionable since it was described based on a specimen with unusually prominent black markings at the apex of the forewings. The case for *hibernica* as a subspecies is more solid, although the distinguishing characteristics are subtle at best and are related to the black markings on the upperside fringes (more heavily marked), the forewing undersides of the male (tinged with yellow) and the hindwing uppersides of the female (strongly tinged with yellow).

DISTRIBUTION

This butterfly is found throughout England, Wales and Ireland, but is somewhat-local further north and especially in Scotland, although the butterfly has spread north rapidly in the last few decades. This butterfly does not form discrete colonies but lives in loose, open populations where it wanders in every direction as it flies along hedgerows and woodland margins looking for a mate, nectar sources or foodplants. On a worldwide basis, the butterfly's distribution extends through Europe and Asia to Japan.

HABITAT

The Orange-tip uses a wide range of habitats that are often damp and includes woodland clearings, rides and glades; hedgerows; meadows and pasture; riverbanks, canals, ditches and fens; arable field margins and set-aside; road verges; and parks and churchyards. The species will also turn up in gardens as anyone that grows Garlic Mustard or Cuckooflower will know.

The white flowerheads of Garlic Mustard [inset PC] **are easy to spot**

STATUS

This is one of the few butterflies whose distribution and abundance are relatively stable or increasing and, as such, the Orange-tip is not a species of conservation concern. The long-term trend shows a stable distribution and a moderate increase in abundance, with the more recent trend over the last decade showing a stable distribution and a significant increase in abundance. Using the IUCN criteria, the Orange-tip is categorised as Least Concern (LC) in both the UK and Ireland.

LIFE CYCLE

The Orange-tip has a single brood each year, with adults flying from the beginning of April, through May and into June. More northern populations emerge later, appearing at the beginning of May in Scotland. In exceptionally early years a small number of adults may appear later in the year in late July, August and September, as was the case as recently as 2018, but this is far from the norm and could represent individuals whose development has been retarded for some reason, a suggestion that is made more plausible given records of individuals in captivity that have remained in the pupal state for two or even three years.

Imago

Early morning is a good time to watch this butterfly as it feeds from a favourite nectar source such as Bluebell, brambles, Bugle, Cuckooflower, dandelions, Garlic Mustard, Greater Stitchwort, Ground-ivy, hawkweeds, Ragged-Robin, Red Campion and vetches. A refuelled male will fly for long periods without stopping to rest or feed as it searches out a mate, much to the chagrin of photographers, although both sexes will drop down to roost on vegetation or a flowerhead as soon as clouds come over.

When a patrolling male encounters another male, the pair will briefly tumble around one another before going their separate ways. When a male locates an already mated female (or rather, one that is not willing to mate, since a female may mate more than once) then she signals her rejection by raising her abdomen, although this posture is also exhibited briefly by virgin females—in the paper 'Courtship and male discrimination between virgin and mated females in the orange tip butterfly *Anthocharis cardamines*', published in *Animal Behaviour* in 1985, Christer Wiklund and Johan Forsberg showed that a receptive virgin female detains a courting male by raising her abdomen for around four seconds before allowing him to mate with her, suggesting that she is determining the suitability of the male by testing his persistence.

The female is most often encountered in the vicinity of foodplants when she is looking to lay her eggs. Preferred plants are those that are relatively large, isolated, unshaded and sheltered, and the female will initially locate a plant by sight before she alights and tastes it with her feet. If the plant is suitable, a single egg is laid on a flower stalk. Eggs are laid singly for good reason—the

A ♀ raising her abdomen often signals rejection, but not always [NF]

A mating pair with the ♂ above

larvae are cannibalistic in their early instars, which ensures that larvae do not outstrip their food supply and perish. In order to reduce losses due to cannibalism, it is thought that the female is able to avoid laying on a plant that already contains eggs by detecting them by both sight (since eggs develop a conspicuous orange colour, which also makes them very easy to find in the field) and smell (since the female leaves pheromones on eggs after they have been laid). We could therefore conclude that most plants would only ever contain a single egg or just a few at the most but, in practice, this is not always the case, especially when foodplants are in short supply. I remember coming across an isolated Cuckooflower plant with 21 eggs on it that were at different stages of development, including some that were newly laid.

Eggs are laid on several species of crucifer, especially Cuckooflower in damp meadows and Garlic Mustard along road verges and ditches, although the butterfly will also use Hedge Mustard, Winter-cress, Turnip, Charlock, Large Bitter-cress and Hairy Rock-cress. It will also lay its eggs on Honesty and Dame's-violet in gardens, but larval survival is thought to be poor on these plants. Cuckooflower is the primary foodplant used in Ireland, simply because Garlic Mustard is much less common.

The egg is greenish-white when first laid

The egg gradually turns bright orange

Ovum
The bottle-shaped, 1.2-mm high eggs are a greenish-white when first laid, but gradually turn orange, before ultimately turning dark brown as the larva develops. They are also extremely easy to find, tucked away on a flower stalk of the foodplant, rarely on a leaf, and are worth examining with a decent hand lens since they are quite beautiful when seen close up, with 18 ribs running from top to bottom. Eggs are sometimes found on the same plants used by the Green-veined White, such as Garlic Mustard, although the two species are not in competition since the Green-veined White larva eats the most tender leaves of the plant, whereas the Orange-tip larva feeds primarily on the developing seed pods.

Larva
After a week or so, the 1.5-mm long yellow-brown **first instar** larva emerges from the egg and eats the shell before moving to feed on the developing seed pod, a diet that will act as its main food source as it grows, although the larva will also eat flowers and leaves once the seed pods have been eaten. The larva has several rows of black warts along the length of its body and each wart bears a long black hair that is adorned with a bead of clear liquid, which is a characteristic also found in later instars. Given its cannibalistic tendencies, the larva will eat any other Orange-tip eggs it encounters as well as other larvae too unless, of course, it is on the receiving end. In *The Butterflies of Britain and Ireland*, Jeremy Thomas estimates that cannibalism accounts for 10% of mortality.

The larva feeds by both day and night in all instars and develops surprisingly rapidly, as anyone who has reared this species in captivity, or who has the good fortune of having this species in their garden, will attest. When fully grown, the first instar is just over 3.5 mm long.

1st instar: a newly emerged larva eating its eggshell

1st instar: the larva has long black hairs that are topped with a bead of clear liquid

2nd instar: a larva feeding on the end of a seed pod [VM]

After only a few days, the larva moults into the **second instar** and is now a more greenish-grey in colour with a pale stripe along each side. The head is a dull olive that contrasts, to some degree, with the colour of the body. When fully grown, the second instar is 6.5 mm long.

After a few days, the larva moults into the **third instar**. The body is now grey-green and remains covered in warts that each bear a hair, and the larva is superbly camouflaged when resting lengthwise along a developing seed pod. The head is now a similar colour to the body, allowing a third instar to be distinguished from a second instar. When fully grown, the third instar is around 10.5 mm long.

After a few days, the larva moults into the **fourth instar**. The larva is now a bluish-green, rather than the grey-green of the previous instar. Another distinguishing feature is that the larva now has clear green mid-abdominal prolegs (the

3rd instar: the larva is well-camouflaged as it rests [VM]

4th instar: a fully grown larva at rest on a seed pod

'claspers'), that lack the dusky black tips found in earlier instars. When fully grown, the fourth instar is 17.5 mm long.

After six or seven days, the larva moults into the final and **fifth instar**. The larva now has a decidedly bluish tinge as the green on its back fades into the bright white stripe that runs along the length of each side. When fully grown, the larva is just over 30 mm long. The differences between a fifth and fourth instar are subtle at best, although the fifth instar has less prominent warts on its body and the accompanying hairs are also less discernible. The last resort when distinguishing fourth and fifth instars is, of course, size—when fully grown, the final instar is over 10 mm longer than the fourth instar.

After nine days or so, the fully grown larva will travel extensively in search of a suitable pupation site (F.W. Frohawk records one larva roaming for 30 hours in captivity). Once a site is chosen, the larva will attach itself head upwards to its chosen surface by a silk girdle around its waist and a silk pad at its tail end.

5th instar: a mature larva will occasionally exhibit this curious posture

5th instar: a close up of the head

5th instar: a larva preparing to pupate

Pupa

The 23-mm long pupa is green when first formed, although many turn light brown to more-closely match their surroundings. As well as having two colour forms, another interesting characteristic of the pupa is that, unlike the pupae of most other species, its abdomen does not flex to any great degree. Also, given that the colour and appearance of the pupa perfectly mimics a dead seed pod, it would be logical to assume that most pupae are formed on the foodplant. However, this does not appear to be the case, which may be because the stems of the most widely used foodplants, namely Garlic Mustard and Cuckooflower, often perish over the winter months. This delicate and delightful butterfly takes to the wing the following spring, after 10 months in the pupal stage.

The green form of pupa

The brown form of pupa

A ♂ expanding its wings after emerging from its pupa

Life Cycles of British & Irish Butterflies

Large White
Pieris brassicae

❖ LARGE WHITE IS SUPPORTED BY GRAHAME & LESLEY HAWKER, KEVIN PENNINGTON & JUDITH WELLS ❖

	January	February	March	April	May	June	July	August	September	October	November	December
IMAGO												
OVUM												
LARVA												
PUPA												

86 *Life Cycles of British & Irish Butterflies*

While both the **Large White** and Small White are known colloquially as 'cabbage whites', the crown must surely go to the Large White whose larvae can reach pest proportions, decimating cabbages and other members of the Brassicaceae family to the point that they become mere skeletons of their former selves. It is unfortunate that this familiar visitor to gardens is met with disdain by those growing vegetables or garden plants such as Nasturtiums, another favourite larval foodplant, for the butterfly brightens up a spring or summer's day wherever it flies. That being said, even early entomologists did not hold this species in high regard. In his *The Larvae of the British Butterflies and Moths* of 1886, William Buckler writes "*The sight of the caterpillars of this species feeding on cabbages, and their unpleasant odour when plucked off and crushed under foot, are among my very earliest recollections; and I suppose the disgust which they inspired has ever since kept me from caring to know much about them*".

All of the early vernacular names refer to the butterfly's size, colour or fondness for cabbages, and the name that we use today was coined by Adrian Haworth in his *Lepidoptera Britannica* of 1803, with James Rennie surely having a lapse in creativity when he named this species, quite simply, 'The Cabbage' in his *A Conspectus of the Butterflies and Moths found in Britain* of 1832. The specific name of *brassicae* refers to the genus *Brassica*, to which cabbages belong.

The female is distinguished from the male by the presence of two black spots on the forewing upperside, together with a black streak along the inner margin. The two sexes are otherwise similar in appearance. This species also exhibits an element of 'seasonal dimorphism'—spring brood adults have less pronounced markings than those of the summer brood.

♂ and ♀ undersides are similar

The ♀ (left) has a wingspan of 63 mm and the ♂ 58 mm [VM]

DISTRIBUTION

This is one of our most widespread species and is found across Britain and Ireland, including Orkney and Shetland, although it is absent from most mountainous regions. On a worldwide basis, the butterfly is distributed throughout the Palearctic and is found across most of Europe, through western Asia as far as the Himalayas, and in parts of North Africa.

This species has a powerful flight, travelling several hundred kilometres on occasion, and it is known to migrate to Britain each year from the continent, sometimes in large numbers, augmenting the resident population in the process. Large Whites have also been seen flying south in the autumn, suggesting that a reverse migration takes place.

Life Cycles of British & Irish Butterflies

There are many stories in entomological literature of these migrations and, in his classic work *Insect Migration* (volume 36 in The New Naturalist series, published in 1958), C.B. Williams recalls one particular migration in early August 1911 that provides an indication of the numbers involved: *"Prof. Oliver … was visiting a small island of about two acres in extent in Sutton Broad, a lake in Norfolk about four miles inland from the coast. As he approached, the whole island was seen to be covered with fluttering white butterflies all of which were caught on the sticky leaves of our insectivorous plant, the sundew. Each small plant had captured four to seven butterflies, mostly still alive, and several counts gave an average of seventy per square foot – or a total of about six million butterflies for the island! There is no doubt that a mass immigration had arrived from over the sea earlier in the day, and had chosen a gigantic fly trap on which to settle".*

HABITAT
This species is found in a wide variety of habitats and can turn up anywhere, including gardens, parks, churchyards; allotments and market gardens; arable fields and set-aside; hedgerows; and coastal vegetated shingle.

Allotments are a major attraction to the Large White

The Large White uses Sea-kale at coastal sites [PC]

STATUS
This widespread and common butterfly is not currently a species of conservation concern with both the long-term and short-term trends showing a stable distribution and a moderate decline in abundance. Using the IUCN criteria, the Large White is categorised as Least Concern (LC) in both the UK and Ireland.

LIFE CYCLE
This butterfly has two broods each year. The first brood emerges in April and peaks in May, and the second brood emerges in July and flies throughout August and into early September. In good years, a third brood emerges at the end of September whose offspring have a protracted development in the cooler weather, with larvae found as late as December in some years—there are even records of larvae overwintering and going on to pupate the following year. The second brood is always larger than the first, often by a considerable margin, and no doubt benefits from an influx from the continent. Immigration, the prevailing weather and the impact of parasitism cause the abundance of this familiar butterfly to fluctuate, sometimes dramatically, from year to year.

Imago
The adult butterfly is a familiar sight across Britain and Ireland and, in good weather, is one of the first species to take to the wing during the day. In overcast conditions, the butterfly can be found with its wings held half open as it directs any light onto its dark body in order to warm itself up so that it is able to fly. Adults are avid nectar feeders and will use a wide variety of flowers including those of Bluebell, Bugle and dandelions in the spring, and thistles, Buddleja, Common Fleabane, Devil's-bit

Scabious, Field Scabious, Hemp-agrimony, knapweeds, Ragged-Robin, ragworts, Wild Marjoram and Wild Teasel in the summer. The butterfly spends a good amount of time looking for a suitable site to roost for the night, eventually settling on a plant that disguises their presence, including those with white flowers, such as White Dead-nettle, and those with pale and variegated foliage.

Males patrol in search of females and are attracted to movement as well as the ultraviolet light that is reflected by the wings of the female. When a potential mate is found, the male will land beside her. A receptive female will keep her wings closed while the male curls his abdomen in order to mate, the pair remaining coupled for around two hours, or longer in cooler conditions. If an already mated female is approached then she will adopt the typical rejection posture found in many of the Pieridae by opening her wings and raising her abdomen, preventing the male from mating with her, and he eventually loses interest and flies off. In their 2003 article 'Antiaphrodisiacs in Pierid Butterflies: A Theme with Variation!', published in *Journal of Chemical Ecology*, Johan Andersson, Anna-Karin Borg-Karlson and Christer Wiklund show that the male Large White passes an anti-aphrodisiac to the female during mating in order to make the female less attractive to other males, and a male may therefore be dissuaded from pairing if she gives off a scent that indicates that she has already paired.

When egg-laying, the female will flutter over the foodplant before alighting on a chosen leaf that she tastes with her feet and, if it is suitable, will close her wings and commence laying at a rate of four eggs per minute, moving her abdomen away from the leaf surface as each egg is laid and before the next is ready to be deposited. Eggs are laid on either surface of a leaf, typically in groups of between 25 and 150, and a female may lay up to 600 eggs in total. Each egg usually abuts its neighbours, resulting in a highly organised yellow egg batch that is very conspicuous against the green of the foodplant. Eggs are laid on wild or cultivated species of the Brassicaceae family, with a strong preference for cultivated varieties of *Brassica oleracea*, such as Cabbage and Brussels-sprout, and varieties of *Brassica napus*, such as Oil-seed Rape. Nasturtium and Wild Mignonette are also used, as is Sea-kale along the coast.

Given the amount of damage that Large Whites can cause, it should come as no surprise that females will avoid plants that are already occupied by another female's offspring. Existing egg batches are detected by both sight and smell—the female leaves a chemical signal after laying her batch of eggs, known as an 'oviposition-deterring pheromone', or ODP. In their paper 'Two related butterfly species avoid oviposition near each other's eggs', published in *Experientia* in 1990, L.M. Schoonhoven, E.A.M. Beerling, J.W. Klijnstra and Y.v. Vugt showed that not only are other Large Whites deterred from laying in close proximity of an existing egg batch, but also Small Whites, and vice-versa. Larvae can also be detected by smell—in his *The Butterflies of the British Isles* of 1906, Richard South writes: "A peculiarity of these caterpillars is that even when not numerous, their presence is indicated by an evil smell that proceeds from them".

Despite all of these clues, there have been numerous occasions when I have found several egg batches not only on the same plant, but on the same leaf. The resulting larvae emerge from each batch at different times, and this might explain why the larvae found in some groups are clearly in very different instars. In the most authoritative tome on this species, *Large White Butterfly – the biology, biochemistry and physiology of* Pieris brassicae *(Linnaeus)*, John Feltwell also suggests that the colour of the eggs may indicate their distastefulness to birds.

The typical rejection posture of the ♀ [NF]

Mating pair with ♂ on the right [VM]

Ovum

The yellow, skittle-shaped egg is 1.2 mm high and has 17 to 19 ribs that run along its side, although not all run the full length of the egg. A day or so before hatching, the egg darkens, with the head of the larva showing as a dark area at the crown.

An egg batch laid on the underside of a cabbage leaf

Larva

The 2-mm long **first instar** larva emerges from its egg after a week or two, depending on temperature, and eats most of its eggshell, leaving only the base. The larva is cylindrical with a black head and a lemon-yellow ground colour, with white hairs that protrude from six rows of black and yellow warts. The larvae spin a thin layer of silk on which they start to feed gregariously alongside one another and remain in the company of their siblings until fully grown. The larvae initially feed only on the leaf cuticle, but soon perforate the leaf when conspicuous holes appear. Unlike the larvae of their close cousin, the Small White, Large White larvae feed on the outer, rather than inner, leaves of the foodplant. Even at this young age, the larvae exhibit aposematic colouring, signalling that they are not palatable to birds. This is no bluff, since the larvae accumulate poisonous mustard oils in their bodies as they feed, and any bird that does not get the message is in for a nasty surprise.

After nine days or so, the fully grown 3.5-mm long first instar larvae moult into the **second instar**. The ground colour is now a greenish-yellow and the larva has a yellow line that runs along its back that separates the numerous black warts found on each side of the body. Each wart bears a hair, although these vary in length. The second instar larva feeds more substantially on the leaves of the foodplant, creating significant holes, and, when fully grown, is around 6.5 mm long.

Larvae can develop very quickly in good weather, and in just under a week, moult into the **third instar**. The larva is similar in appearance to the second instar, although the black markings all over its body are more significant and this results in a more clearly defined yellow line along its back. The fully grown third instar larva is 8.5 mm long.

After another five days or so, the larva moults into the **fourth instar**. It is now so similar in appearance to the third instar that the only distinguishing feature is size—when fully grown, the fourth instar is 19 mm long and twice the length of a third instar.

After another five days, the larva moults for the last time into the **fifth instar**. The larva is similar in appearance to the previous instar but is much larger, growing up to 42 mm in length, and the head

1st instar: larvae emerging from their eggs

1st instar: fully grown larvae preparing to moult [VM]

2nd instar: larvae showing a clear yellow stripe along their backs

3rd instar: larvae continue to feed side-by-side

4th instar: larvae are similar in appearance to third instar larvae [VM]

5th instar: fully grown larvae are 42 mm long

is now comprised of two grey lobes with a whitish plate at the front (known as the 'clypeus') that turns yellow as the larva matures (Vince Massimo, pers. comm.). As in earlier instars, the seemingly random placement of black blotches on each of the segments (each of which is divided into six 'sub-segments') is actually a repeated pattern. Behaviourally, larvae no longer spin silk on the leaves and no longer feed in groups, appearing less gregarious.

After wandering for some time, the fully grown larva finds a suitable pupation site that is typically away from the foodplant and that affords some protection from the weather, such as a fence, tree trunk, or an overhang on a building, such as its eaves. It then holds itself in its pre-pupation pose, often under a horizontal surface, with a silk pad at its tail end that is gripped by the claspers, and a silk girdle around its waist.

Large White larvae are particularly vulnerable to the parasitic wasp *Cotesia glomerata*, which deposits between 15 and 80 eggs inside each young larva that it parasitises. The wasp larvae feed on the layer of fat beneath the skin of their host, thereby avoiding vital organs, and, when their host is fully grown, break through the skin and pupate within yellow cocoons on or near their host, with the wasps emerging after a week or so. In some years, the wasp can have a dramatic effect on the number of Large White larvae that reach adulthood.

5th instar: the markings on the middle segments exhibit a repeating pattern

Budding entomologist Edward Eeles, helping his grandad find larvae

5th instar: the two lobes of the head are now grey [VM]

It is clearly important for the wasp to locate a young larva if its offspring are to successfully develop. As well as locating potential hosts based on the chemical signals given off by the plant following feeding damage, the wasp is also able to distinguish between the first and fifth instars, as discovered by Letizia Mattiacci and Marcel Dicke in their aptly titled paper 'The parasitoid *Cotesia glomerata* discriminates between first and fifth larval instars of its host *Pieris brassicae*, on the basis of contact cues from frass, silk, and herbivore-damaged leaf tissue', that was published in *Journal of Insect Behavior* in 1995.

5th instar: a larva ready to pupate [VM]

5th instar: the yellow cocoons of *Cotesia glomerata* [ABr]

Pupa

The larva moults for the final time two days after it has fixed itself for pupation, revealing a 25-mm long pupa that is held in position by the silk girdle and by the cremaster to the silk pad. The pupa has a 'beak' on its head and a plain ground colour that is speckled with black markings. The colour of the pupa varies considerably and is determined, to some extent, by the substrate on which it is formed, with 'dull white' seemingly the most common ground colour. This stage lasts around two weeks for pupae that go on to produce another brood in the same year, and around eight months for pupae which overwinter and that produce the spring brood.

Dull white form of pupa

Pale green form of pupa [VM]

Small White
Pieris rapae

❖ SMALL WHITE IS SUPPORTED BY SUE COLLINS, BROOKE NICOLSON & STUART WOODLEY ❖

It is unfortunate that the Small White, arguably the less destructive of our two 'cabbage whites', is considered 'guilty by association', since this butterfly is one of the first to grab the attention of children all over Britain and Ireland and is often their first introduction to the wonders of the butterfly world. While the feeding damage of the Large White can decimate cabbages and other members of the Brassicaceae, the Small White lays its eggs singly and, unless foodplants are in short supply when many eggs may be laid on the same plant, the damage is usually less devastating. That being said, this butterfly can be a serious pest on occasion.

The butterfly exhibits an amount of both sexual and seasonal dimorphism—the male has a single spot on the forewing whereas the female has two (and a mark that runs along the inner margin), and adults of the spring brood have lighter markings than those of the summer brood—although all adults have a wingspan of 38 to 57 mm, irrespective of the sex and brood. The differences between sexes and seasons resulted in James Petiver giving this butterfly four different names in his *Papilionum Britanniae* of 1717—the spring brood male and female were named the 'Lesser white unspotted Butterfly' and the 'Lesser white double spotted Butterfly' respectively, and the summer brood male and female were named the 'Lesser white Cabbage Butterfly' and the 'Lesser white treble spotted Butterfly'. The name that we use today was given to the butterfly by Adrian Haworth in his *Lepidoptera Britannica* of 1803 and James Rennie was being rather specific when he gave this butterfly the name 'The Turnip' in his *Conspectus of Butterflies and Moths* of 1832, and the butterfly's specific name, *rapae*, is also a reference to the Turnip, *Brassica rapa* subsp. *rapa*, one of the butterfly's many larval foodplants.

Both ♂ and ♀ have plain undersides and a wingspan of 38 to 57 mm

♂ spring brood [NF]

♀ spring brood [VM]

♂ summer brood [VM]

♀ summer brood [NF]

Life Cycles of British & Irish Butterflies

DISTRIBUTION
This is one of the most widespread species found in Britain and Ireland and can turn up almost anywhere and, while it is relatively scarce in northern Scotland, it has been recorded from the Outer Hebrides, Orkney and Shetland. On a worldwide basis, the butterfly is found throughout Europe and eastwards across Asia to China and Japan. The butterfly was accidentally introduced to Quebec in 1860, possibly as diapausing pupae, and has since spread across most of North America. The butterfly was also accidentally introduced to New Zealand in 1930 and Australia in 1939 where the butterfly has become a serious pest.

This species is also known to migrate to our shores from the continent, sometimes in great swarms, where it augments the resident population. An extraordinary migration was witnessed on 5 July 1946 and reported in *The Canterbury Journal, and Farmers' Gazette* on 11 July: "*Such was the density and extent of the cloud formed by the living mass, that it completely obscured the sun from the people on board the continental steamers on their passage, for many hundreds of yards, while the insects strewed the decks in all directions. The flight reached England about twelve o'clock at noon, and dispersed themselves inland and along shore, darkening the air as they went. During the sea passage of the butterflies the weather was calm and sunny, with scarce a puff of wind stirring, but an hour or so after they reached terra firma, it came on to blow great guns from the S.W., the direction whence the insects came*". It is also thought that some individuals attempt a reverse migration in late summer.

Evidence of the butterfly's mobility comes from the accidental introduction to Melbourne, Australia in 1939. Three years after the introduction, the species had reached the west coast of Australia some 3,000 km away in only 25 generations. While this could be put down to immature stages being transported, it may also suggest that adults are able to fly over 120 km in their lifetime, although most butterflies will undoubtedly only travel a kilometre or two.

HABITAT
This species is found in a wide variety of habitats and can turn up almost anywhere, including open grassland; hedgerows; arable crops; coastal vegetated shingle; gardens, parks and churchyards; allotments and market gardens. Although this butterfly is not colonial, it will congregate in areas that are rich in flowers and larval foodplants, including gardens and fields with crops such as Oil-seed Rape.

The butterfly can be found in just about every allotment in Britain and Ireland

STATUS
The long-term trend of this common and widespread butterfly shows a stable distribution and a moderate decline in abundance, with the more recent trend over the last decade showing both a stable distribution and abundance. Using the IUCN criteria, the Small White is categorised as Least Concern (LC) in both the UK and Ireland.

LIFE CYCLE

The butterfly has two broods in most years throughout its range, with three generations in the south. Overwintering pupae give rise to the adults of the first brood, which emerge from mid-April, peak around the middle of May and gradually tail off through June. The second brood starts to emerge in late June and early July that, depending on location, may give rise to a third brood that emerges in early September. Any summer broods are always stronger than the spring brood, since they are bolstered by migrants from the continent.

Imago

This highly mobile butterfly is a familiar sight in gardens where it is attracted to a wide variety of nectar sources, including Buddleja and lavenders. In the wider countryside, it is particularly fond of thistles, but will also nectar on the flowers of other plants such as Bluebell, Bugle, Common Bird's-foot-trefoil, Common Fleabane, Daisy, dandelions, Greater Stitchwort, hawkweeds, Hemp-agrimony, knapweeds, Ragged-Robin, ragworts, Red Campion, Red Clover, Sainfoin and Wild Marjoram. When roosting and during dull weather, the butterfly will settle on plants with white flowers or variegated foliage, where it is surprisingly well camouflaged.

The male will pursue any female he encounters and, when she has landed, will land alongside her, releasing pheromones from the androconia (specialised scent scales) that are scattered over his wing uppersides. An encounter with an unreceptive female results in her spreading her wings and raising her abdomen, making it physically impossible for the male to mate with her. If the female is receptive, however, then the pair mate and remain coupled for around two hours. Like the Large White male, the Small White male passes an anti-aphrodisiac to the female during mating that makes her less attractive to other males, although this also attracts the attention of the tiny *Trichogramma brassicae* wasp that hitches a ride on the female and then goes on to parasitise her eggs as they are laid. The effect of the anti-aphrodisiac clearly wears off since both sexes may mate more than once, and one study showed that females mated 2.4 times on average. The benefit to the female is not merely the transfer of sperm to fertilise her eggs, but also a 'nuptial gift' of nutrients that she absorbs and that extends her life.

A mating pair [VM]

Eggs are laid singly on the foodplant and normally on the underside of a leaf, with a preference for plants growing in warm and sheltered areas. Favoured plants often receive eggs from several females and it is not uncommon for a female to lay a single egg, fly away from the plant and return almost immediately to lay another egg, a process that she may repeat several times before eventually departing. This is somewhat surprising since, like the Large White, the Small White female leaves an 'oviposition-deterring pheromone' after she has laid that deters both Small Whites and Large Whites from laying on the same plant. Larval foodplants include both cultivated and wild species of the Brassicaceae family, especially cabbages in gardens and allotments, and plants such as Charlock, Garlic Mustard, Hedge Mustard, Hoary Cress, Oil-seed Rape, Sea-kale and Wild Mignonette in the wider countryside. Nasturtium is also used in gardens.

The egg is a very pale green when it is first laid

The egg turns a primrose-yellow after 24 hours

Ovum

The bottle-shaped egg is typical of the Pieridae family, with 12 ribs that run the length of the egg from top to bottom. The egg is a very pale green when it is first laid, but soon develops a primrose-yellow colour, ultimately turning grey just prior to hatching.

Larva

After around a week, depending on temperature, the 1.8-mm long **first instar** larva emerges from the egg, eats the remainder of the eggshell and then moves to the underside of the leaf to feed. On cabbages and other brassicas, the larva will move into the heart of the plant where it can cause considerable damage to the developing leaves. The shiny larva has a pale brown head and a pale yellowish-brown ground colour, although it becomes tinged with green as it starts to feed on the foodplant. It is covered in several rows of white tubercles that each bear a number of white hairs, many of which exude a bead of clear liquid. The larva feeds on the leaf surface rather than the leaf edge, and always rests nearby, returning to the same spot to feed. The leaf is eventually perforated, and the larva creates tell-tale holes in the leaf which increase in size as the larva grows. When fully grown, the first instar is 3 mm long.

The Small White larva's defence strategy is one of camouflage and evasion, unlike its close cousin, the Large White, whose deterrent is based on aposematic colouring (indicating that it is distasteful) and 'safety in numbers'. As well as avian predators, the Small White larva is predated by harvestmen, spiders and beetles.

1st instar: a fully grown larva

2nd instar: the larva is now covered in black hairs

2nd instar: a larva with typical feeding damage

After only a few days, the larva moults into the **second instar**, eating its cast skin as it does after every moult. The cylindrical and shiny larva has a yellowish-green ground colour and head, and remains covered in white tubercles, although these now bear black hairs, rather than the white hairs found in the first instar, and the hairs continue to exude a clear bead of liquid. The body is also covered in shorter black hairs, with white hairs at the base of each side, and the larva now has a very pale line that runs along its back. When fully grown, the second instar is 5.5 mm long.

After as little as two days, the larva moults into the **third instar** and now has a more uniform green ground colour, but is otherwise similar to the second instar. The main distinguishing feature is size—when fully grown, the third instar is 8 mm long.

After five days, the larva moults into the **fourth instar** and now has a velvety appearance and is noticeably less shiny than earlier instars. The larva retains the white tubercles and associated black hairs along its back, as well as the white hairs found on each side of the body. The larva's consistent green ground colour accentuates a subtle yellow line that runs along the centre of the back, as well as a series of yellow markings that are positioned in line with the spiracles on each side of the body. There is one yellow

3rd instar: the larva is now a more uniform green

4th instar: the larva has a velvety appearance

5th instar: the yellow markings on a Small White larva

5th instar: male larvae develop testes that may be visible as pale patches

5th instar: a pre-pupation larva [VM]

marking on the first, second and third segments, none on the last two segments, and two markings on the remaining eight segments—one before each spiracle, with a fainter yellow marking after each spiracle, when looking towards the head. Behaviourally, the day-feeding larva now rests on the upperside of the leaf along the midrib and, when fully grown, the fourth instar is around 13 mm long.

After four or five days, the larva moults for the last time into the **fifth instar**. As in the previous instar, the larva has a velvety green appearance, with a thin yellow line along its back. The yellow markings either side of each spiracle are more substantial than in the fourth instar, but have the same arrangement. Another interesting characteristic of this species is that a mature larva will often develop two pale patches on the eighth segment, which are an indication that testes are developing and that the larva is male.

In less than three weeks after emerging from the egg, having spent seven days in this instar, the larva is fully grown and around 25 mm long. The larva wanders in search of a pupation site and, while some spring brood larvae may pupate on the plant or other vegetation, summer brood larvae invariably use a more permanent surface such as a fence, tree trunk or building. Once found, the larva will select a horizontal or vertical surface before spinning a girdle around its waist and a silk pad at its tail end. The larva may even make its way into a greenhouse or conservatory, and early sightings of this species each year are almost certainly the result of a pupa that has formed in one of these sheltered and warm locations. Like the Large White, the Small White larva is attacked by several parasitoid flies and wasps, including the *Cotesia rubecula* wasp that is closely related to *Cotesia glomerata* that attacks Large White larvae.

Pupa

The 19-mm long pupa is held in place by the silk girdle and the cremaster and has several protuberances on its body at the base of the wing cases and on the thorax, and a noticeable 'beak' on the front of its head. The pupa has two main colour forms—light green and light brown—and if it goes on to produce adults in the same year then it will do so in three weeks, otherwise it will overwinter for seven months, with adults emerging the following spring.

The light green form of pupa [VM]

The light brown form of pupa

Life Cycles of British & Irish Butterflies

Green-veined White
Pieris napi

❖ GREEN-VEINED WHITE IS SUPPORTED BY DEBBIE COOPER, BILL HIGGINS & BRIAN & JACKIE MATTHEWS ❖

	January	February	March	April	May	June	July	August	September	October	November	December
IMAGO												
OVUM												
LARVA												
PUPA												

The **Green-veined White** is a common butterfly of damp grassland and can be found fluttering near hedgerows, ditches, river banks, meadows, woodland rides, parks and gardens—wherever nectar sources and lush growth of its larval foodplants abound. The butterfly's familiarity is also due to its broad distribution in both Britain and Ireland, as well as its multi-brooded nature—adults can be found flying from the start of spring to the end of autumn.

The butterfly is often mistaken for its close cousin, the Small White, but can be distinguished by the prominent veins found on its undersides, from which the butterfly gets its name, although the 'green' is, in fact, an illusion created by a subtle combination of yellow, black and grey scales. The underside can also distinguish this species from a female Orange-tip, whose undersides have a mottled appearance. Like the Small White, the butterfly exhibits both sexual and seasonal dimorphism—the male has a single spot (or no spot) on each forewing whereas the female has two, and adults of the summer brood have bolder markings than those of the spring brood, despite the female of the spring brood usually having a more extensive dusting of grey scales on her forewings. Adults have a wingspan of between 40 and 52 mm and, in both sexes, summer brood adults are generally larger than those of the spring brood. The undersides of male and female are similar, and those of the spring brood female can be particularly yellow.

♂ spring brood

♀ spring brood

♂ summer brood

♀ summer brood [VM]

♂ and ♀ have similar undersides

The spring brood ♀ may have particularly yellow undersides

Life Cycles of British & Irish Butterflies

As with the Small White, James Petiver gave this butterfly several names in his *Papilionum Britanniae* of 1717, presumably thinking that these were different species—the spring brood male was named the 'Lesser, white, veined Butterfly', the summer brood male the 'Lesser, spotted, white, veined Butterfly' and the female the 'Common white, veined Butterfly'. In his *A Natural History of English Insects* of 1720, Eleazar Albin used the name 'Green vein'd Butterfly' which ultimately led to the name we use today that was provided by William Lewin in his *The Papilios of Great Britain* of 1795. The specific name *napi* is a reference to Oil-seed Rape, *Brassica napus*, one of the larval foodplants.

This is a variable species and there are three named subspecies on our shores, although this designation is questioned by some. The subspecies found in England and Wales is known as *sabellicae*. The subspecies *thomsoni* is found in Scotland whose female is more heavily suffused with black on its uppersides, and has a darker underside. Unsurprisingly, there is no clear dividing line in the north of England and southern Scotland between the distribution of subspecies *sabellicae* and subspecies *thomsoni*. The subspecies *britannica* is found throughout Ireland and has pronounced dark markings and particularly yellow hindwing undersides.

DISTRIBUTION

This is the most widespread of our 'whites' and can be found almost everywhere, although it is absent from Shetland and areas of the Scottish Highlands. On a worldwide basis, the butterfly has a Holarctic distribution and is found across Europe and temperate Asia as far as Japan, in the Atlas Mountains of North Africa, and in North America where it is known as the Mustard White.

The butterfly flies widely across the countryside but can be found concentrated in suitable areas of habitat where nectar sources and larval foodplants grow in profusion, although it does not show colonial tendencies, except in more northern areas, where it is more sedentary. The butterfly is known to undertake a modest level of migration from the continent on occasion, but this is nowhere near the scale of either the Small or Large Whites.

A path cutting through a favoured damp area next to the River Kennet at Woolhampton in Berkshire

HABITAT

The butterfly can be found in a wide variety of locations, wherever foodplants and nectar sources abound and where conditions are warm, sheltered and relatively humid, including hedgerows; lakes, ponds, ditches, canals, rivers and streams; gardens, parks and churchyards; woodland clearings, rides and glades; meadows and pasture; road verges; upland heathland; lowland fen; and arable field margins and set-aside. This species favours damp areas, but can also be found in small sheltered pockets in otherwise dry and open habitat, such as patches of mixed scrub within a chalk grassland landscape.

STATUS

The butterfly's widespread distribution, mobility and use of a wide range of larval foodplants has dampened the negative effects of habitat loss and poor seasons and this species is not, therefore, a species of conservation concern. The long-term trend shows both a stable distribution and abundance, with the more recent trend over the last decade showing a stable distribution and a significant increase in abundance. Using the IUCN criteria, the Green-veined White is categorised as Least Concern (LC) in both the UK and Ireland.

LIFE CYCLE

First-brood adults start emerging from mid-April at southern sites, peak around the middle of May and gradually tail off through June. The second brood, which is always stronger than the first, starts to emerge in early July and, in good years, may emerge earlier in late June and give rise to a third brood that flies from early September until early October. The butterfly can therefore be found continuously at southern sites from mid-April through to early October. Adults emerge later at more northern sites and those furthest north and at high altitude produce only a single brood that emerges in early June and flies until the end of July.

The larval stage is incredibly short, lasting less than three weeks, and the butterfly always overwinters as a pupa. In areas where this species is multi-brooded, the offspring may differ somewhat in their developmental paths, with some going on to produce another brood, while others remain as a pupa until the following spring.

Imago

Adults feed on a wide variety of flowers, including Betony, Bluebell, Bugle, buttercups, Common Fleabane, Cuckooflower, Greater Stitchwort, Hemp-agrimony, knapweeds, Ragged-Robin, ragworts, Red Campion, thistles, vetches and Wild Marjoram. Males will also congregate on wet mud (a phenomenon known as 'mud-puddling') and animal dung where they obtain essential nutrients that are passed to the female during mating and that prolong her life, allowing her to lay more eggs.

Four ♂s on dung, with a ♂ Small White (on left) [NF]

In-depth studies on this species were started in the 1980s by Johan Forsberg and Christer Wiklund, and their work has told us much about the ecology of the Green-veined White. Like many species, the males emerge a few days earlier than the females, and spend much of their time searching for a mate when they are quite conspicuous as they flutter along hedgerows, woodland rides, scrub edges or a water's edge, investigating any white object they encounter, in the hope of finding a receptive female. Females, on the other hand, stay perched on grasses or other vegetation after they have emerged, where they await a male, although they will take to the air and even approach males on occasion.

Once the male finds a female, then he will flutter around her, showering her with pheromones from the scent scales on his wings. An unresponsive female will raise her abdomen to reject the male, as is typical in the Pieridae. Forsberg and Wiklund found that a virgin female may also initially reject the male by raising her abdomen, before taking to the wing with the male in tow, presumably to assess his quality as a mate. If the female does not take flight, or after she had landed, she will close her wings, allowing the male to mate with her. After several seconds the male will then take flight, with the female dangling below him, as he finds another spot. The pair remain coupled for between two and seven hours, before they go their separate ways.

As in the Large and Small Whites, the male passes an anti-aphrodisiac to the female during mating that makes her less attractive to other males, although this can only have a temporary effect since females are known to mate up to five times, although twice is more typical. Forsberg and Wiklund found that initial pairings tended to occur in the morning, with secondary pairings taking place in the afternoon. Secondary pairings are completely unnecessary for the fertilisation of the eggs and it is thought that the primary benefit to the female is that, in addition to sperm, she receives nutrients from the male that she absorbs

A mating pair from the spring brood (♀ on left) [WL]

to extend her life. The combined packet of sperm and nutrients may be as much as 15% of the male's body weight. Helena Larsdotter-Mellstrom and colleagues have also determined that the male will increase the size of the 'nuptial gift' by as much as 26%, based on the female's mating history (signalled by the anti-aphrodisiac), in order to increase his chances of success when sperm competition is a factor.

Egg-laying females are quite easy to spot, with a flight that appears slow and clumsy as she flits from plant to plant, tasting each with receptors on her feet to test for suitability, especially the presence of mustard oils that make the larva, pupa and adult distasteful to birds when the foodplant is consumed. The female will also alight on non-foodplants, suggesting that she locates foodplants primarily by taste rather than sight.

Eggs are laid singly, usually on the underside of a leaf of the foodplant, occasionally on a leaf stem or seed pod, with small and young plants preferred. The primary larval foodplants are Charlock, Cuckooflower, Garlic Mustard, Hedge Mustard, Large Bitter-cress, Water-cress, Wild Cabbage and Wild Radish. Unlike the Small and Large Whites, the Green-veined White does not normally lay on cultivated members of the Brassicaceae family and is not therefore a pest of garden crops, although it will occasionally lay on Nasturtiums. Also, eggs are sometimes found on the same plants as those used by the Orange-tip, although the two species are not in competition since the Green-veined White eats the leaves of the plant, whereas the Orange-tip eats the flowerheads and developing seed pods.

An egg laid on the underside of a Water-cress leaf

Ovum
The 1-mm high and bottle-shaped egg hatches in as little as three days, although this depends on temperature. It has 14 or 15 ribs that run along its sides with most running from top to bottom. The egg is a very pale yellow when it is first laid, but grows whiter just prior to hatching.

Larva
The 1.3-mm long **first instar** larva emerges from the egg in four or five days and eats most of its shell before starting to feed on the leaves of the foodplant, where it creates small holes as it breaks through the leaf surface. The larva is a very pale brown when it first emerges from the egg, but soon turns green as it ingests food. The body is covered in short hairs and each segment contains six white tubercles that each bear a long dark hair that is topped with a bead of clear fluid. The six tubercles on the first three segments are positioned in a line at the front of each segment, whereas the remaining segments have four tubercles at the front of each segment, with two further back.

1st instar: the newly emerged larva is very pale brown

1st instar: the larva turns green as it ingests food

2nd instar: the pale green larva is mottled with dark green blotches

The development of the larva is extremely rapid and, after only three days, the fully grown 3.5-mm long first instar moults into the **second instar**, the larva eating its cast skin as it does after every moult. The head is pale greenish-brown, and the body is pale green and mottled with darker green blotches. Each segment has six white hair-bearing tubercles as in the previous instar, although these are now much more prominent. When fully grown, the second instar is 6.5 mm long.

After three more days, the larva moults into the **third instar** and is similar in most respects to the previous instar, although the mottling is less marked and there is a very faint yellow line just below the spiracles along each side. When fully grown, the third instar is 10.5 mm long.

3rd instar: the larva has reduced mottling

4th instar: a faint yellow line runs along the back and the base of each side

5th instar: the larva now has a completely green head | 5th instar: a pre-pupation larva

After a few more days, the larva moults into the **fourth instar** and now has a greener head. The longest hairs emerge from the white tubercles as in previous instars, although the largest of the short hairs that cover the remainder of the body now emanate from black bases. The larva now has a faint yellow line running along its back (although this is nowhere near as prominent as that in a Small White larva), and along each side in line with the spiracles, which are themselves encircled with yellow. The fully grown fourth instar is 16 mm long.

4th instar: the white tubercles on the 3rd (left) and 4th (right) segments

After another three days, the larva moults into the final and **fifth instar** and is similar to the previous instar, although the head is now green, there is no yellow line along the back, the yellow markings encircling each black spiracle are more substantial and the larva is, of course, larger. After four days, and after only two and a half weeks in this stage, the larva is fully grown and between 24 and 27 mm long. It prepares for pupation by locating a suitable surface where it holds itself upright with a silk girdle and a silk pad at its tail end. Larvae that go on to produce another brood occasionally pupate on the foodplant but the majority, and those that go on to overwinter, usually pupate away from the foodplant and low down in vegetation, although they will pupate on tree trunks and even fences on occasion.

5th instar: the spiracles are encircled with yellow

Pupa

The 19-mm long pupa is held in place by the silk girdle and the cremaster and, like the pupa of the Small White, has several protuberances on its body at the base of the wing cases and on the thorax, with a noticeable 'beak' on the front of its head. The pupa comes in two colour forms—green and light brown—although intermediate forms are also produced, and it is thought that the colour may be influenced by the substrate on which the pupa is formed. There is also an amount of variation in terms of markings—some pupae have no markings, while others are heavily marked. There is also some variation between those pupae that go on to produce another brood, and those that overwinter. Those pupae that produce adults in the same year have a slight shine with smooth wing cases, and thinner shells that give a semi-translucent appearance, whereas those that overwinter have thicker and waxy shells, with raised veins on the wing cases (Vince Massimo, pers. comm.). If the pupa is not overwintering then, after around 10 days, this delightful butterfly of damp grassland emerges.

The green form of pupa [VM] | The light brown form of pupa

Life Cycles of British & Irish Butterflies 103

Clouded Yellow
Colias croceus

❖ CLOUDED YELLOW IS SUPPORTED BY KEVIN DOYLE, MIKE GIBBONS & MR MARK RICKUS ❖

	January	February	March	April	May	June	July	August	September	October	November	December
IMAGO												
OVUM												
LARVA												
PUPA												

A small number of species found in Britain and Ireland are noted for their migrations and the Clouded Yellow, with its powerful and rapid flight, is such a butterfly. The vast majority of butterflies found on our shores originate from North Africa and southern Europe since, while small numbers manage to successfully overwinter on the south coast of England each year, both larva and pupa of this continuously brooded species are easily killed by damp and frost and perish over the winter.

Since we are dependent on the strength of the influx from the continent, numbers of this distinctive orange-yellow butterfly vary significantly from year to year, with no discernible pattern. Occasionally we are on the receiving end of a huge migration that results in a 'Clouded Yellow year' where butterflies may be found across Britain and Ireland, including the Orkneys. The first record of such an influx was in 1804 and the most recent in 2013. It is a sad fact that such a spectacle is typically followed by a paucity of records the following year, confirming that this species is largely unable to survive our winter.

The butterflies seem to be somewhat gregarious when migrating and there are several records of groups of butterflies flying northwards or coming in like a 'cloud' off the sea. In his *The Complete Book of British Butterflies* of 1934, F.W. Frohawk shares the content of a letter he received from a Rev. D. Percy Harrison: *"My greatest experience was in Cornwall as far back as 1868, when I was only 11, and sat on a cliff near Marazion, and saw a yellow patch out at sea, which as it came nearer showed itself to be composed of thousands of Clouded Yellows, which approached flying close over the water, and rising and falling over every wave till they reached the cliffs, when I was surrounded by clouds of C. (= croceus) edusa, which settled on every flower ... the nearest coast of France would be Cherbourg. They swarmed in the district for a space of some three weeks and were good specimens when they arrived"*.

A rare upperside shot of a ♀ that is in the process of taking flight [CK]

Many migrant species are first seen wherever the distance between the continent and our shores is the shortest, with many first arrivals being sighted in Kent and Sussex. However, the Clouded Yellow is the exception to the rule, with many first sightings coming from Dorset and neighbouring counties, suggesting that large expanses of water are no barrier to this species.

The ♂ has a wingspan of between 52 and 58 mm [IL]

The ♀ has a wingspan of between 54 and 62 mm [MC]

Life Cycles of British & Irish Butterflies

Southerly migration in autumn has also been observed on occasion, but never in the same numbers as the immigration. In the book *Insect Migration* of 1958 (number 36 in The New Naturalist series), C.B. Williams describes an experience of the return migration at the western end of the English Channel: *"After arriving in this country, usually in early June, the immigrants lay eggs in the clover fields, and about the first week in August there is an emergence of home-bred adults. These fly round for a few weeks and then there appears to be a return flight to the south. For example, on 14th October, 1947 – one of the years of great abundance – J. Blake was on a steamer sailing up the Channel from off Ushant to Start Point in Devon. For many miles he saw Clouded Yellows over the sea moving steadily to the S.S.W. He considered that the flight was on a front of about fifty miles and that there must have been well over a hundred thousand butterflies taking part"*.

Because of its similarity to some of the whites, especially the Brimstone, this butterfly is often confused with them when seen in flight. However, experienced observers will immediately latch on to a different hue than that of other species, the Clouded Yellow appearing an orange-yellow as the orange uppersides contrast with the yellow undersides, and it is only the latter that you can expect to see when the butterfly has settled since this species always keeps its wings firmly closed once it has landed.

The name 'Clouded Yellow' was given to this butterfly by Benjamin Wilkes in his *Twelve New Designs of English Butterflies* in 1742, which recognises the dark borders found on the uppersides of both male and female, although some believe that it may refer to the 'clouds' of migrants that have been seen in numbers coming in off the sea. Various authors, both before and after Wilkes, have given this butterfly alternative names, but each pays homage to the butterfly's distinctive colour and pattern, including Saffron Butterfly (for the male), Spotted Saffron Butterfly (for the female), Clouded Orange, and Clouded Saffron. In fact, the specific name of this species, '*croceus*', is Latin for 'saffron'.

A ♀ of the form *helice*

There are two butterfly species found in Britain and Ireland where the female exhibits a particular form—the Clouded Yellow produces a form known as '*helice*' and the Silver-washed Fritillary produces a form known as '*valesina*'. The *helice* form, which occurs in around 10% of females, has a creamy white ground colour, rather than orange, and this makes identification when in flight a little more difficult due to the similarity with the Brimstone and some of the whites, although the dark borders usually give the game away, even in the absence of the orange-yellow colouring.

DISTRIBUTION

The Clouded Yellow has a distribution befitting a highly migratory species and can be found just about anywhere in Britain and Ireland although its distribution is, as we might expect, concentrated in the south coast of England where many of the migrants from the continent remain to feed, mate, and lay eggs. The resulting offspring may venture further afield, and any inland concentrations are almost certainly the result of a local emergence. On a worldwide basis, the heartland of this butterfly is in the Mediterranean region and in North Africa, from where it begins its migration, ultimately spreading into most of Europe and western Asia.

HABITAT

This butterfly can be found in just about any open habitat in the countryside, especially coastal cliffs and calcareous grassland in southern England. It is also one of the few species that can tolerate arable farmland, often because there is a supply of clovers nearby that are not only cultivated as fodder (that

Chimney Meadows in Oxfordshire [WT]

is used to feed livestock over the winter) but are also a key larval foodplant of this species. Other larval foodplants include Common Bird's-foot-Trefoil and Horseshoe Vetch.

STATUS
The Clouded Yellow is one of the most widespread species in Europe and is not a species of conservation concern since any trends are largely a reflection of the extent of the immigration from continental Europe. The long-term trend shows a significant increase in both distribution and abundance, with the more recent trend over the last decade showing a moderate decline in distribution and a significant decline in abundance. Using the IUCN criteria, the Clouded Yellow is categorised as Least Concern (LC) in both the UK and Ireland.

LIFE CYCLE
The butterfly can put in an appearance as early as March in good years, although these early sightings usually represent a local emergence from larvae that have managed to survive the winter at coastal undercliffs. These early individuals are soon followed by migrants that arrive in late May and early June, resulting in a much larger overall population by July and August as the offspring of the first arrivals mingle with new immigrants. In good years, this species can produce up to three broods in the south of England, with the third brood emerging in late September and October. In his *Natural History of British Butterflies* of 1924, F.W. Frohawk provides an account of the rapid development of this species: *"The first brood, from eggs laid in June, only occupies about fifty-four days from the deposition of the eggs to the emergence of the imagines"*.

Imago
Given its ability to migrate over long distances, it goes without saying that this butterfly has a powerful flight and is best sought out wherever it is able to refuel on favourite nectar sources such as Common Fleabane, dandelions, knapweeds, ragworts, thistles, vetches and Wild Marjoram. Much like the equally migratory Painted Lady, this is the best time to get close to this butterfly since it will remain in an area to feed on its chosen flowers, even if it is disturbed.

Once settled in a favourable location, the male will adopt a strategy of patrolling, flying up and down a stretch of land, in order to find a mate. I encountered a male doing just this along an area of Bokerley Dyke at Martin Down, on the border

A mating pair [PB]

between Dorset and Hampshire, and, as soon as a female flew into the area, she was immediately met with. She then landed in long grass and was mated without any discernible courtship, the pair remaining 'in cop' for just over an hour. The female is subsequently able to lay an extraordinary number of eggs—up to 600 have been recorded from a single female.

Ovum

The skittle-shaped eggs are laid singly on the upperside of a leaf of the foodplant and are just over 1 mm high, each egg possessing a number of longitudinal ribs that run its length. The colour of the egg changes as it develops—when first laid it is a pale yellow but it soon develops a rose-orange hue before ultimately deepening to a purple-grey just before the larva emerges.

The newly laid egg is a pale yellow

The egg develops a rose-orange hue a day or so after being laid

The egg is a purple-grey just prior to hatching

Larva

The 1.5 mm-long **first instar** larva emerges from the egg after between one and two weeks, depending on temperature, when it proceeds to consume the majority of its eggshell. The first instar is a dull green in colour and is distinguished from all other instars by its black head. It feeds during the day primarily on the epidermis of the leaf but does not perforate it and, when preparing to moult, fixes itself to a silk pad spun on the underside of the leaf. When fully grown, the first instar is just under 3 mm long and is now a yellowish-green in colour.

1st instar: this is the only instar with a black head

After around 10 days, the larva moults into the **second instar** and eats the old skin, which it does after each moult. The larva is now dull green and, unlike the first instar, has a dusky brown head as well as a very faint white line that runs along each side, in line with the spiracles. The larva rests in a straight position on the upperside of the leaf midrib, facing the leaf base. Unlike the first instar, the second instar will now perforate the leaf while feeding and, when fully grown, is just over 4.5 mm long.

2nd instar: the larva has a dusky brown head and rests facing the leaf base

After only five days or so, the larva moults into the **third instar**. The larva now has a velvety appearance thanks to numerous black warts and whitish hairs that can be seen at high magnifications. The main distinguishing features with earlier instars are the greener head and the presence of a more prominent white line that runs the length of each side of the body, in line with the spiracles. When fully grown, the third instar is just under 9.5 mm long.

In under a week, the larva moults into the **fourth instar** and is around 16 mm long when fully grown. It is similar in appearance to the previous instar, although the white lateral lines are much more prominent, and a pale yellow-orange spot is present next to the spiracle on each segment. Behaviourally, the day-feeding larva also rests with its front segments slightly raised, which is reminiscent of the behaviour of a Brimstone larva that many observers will be familiar with.

3rd instar: the larva now has a greenish head

4th instar: a larva resting with its front segments raised

4th instar: there is now a pale yellow-orange spot next to each spiracle

In under a week, the larva moults for the last time into the **fifth instar**. The larva is similar in appearance to the previous instar and, apart from the obvious difference in size, can be distinguished by the beautiful pair of markings found on each segment along the lateral line, where each white spiracle is bordered with a yellow marking to the front and an orange marking to the rear. When strung out along the side of the larva, the resulting pattern has the appearance of decorative bunting. When fully grown, the larva is around 33 mm long and, when preparing to pupate, finds a suitable surface before constructing a silk pad at its tail end, and a silk girdle around its waist, to hold itself upright. The effect of temperature on larval development is significant, with some authors stating that the larva will complete this stage in three weeks in relatively high temperatures, but will take six weeks (or longer over the winter) when temperatures are cooler.

5th instar: a fully grown larva

5th instar: the lateral line is now adorned with a beautiful pattern of yellow and orange markings

5th instar: the yellow and orange markings fade as the larva prepares to pupate

Pupa

The larva moults for the last time in just over 24 hours. The resulting pupa is 23 mm long and is held in position by the silk girdle and the cremaster, whose hooks are attached to the silk pad previously constructed by the larva. Prior to emergence, the fully formed butterfly can be seen quite clearly through the pupal case and, after two or three weeks in this stage, this orange-yellow beauty takes to the skies.

The pupa is attached to a suitable surface by a silk girdle and the cremaster

The fully formed butterfly is clearly visible through the pupal case

Life Cycles of British & Irish Butterflies

Brimstone
Gonepteryx rhamni

❖ BRIMSTONE IS SUPPORTED BY CLARE MORTIMER, DAVID OAKLEY-HILL, CLIVE PRATT & CHRIS ROWLAND ❖

	January	February	March	April	May	June	July	August	September	October	November	December
IMAGO												
OVUM												
LARVA												
PUPA												

If there is one sight that epitomises the end of winter, it is surely the flash of yellow of a male Brimstone as it flits through woodland or a country lane, since this butterfly overwinters as an adult and is one of the first to appear, in February and March. There are several suggestions regarding the origin of the word 'butterfly', but the thought that it is derived from 'butter-coloured fly', as characterised by the male Brimstone, has to be my favourite. The female, on the other hand, is a whitish green, and both sexes have an exquisitely scalloped wing shape that perfectly matches

The ♂ is yellow, with a wingspan of 60 to 74 mm [IL]

The ♀ is a whitish green and has a similar wingspan to the ♂

a leaf when the butterfly is roosting or hibernating within foliage such as an Ivy bush. The specific name *rhamni* is a reference to *Rhamnus*, the genus of Buckthorn, one of the larval foodplants of this species.

We have two subspecies in Britain and Ireland—the nominate subspecies, *Gonepteryx rhamni rhamni*, is found throughout its distribution in Britain and the subspecies *gravesi* is found throughout Ireland. The latter is deceptively pale, and I remember watching a male nectaring in the Burren in County Clare and convinced myself that it was a female, before it took flight when I caught my first glimpse of yellow that confirmed that it was, in fact, a male.

A ♂ of the subsp. *gravesi* has a very pale yellow underside

DISTRIBUTION
The distribution of this species is confined to its two larval foodplants, Buckthorn and Alder Buckthorn, which are found throughout England, Wales and in central Ireland in a belt that stretches from Counties Clare, Galway and Mayo in the west to County Kildare in the east. Vagrants turn up in Scotland now and again, not least in the Borders where it has been seen for the last several years, possibly helped by the proactive planting of buckthorns in the area. On a worldwide basis, the butterfly is found throughout Europe, through Asia to the far east, and in North Africa.

HABITAT
The Brimstone is not colonial but is a great wanderer and can be found in almost any habitat where its foodplants grow nearby, including

Alder Buckthorn grows at ride edges at Pamber Forest in Hampshire

broadleaved woodland; mixed scrub; hedgerows; lowland fens; arable field margins and set-aside; gardens, parks, churchyards; and road verges.

STATUS
Both long-term and short-term trends show a moderate increase in distribution and a stable abundance. Using the IUCN criteria, the Brimstone is categorised as Least Concern (LC) in both the UK and Ireland.

LIFE CYCLE
The Brimstone is single brooded and may be found in every month of the year, although peak flight times are in April and May as the hibernating adults emerge, feed and mate, and again in July and August when their offspring reach adulthood, with males emerging a week or so before females in both cases. Autumn is a good time to see this species when the adults gorge themselves on various nectar sources as they build up their fat reserves in preparation for hibernation.

Given that the Brimstone overwinters as an adult, it is one of a handful of species that spends most of its life as the adult butterfly and many observers will have come across battered individuals in mid-summer that overlap with the new brood. These worn individuals will have survived the best part of a year as an adult, making this our longest-lived of butterfly.

Imago
Adults spend much of their time feeding, where they always settle with their wings closed, showing a preference for purple and nectar-rich flowers, especially thistles. Other nectar sources include Betony, Bluebell, Bugle, Common Fleabane, Cowslip, dandelions, Devil's-bit Scabious, knapweeds, Primrose, Ragged-Robin, Red Campion, Selfheal, vetches and Wild Marjoram. The long proboscis of the Brimstone also allows it to take nectar from flowers, such as Wild Teasel, that are beyond the reach of many other butterfly species. With the approach of autumn, the butterfly settles down to hibernate—often under leaves of Ivy, Holly or bramble.

Males are the first to be seen in the spring and, when warming their bodies, can often be found resting with their wings at right angles to the sun in order to get its full effect, which may result in them lying almost flat on the ground. Males are also particularly visible as they patrol woodland edges, hedgerows and other habitats looking for a mate and they are best sought out in the middle of the day when it is at its warmest, since the butterfly goes to roost much earlier than other species and has generally stopped flying by 2.30 pm. Females emerge a little later in the spring than males, usually when the temperature is consistently warm and does not fluctuate between warm and cold spells.

When a virgin female is found, the pair will spiral around one another, eventually climbing high into the air, often out of sight, before tumbling back down into a bush where they then mate, remaining coupled for a few hours. However, mating may be a protracted affair and, in a few exteme cases, pairs have remained 'in cop' for a number of days, although such lengthy pairings do not seem to affect the adult butterflies when they finally go their separate ways. When an already mated female is located, however, she will try to shake off the male by maintaining a somewhat static position in the air, with the pair simply bobbing up and down in almost the same spot and this is the perfect time to capture this species in flight. If the male is particularly persistent then the female will drop down to the ground and exhibit the typical rejection posture found in many Pierids, by opening her wings and raising her abdomen. This is normally enough for the male to eventually lose interest although I have, on occasion, seen a female surrounded by several males that won't take no for an answer, when they can be found walking over her open wings.

A ♂ pursuing two ♀s

An already mated ♀ surrounded by ♂s

A mating pair with ♀ on the left [PCh]

Females are particularly choosy about the plants on which they lay their eggs—even on sites with many buckthorns present, only a very small proportion are used by females, who typically lay on plants that are isolated, sheltered and growing in sunny areas, such as at the edge of a woodland ride. In more open areas, the female will select much smaller and less mature bushes that are under 1 m tall and I have spent many a time following a female through a patch of scrub where she is able to pick out a buckthorn seemingly at ease before laying an egg. Egg-laying is a slow and deliberate affair as she carefully chooses the leaf on which to lay, and this is one of the best times to get a photo.

A ♀ in the process of egg-laying [JA]

The primary larval foodplants are Alder Buckthorn on moist acid soils, such as in woodlands, heathland and wetlands, and Buckthorn on the calcareous soils of chalk and limestone grassland, and it is worth growing a buckthorn in your garden if you are interested in finding the immature stages close to home. I have an Alder Buckthorn that had no eggs on it for three years but, in its fourth year, in 2017, clearly became suitable since no less than 140 eggs were laid on it. It has played host to Brimstone eggs and larvae every year since.

Ovum

The bottle-shaped eggs are laid singly on the youngest buckthorn leaves, often those that have yet to fully unfold, at all heights on the plant and almost always on its sunny side. Eggs are typically laid on the underside of the leaf, although eggs are occasionally laid on the leaf upperside or on the end of the twig next to the unfurling leaves. Several eggs may be found together, and this is usually the result of different females using the same leaf, or the same female revisiting the same spot at different times. On occasion, the female will lay an egg, flutter around the vicinity for a moment, and then return to exactly the same spot to lay another egg, and this may be repeated several times, with several eggs laid on the same leaf.

Three bottle-shaped eggs laid on the same Alder Buckthorn [MC]

The newly laid egg is a pale green

The coloured-up egg is a deep yellow

Newly laid eggs are pale green, but soon turn yellow and eventually grey as the larva develops. Eggs are approximately 1.3 mm high and have 10 longitudinal ribs running top to bottom, although this number does vary to some degree. Eggs are occasional asymmetrical and bend to one side, which is a characteristic found in other species in the Pieridae family also.

Larva

After one or two weeks, the 2-mm long **first instar** larva eats an untidy hole at the top or side of the egg before making its exit, leaving the remainder of the eggshell uneaten. The larva typically takes up position next to a rib on the underside of the leaf, remaining on the underside for its first two instars. This, however, is a rule of thumb and it is not unusual to find larvae feeding on the upperside of the leaf in the early instars on occasion. The larva feeds by nibbling away at the centre of the leaf, leaving tell-tale holes where the leaf has been perforated that can give away the presence of an early instar larva. The larva is a mottled olive-brown when it first emerges and is covered in short hairs that each end in a pale yellow patch at their base. The larva becomes a more consistent yellow-green as it grows with the pale patches becoming less noticeable. When fully grown, the first instar is around 5 mm long.

After eight days or so, the larva moults into the **second instar** and eats its old skin, as it does in all instars. The old skins must be relatively nutritious, with at least one instance of a larva eating another's cast skin (Vince Massimo, pers. comm.). The second instar is a uniform green, with short black hairs running in two longitudinal lines along the back, each ending in a pale patch at its base that becomes less noticeable as the larva grows, as in the first instar. The larva is also sprinkled with small black spots all over its body and continues to position itself against a rib on the underside of the leaf from which it feeds, where the characteristic holes that it creates are much more substantial. The fully grown second instar is around 8 mm long.

After only four or five days, the larva moults into the **third instar** and is similar to the second instar with only minor differences in appearance. The main distinguishing feature with the previous instar is, in fact, its behaviour—the larva now always rests along the midrib of the upperside of the leaf rather than the underside, and it also feeds on the leaf edges rather than the central parts of the

1st instar: the newly emerged larva is olive-brown

1st instar: the larva develops a greener hue as it feeds

2nd instar: the larva continues to rest alongside a rib on the leaf underside

3rd instar: the larva now rests on the leaf upperside

4th instar: the larva lacks the pale yellow patches of previous instars

4th instar: a larva eating a developing Alder Buckthorn flower bud [VM]

leaf. The larva also has a curious habit of raising the front part of its body off the leaf when at rest, presumably to reduce the amount of silhouette when viewed from below that might otherwise alert a bird to its presence. When fully grown, the third instar is 13 mm long.

After five days or so, the larva moults into the **fourth instar** and is similar in appearance to the third instar but is now devoid of any pale yellow patches running along its back. The body and head are also now heavily sprinkled with black spots, with the body also showing a graduation of colour from green on the back to greenish-blue on the sides and finally to a white stripe. The larvae are now quite visible to the naked eye and there have been many occasions where I have found several larvae in close proximity on a favoured bush. Vince Massimo, who has studied this species in some detail, has found that later instars will also eat developing flower buds. The fully grown fourth instar is 20 mm long.

After just under a week, the larva moults into the final and **fifth instar** and, when fully grown, is around 35 mm long. It is similar in appearance to the previous instar, with a consistent green along the length of its body and a similar graduation of colour down the sides of the body, going from green on the back to greenish-blue on the sides and then to a prominent white stripe. In these later instars, the larva exudes an amber liquid from the tips of the fine hairs that cover its body, and this is especially noticeable in the final instar. This liquid may be distasteful to birds and therefore act as a deterrent. Prior to pupation, the larva typically leaves its host plant before securing itself to the underside of a leaf or plant with a pad of silk at its tail and a silk girdle around its waist. After six days or so, the larva becomes noticeably paler before pupating.

5th instar: a larva in a typical resting pose

5th instar: a larva preparing to pupate [VM]

Pupa

The 25-mm long pupa has a curious shape that gives it the appearance of a curled leaf. Before the adult butterfly emerges, the orange spot found in the centre of the forewing can be clearly seen through the pupal case. After approximately two weeks as a pupa, this long-lived butterfly emerges.

A pupa fastened to the underside of a leaf by a silk girdle and the cremaster

A ♂ pupa with the orange spot on the forewing upperside clearly visible

Life Cycles of British & Irish Butterflies

Nymphalidae

The **Nymphalidae** is the largest of the butterfly families, representing 6,000 or so species that are found in all zoogeographical regions of the world. The species found in Britain and Ireland are grouped into five subfamilies—the Satyrinae (the browns), with all others spread across the Heliconiinae, Limenitidinae, Apaturinae and Nymphalinae subfamilies. In earlier classifications, the browns have been placed in their own family, the 'Satyridae'.

Family	Subfamily	Genus	Species	Vernacular name
Nymphalidae	Satyrinae	Lasiommata	megera	Wall
		Pararge	aegeria	Speckled Wood
		Coenonympha	tullia	Large Heath
			pamphilus	Small Heath
		Erebia	epiphron	Mountain Ringlet
			aethiops	Scotch Argus
		Aphantopus	hyperantus	Ringlet
		Maniola	jurtina	Meadow Brown
		Pyronia	tithonus	Gatekeeper
		Melanargia	galathea	Marbled White
		Hipparchia	semele	Grayling
	Heliconiinae	Boloria	euphrosyne	Pearl-bordered Fritillary
			selene	Small Pearl-bordered Fritillary
		Argynnis	paphia	Silver-washed Fritillary
			aglaja	Dark Green Fritillary
			adippe	High Brown Fritillary
	Limenitidinae	Limenitis	camilla	White Admiral
	Apaturinae	Apatura	iris	Purple Emperor
	Nymphalinae	Vanessa	atalanta	Red Admiral
			cardui	Painted Lady
		Aglais	io	Peacock
			urticae	Small Tortoiseshell
		Polygonia	c-album	Comma
		Euphydryas	aurinia	Marsh Fritillary
		Melitaea	cinxia	Glanville Fritillary
			athalia	Heath Fritillary

IMAGO

Most of the members of this family are of a medium or large size when compared with those from other families. The forelegs in both sexes are vestigial and useless for walking, and this family is sometimes referred to as the four-footed butterflies. The brush-like appearance of the forelegs has also resulted in the other common name for this family—the brush-footed butterflies.

A close-up of the vestigial forelegs of a Red Admiral

A ♂ Gatekeeper showing the distinctive sex brands on his forewings

A ♂ Silver-washed Fritillary with prominent sex brands on the veins of his forewings

The members of the Satyrinae subfamily are predominately brown, with the obvious exception of the Marbled White. All other species in the Nymphalidae family—the fritillaries and aristocrats—are more brightly coloured. The members of the Satyrinae also have androconia grouped into a sex brand on the male forewing, and this is especially noticeable in the Gatekeeper. A sex brand is also present in members of the *Argynnis* genus (the Silver-washed, Dark Green and High Brown Fritillaries). The uppersides and undersides of the adult also differ significantly in this family, especially in those species that are members of the Heliconiinae, Limenitidinae, Apaturinae and Nymphalinae subfamilies.

The barrel-shaped egg of the Meadow Brown, a member of the Satyrinae

The oval and ribbed egg of the Purple Emperor

The egg of a White Admiral is made up of a series of hexagonal pits

OVUM
Eggs of the Satyrinae subfamily are usually barrel-shaped and may have discernible ribs. The eggs of the remaining subfamilies vary in shape but are often oval and strongly ribbed. The egg of the White Admiral is unique in our Nymphalidae for being spherical but made up of a series of hexagonal pits.

LARVA
The larvae of the Satyrinae subfamily are covered in short hairs, tend to taper at both ends, and have an anal segment that ends in two points in later instars. All of the Satyrinae in Britain and Ireland feed on grasses. The larvae of the remaining subfamilies have branched spines in later instars, with the exception of the Purple Emperor that has a smooth surface in all instars.

The anal segment of a Speckled Wood larva ends in two points

A spiny Peacock larva

A Purple Emperor larva

PUPA
The pupae of the Satyrinae subfamily are either suspended head-down, attached by the cremaster or old larval skin to a silk pad, or simply lie unattached at the base of the larval foodplant. The pupae of all other subfamilies always pupate head-down.

A Meadow Brown pupa attached to the old larval skin

A Marbled White pupa formed at the base of a grass tussock

A Glanville Fritillary pupa

Life Cycles of British & Irish Butterflies

Wall
Lasiommata megera

❖ WALL IS SUPPORTED BY VIVIAN RUSSELL & BILL STONE, SUFFOLK BUTTERFLY RECORDER ❖

	January	February	March	April	May	June	July	August	September	October	November	December
IMAGO												
OVUM												
LARVA												
PUPA												

The Wall was given its name by the legendary entomologist Moses Harris in his *The Aurelian* of 1766, where he describes a patrolling male: *"This Fly is very common in Fields, and by Roadsides. It delights to fly along very low in dry Ditches, seldom straying from the Bank, or Field, where it was bred; but, when it comes to the End of the Bank, will return back again, frequently settling against the Bank, or perhaps against the Side of a Wall; and is, for this Reason, called The Wall Flie".* An earlier name for this species was much more descriptive—in his *Musei Petiveriani* of 1699, James Petiver called this butterfly 'The golden marbled Butterfly, with black eyes' although this is, admittedly, a bit of a mouthful and he later changed the name to 'The London Eye'. Male and female are easily distinguished since the male has a prominent sex brand on each forewing.

The ♂ has a 45–53 mm wingspan and conspicuous sex brands on its forewings

The ♀ is only slightly larger than the ♂ and lacks the sex brands

This butterfly has a characteristic behaviour of resting with wings two-thirds open on any bare surface (not just walls), including bare ground, in order to raise its body temperature to 25–30°C before it takes flight. This basking behaviour allows the butterfly to get the full effect of the sun both directly and indirectly as sun rays get reflected back onto the butterfly from whichever surface it is on, allowing it to raise its temperature by 8–10°C higher than the surrounding area. In particularly hot weather such basking is avoided, and the butterfly may even retreat to a suitably shaded spot to avoid overheating.

The underside is similar in both sexes with a lovely ring of eyes on the hindwing

DISTRIBUTION

This species was once found throughout England, Wales, Ireland, parts of Scotland, the Isle of Man and Channel Islands, and anyone who has been interested in butterflies for a while will remember just how common this species used to be when it would crop up in many different habitats. Today, however, is a very different picture, with this species suffering severe declines over the last several decades. It is now found primarily in coastal regions and has been lost from many sites in central and eastern England and, in Scotland, it is now confined to coastal areas in the south-west and in the Scottish Borders on the east coast. On a worldwide basis, the butterfly is found throughout most of Europe, North Africa and in temperate regions of western and central Asia.

Where the butterfly does occur, it is found in relatively small colonies of no more than 20 or 30 adults at peak. Colonies are self-contained, although some individuals will wander, allowing the species to colonise nearby sites if they are suitable. Although no formal study has taken place regarding the mobility of this species, Jeremy Thomas, in *The Butterflies in Britain and Ireland*, says that *"There are at least two records of adults reaching the Outer Dowsing light vessel, 50 km off the Norfolk coast".*

HABITAT

The Wall is primarily a grassland butterfly but needs areas of bare and stony ground, such as farm tracks, where the adult butterfly can bask as well as find suitable sites for its eggs. Other habitats include limestone pavement, disused quarries, sand dunes, cliff edges, brownfield sites and disused railway lines. It may also be found in open woodlands in areas that receive direct sunlight, thereby providing the shelter and sun that this species needs.

Ballard Down, near Swanage on the Dorset coast, has patches of bare ground favoured by the Wall

The butterfly can be found in sheltered areas of the Burren in County Clare, Ireland

STATUS

The long-term trend shows a significant decline in both distribution and abundance, with the more recent trend over the last decade showing a moderate decline in both distribution and abundance. Using the IUCN criteria, the Wall is categorised as Near Threatened (NT) in the UK and Endangered (EN) in Ireland.

Many theories have been put forward for this drastic decline, including habitat loss and agricultural intensification, but a compelling argument is based on research by a team at Louvain University in Belgium, led by Professor Hans Van Dyck. The team suggests that the Wall has become a victim of climate change, where warmer temperatures in autumn have encouraged the butterfly to progress to adulthood and then attempt to fit in another brood. This has resulted in a 'developmental trap' where some of the offspring of this new brood are unable to reach their third larval instar, the only instar in which they can successfully overwinter, resulting in a 'lost generation'. The research also showed that captive-bred Wall larvae placed at inland sites quickly developed, resulting in a precarious third generation, but at cooler coastal sites only 42.5% of larvae went on to produce another generation, and this may explain why this butterfly is confined to primarily coastal regions.

LIFE CYCLE

The first generation of adults emerges in late April and early May, peaking at the end of May and early June, with a slightly later emergence in the north of England and Scotland. This first brood gives rise to an often-larger second brood that emerges at the end of July, or in August further north. There are typically two generations each year although a smaller third generation may appear from early September, at least in the south of its range, as was the case in the hot summer of 2018 (Bob Eade, pers. comm.).

Imago

Given the decline of this species at inland sites, enthusiasts are most likely to encounter this butterfly at coastal locations where the butterfly is a familiar sight on footpaths, flying up when disturbed, before landing a few metres ahead. In his *The Lepidoptera of the British Islands* of 1893, Charles Barrett wrote "*No butterfly introduces itself more constantly to the notice of passers along a country road, since its habit is to start suddenly and swiftly up, and fly on a little way, then settle on a bare spot, spring up again as soon as approached, and keep a short distance in advance, in a most persistent manner, so that as a wayside companion it is hardly to be surpassed*". The butterfly is also one of the wariest of all our butterflies and would appear to detect the slightest movement or noise, including the opening of a camera shutter, making it extremely difficult to photograph as it annoyingly takes to the air just as the picture is being taken.

The male of this species is territorial and will inhabit a particular sunny edge, such as a path, hedgerow or roadside verge, waiting for a passing female. Males will typically perch in a favoured position but will, in sunny and warm conditions, adopt a second strategy of patrolling in order to find a mate. Several studies in various parts of Europe show that the butterfly will also undertake 'hill-topping' where the male uses a high point in the landscape that females visit when looking for a mate. Whatever the strategy, the male will investigate all passing insects, and rival males will spiral high into the air before shortly coming back to the ground. Any virgin female that is encountered is pursued until she lands on a patch of bare ground, when the male lands behind her before fluttering around her in a semi-circle so that he is facing her head on. He then undertakes a crude form of courtship, certainly cruder than the elegant courtship of the Grayling, although the outcome is the same—the male bows to the female several times, with heads touching on occasion (some suggesting that the male headbutts the female) and with the male's antennae clubbing the female, while he showers her with pheremones. With the female seduced, the pair mate before disappearing into surrounding vegetation.

The female is much more sedentary and less conspicuous than the male. Once mated she moves away from male territories to feed and lay her complement of eggs. Both sexes are avid nectar feeders and will visit a variety of flowers, especially Common Fleabane, Daisy, hawkweeds, knapweeds, Ragged-Robin, ragworts, thistles, Water Mint, Wild Marjoram and Yarrow. As dusk approaches, both sexes will roost upside-down beneath leaves, tree boughs, fences or some other suitable surface in the sunniest parts of the site.

Sites for egg-laying are typically sheltered and warm compared to their surroundings, and include grass clumps, rabbit scrapes, hoof prints and other hollows, often in the side of a bank, where eggs are laid low down on the plant on leaf tips and stems, or on bare grass roots, usually on the host plant, but not always. Eggs may also be laid on grasses besides fences or under shrubs such as Gorse and brambles. An egg-laying female will crawl among the vegetation to find a suitable spot and the resulting egg can be quite easy to find when attached to an exposed bare root, especially when using a pair of close-focusing binoculars. Various grasses are used by the larva, including bents, Cock's-foot, False Brome, Tor-grass, Wavy Hair-grass and Yorkshire-fog.

A mating pair with ♂ on the right [BE]

This hoof print has two eggs laid on the exposed grass roots, with a closer view (inset) of one of the eggs

Ovum

The 1-mm high spherical eggs are laid singly, or occasionally in twos and threes, and are an ivory white, gradually becoming more translucent as the larva develops, when the markings on the head of the larva become quite visible through the eggshell. The egg looks smooth to the naked eye, but actually has a very subtle network pattern over most of its surface.

The egg turns an ivory white after a couple of days

The head of the larva is visible through the eggshell prior to hatching

Larva

After around 10 days, the 2.5-mm long **first instar** larva emerges from the egg by eating away at the crown until it is able to push its way out of the hinged top. It then eats its eggshell before feeding, primarily at night, on the leaf on which the egg was laid. If the egg was laid on a grass root, then the larva will move to a grass blade to feed although there are records of larvae feeding on fresh grass roots immediately after emergence. Larvae become more mobile as they mature and will move to feed on fresh leaves, and even move to an adjacent grass tussock. The key characteristic of the first instar is that the head and body are covered in olive-grey tubercles that are not present in later instars, with each bearing a relatively long black or white hair. When fully grown, the first instar is just over 6 mm long.

1st instar: the larva is covered in relatively long hairs

1st instar: the head and body are covered in olive-grey tubercles

1st instar: a fully grown larva prior to moulting

2nd instar: a pattern of white and dark lines run the length of the body

After around 10 days, the larva moults into the **second instar**. The olive-grey tubercles found in the first instar are no longer present and the larva now has an overall green ground colour, with several longitudinal stripes running the length of the body, ranging from a dark stripe along the middle of its back to a white stripe that runs laterally on each side. These result in a striped pattern of alternating white and dark lines that is not found in later instars to the same degree. The larva also has two fine points on the anal segment that become more prominent in later instars. The fully grown second instar is around 10 mm long.

After around a week, the larva moults into the **third instar** and will now feed by day as well as by night. The larva retains the dark line down its back but is now a more uniform pale green and the few white stripes that remain are spaced further apart than in the second instar, with no resulting pattern of alternating white and dark lines. It also differs from the fourth and final instar by having almost translucent, rather than green, points on the anal segment. If the larva overwinters then it will do so in the third instar, although it does not become fully torpid and will feed during the winter if the weather is mild enough. The fully grown third instar is around 14 mm long.

3rd instar: the larva lacks a clear pattern of alternating white and dark lines

3rd instar: a larva preparing to moult, showing translucent points on the anal segment

If the larva does not overwinter then, after around a week, it moults for the final time into the **fourth instar**. It is very similar in appearance to the third instar but now has a bluish tinge and the anal points are no longer translucent, but light green. The spiracles are also a pale orange and, although very difficult to discern in the field, this can be a useful diagnostic when examining photos since this feature is unique to this instar. The fully grown larva is around 30 mm long and, prior to pupation, spins a silk pad on a blade of grass or other vegetation from which it hangs head-down in a compact J-shape where, after a couple of days, it pupates.

4th instar: the larva is almost invisible when placed in the context of its surroundings [BE]

4th instar: the larva has a bluish tinge

4th instar: a pre-pupation larva in a typical J-shape pose [BE]

Some observers have difficulty distinguishing Wall and Speckled Wood larvae since they both feed on grasses and have a similar appearance. However, the Speckled Wood prefers much shadier conditions than the Wall, which prefers much sunnier sites.

Pupa

The pupa hangs head-down, attached by the cremaster to the silk pad that was laid down by the larva. The pupa varies in colour from grass green, when it is extremely well camouflaged, through to black, at its most extreme, although this form seems to be particularly rare. The colour of the pupa does not affect the colour of the resulting adult in any way. The pupa has two rows of white tubercles on its back that are particularly conspicuous on a black pupa. The pupa grows darker as the adult develops within, and the adult butterfly is quite visible through the pupal case when it has fully formed, the adult emerging after around two weeks. Unsurprisingly, this 'colouring up' is less visible in a black pupa.

The Speckled Wood is unique among our butterflies in being able to pass the winter as both a larva and a pupa. However, in his *The Complete Book of British Butterflies*, published in 1934, F.W. Frohawk says of the Wall: "*The pupal stage varies according to the broods; in the summer it averages about 14 days, and seven months in the case of those that pass the winter as pupae*". However, with no records of this species overwintering as a pupa, this statement could simply be an error on the part of the author, especially since it is not mentioned in his other works.

The grass green form of pupa

On rare occasions, the pupa is almost black

A ♀ pupa with the adult just hours from emerging

Life Cycles of British & Irish Butterflies

The Speckled Wood is a familiar sight within its woodland home since it is not only relatively common, but is also particularly robust, flying in overcast conditions as well as in heavy shade—I always consider this butterfly a constant companion whenever I'm walking in or near woodland when other species are nowhere to be seen. The adult butterfly has a beautiful mix of creams and browns on its uppersides, with a very attractive series of eyespots that run along the edge of the upperside of the hindwing, with a single eyespot on the forewing. These eyespots were the inspiration for an earlier name of 'Wood Argus' ('Argus' is the many-eyed shepherd of Greek mythology) given to the butterfly by Benjamin Wilkes in *The English Moths and Butterflies* of 1749.

The ♂ has a wingspan of 46 to 52 mm

The slightly larger ♀ has a wingspan of 48 to 56 mm [NF]

♂ and ♀ undersides are similar

This ♀ of subsp. *oblita* from north-west Scotland has prominent white markings

This ♀ of subsp. *insula* from the Isles of Scilly has orange and cream markings [MC]

This species is one of a handful whose physical appearance changes from north to south of its range, forming a 'cline'. Those in the north have a darker brown ground colour and whiter markings than those in the south, especially those found in the Scilly Isles, which have a lighter ground colour with cream and orange markings. Travelling even further south across France and into Spain and southern Europe, the butterfly has markings that are of such a bright orange that this species is often mistaken for its close cousin, the Wall. The differences between regions has given rise to a number of named subspecies, even within Britain and Ireland, although there is no clear-cut demarcation from one subspecies to the next from region to region, such is the nature of clines.

The subspecies *oblita* represents the dark and white-marked individuals found in north-west Scotland, the subspecies *insula* represents the individuals with orange and cream markings found in the Isles of Scilly, and the subspecies *tircis* is found in-between and is the most widespread. There are also minor differences between the spring and summer broods, where the markings on spring brood individuals are normally larger than those of the summer brood, and the butterfly has a lighter appearance as a result.

Life Cycles of British & Irish Butterflies

DISTRIBUTION

The butterfly is widely distributed throughout Britain and Ireland, but is noticeably less common in northern England, southern and eastern Scotland, the Outer Hebrides, and the exposed higher ground in central Wales. It is absent from Orkney and Shetland. However, this is a species that is expanding its range and it is anticipated that, over time, it will fill in some of these gaps. On a worldwide basis, the Speckled Wood is found from Spain in the west to Russia and the Balkans in the east, and also in parts of North Africa.

HABITAT

As its names suggests, this butterfly is found primarily in woodland, although it can also be found anywhere there is sufficient scrub to provide the shaded conditions and dappled sunlight that this butterfly relishes, such as gardens, parks, churchyards and hedgerows. The butterfly also prefers damp areas where the grasses used as larval foodplants grow tall and lush.

A mix of sun and shade near Hazeley Heath in Hampshire provides perfect habitat

STATUS

Despite an unexplained collapse in numbers in the late 19th century, this is one of our few species that is doing well, with a sustained expansion of its range and, as such, is not a species of conservation concern. It is thought that this butterfly has benefited from a reduction in coppicing, resulting in the shadier woodland that this butterfly favours. The long-term trend shows a significant increase in both distribution and abundance, with the more recent trend over the last decade showing a stable distribution and abundance. Using the IUCN criteria, the Speckled Wood is categorised as Least Concern (LC) in both the UK and Ireland.

LIFE CYCLE

This species is unique among the butterflies of Britain and Ireland in that it can overwinter in two different stages—as both a third instar larva and as a pupa. As a result, there is a mixed emergence and adults can be seen anytime from April through to October, with a few adults often seen as early as March or as late as November, especially at southern sites. Populations in the south are the earliest to emerge and the fastest to develop, with three broods, whereas those further north emerge later, develop more slowly and are generally restricted to two broods.

Imago

Both sexes feed from aphid honeydew in the tree canopy (and sap when available) but will also take nectar from a variety of plants if honeydew is in short supply, such as brambles, Common Fleabane, Cuckooflower, dandelions, ragworts and Wild Privet.

The male is the most conspicuous of the two sexes due to its territorial and patrolling behaviours, whereas the female spends most of her time high up in bushes and trees where both sexes roost on the undersides of leaves. With the onset of morning, the adults will manoeuvre themselves with outstretched wings to get the full benefit of the sun and, when they fly down to bask on shrubs closer to ground level, provide the perfect photo opportunity as they continue to soak up the sun.

The male has two strategies for finding a mate. The first is to set up a territory in a particular clearing or hedgerow early in the morning, where he will rest on a prominent sunlit perch waiting for a passing female. If a male encounters another male then the pair spiral around one another, often for some time (I measured over a minute on one occasion), before the defending male returns to his favourite perch.

If no suitable territory can be found, or when there is a large number of males, the male will adopt the second strategy of patrolling. Research has also shown that a male that is lighter in colour (and that takes longer to warm up than a darker male) tends to be less mobile and more territorial, essentially waiting for a female to come to him, whereas a male that is darker tends to be more active and patrols in search of a female. The territorial and less-mobile males also tend to have four eyespots on the hindwing upperside, unlike the three eyespots of the patrolling males, presumably because they are more susceptible to predators and the additional eyespot helps divert the attention of an attacking bird away from the essential body parts.

A mating pair [NM]

The ♀ plays dead when rejecting the advances of a ♂ [NF]

A ♀ depositing a single egg on the underside of a grass blade [NF]

When a male encounters a receptive female, she will fall to the ground or a nearby leaf where, after a brief courtship, they fly off to a suitable platform and mate. If the female is unreceptive then she will simply 'play dead' until the male loses interest and flies off, when she miraculously 'comes back to life'.

The female has a more laboured flight than the male and, when egg-laying, is quite conspicuous as she flies slowly and deliberately over the larval foodplant before alighting on a grass blade and curling her abdomen to lay a single egg, occasionally two, on the blade underside. The primary larval foodplants are Cock's-foot, Common Couch, False Brome and Yorkshire-fog, although over 15 species of grass were recorded as larval foodplants at one Buckinghamshire site. It is thought that temperature is a major factor when a female is choosing a plant on which to lay—in spring and autumn eggs are laid on plants in more open positions, with more shaded plants used in summer.

Ovum

The spherical eggs are just under 1 mm tall and are a translucent light green, remaining this colour until the larva is close to emerging, when its black head can be seen through the shell near the crown. The eggshell is covered in a fine honeycomb pattern with subtle keels that run longitudinally from top to base.

Two eggs were laid on this grass blade [VM]

The egg is a translucent light green

Larva

After one or two weeks, depending on temperature, the 2.5-mm long primrose-yellow **first instar** larva emerges from the egg and usually, but not always, eats its eggshell after emerging. The larva has a shiny black head that distinguishes it from all other instars and its body is covered in tubercles that run longitudinally along the length of the body and that each bear a long hair. The larva feeds on the leaf edges by both day and night and, when fully grown, the first instar is now a yellowish-green and just over 4 mm long.

After around nine days, the larva moults into the **second instar** and now has a green head, a light green ground colour and two translucent points on the anal segment. The larva is also adorned with a series of longitudinal stripes that run the length of its body. The fully grown second instar is just over 8 mm long.

In just under a week, the larva moults into the **third instar**. Having spent hours poring over photographs and specimens of second and third instar larvae, I can confirm that they are very difficult to tell apart. In addition to a small difference in size, the third instar is slightly darker, the stripe along the middle of the back is more pronounced and the length of the hairs on the body are proportionally much shorter.

1st instar: the larva is distinguished from other instars by its black head

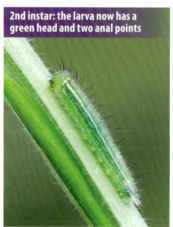

2nd instar: the larva now has a green head and two anal points

3rd instar: the larva has proportionally shorter hairs than a 2nd instar larva

The development of a late-summer larva has five possible outcomes: the larva attempts to hibernate before the third instar but will perish (usually with the onset of extremely cold weather in the first couple of months of the following year); the larva hibernates in the third instar (the only instar in which a larva can successfully overwinter); the larva attempts to hibernate in the fourth instar but will perish; the larva develops into a pupa that overwinters; and the larva goes on to pupate with the adult emerging later in the same year and whose offspring will have to navigate the same tricky path in order to overwinter successfully. Studies have shown that the development path taken by any given individual is driven by both temperature and photoperiod, with some individuals developing very quickly and others relatively slowly, resulting in some individuals overwintering as a third instar larva and others as a pupa. The different rates of development also result in overlapping broods and it is possible to see this butterfly at any time from April to October (and even March and November in good years). It has also been suggested that the variable rates of development can help 'spread the risk' in terms of larval parasitism or unfavourable weather conditions when adults emerge.

Larvae that overwinter will do so shortly after moulting into the third instar (when just over 10 mm long) by fixing themselves to a grass stem low down on the foodplant with a silk pad. Larvae will also feed during the winter months if the weather is mild enough and may wander between plants. Those larvae that do not overwinter become fully grown in this instar in around a week, when they are just under 13 mm long.

The time taken for a third instar to moult into the final and **fourth instar** is highly variable—in early spring then those larvae that have overwintered can take up to a week before they moult, whereas those of subsequent generations take a couple of days at most. The fourth instar is a much brighter green than earlier instars and, in some cases, can appear almost fluorescent against the grasses on which it feeds. When fully grown, the larva is around 27 mm long and this is clearly a distinguishing characteristic between the fourth instar and earlier instars.

4th instar: the larva appears almost fluorescent against its foodplant

An interesting characteristic of this species is that those larvae that are female and that are developing at low temperatures will add an extra instar and go through five instars and not four, a phenomenon known as 'plasticity', although this is the exception to the rule. Additional instars are produced in other species too, usually when a larva is under-developed.

4th instar: a larva preparing to pupate

When preparing to pupate, the larva spins a silk pad and hangs head-down, low down on a grass blade or other vegetation, where it gradually grows paler, pupating after a couple of days. For those fortunate individuals that have Speckled Wood breeding in their garden, it is well worth checking out fences and other garden objects, such as plant pots, where pupae of this species have also turned up on occasion.

Pupa

The 13-mm long pupa is formed head-down, attached to the silk pad by the cremaster. The colour of the pupa varies somewhat, from a pale dull olive to a bright green. If the butterfly does not overwinter in this stage then, after a couple of weeks, the adult butterfly emerges, taking to the wing in its woodland home.

A dull olive pupa showing subtle white markings on the abdomen

A bright green pupa [VM]

A pupa that is colouring up as the adult develops [VM]

Life Cycles of British & Irish Butterflies

Large Heath
Coenonympha tullia

❖ LARGE HEATH IS SUPPORTED BY JANE JONES, HANNAH LEHNHART-BARNETT & TAM STEWART ❖

	January	February	March	April	May	June	July	August	September	October	November	December
IMAGO												
OVUM												
LARVA												
PUPA												

The **Large Heath** is what many would consider a 'northern' species, given its distribution, and it is also rather unique in that it is more or less confined to blanket bogs and raised bogs where its larval foodplants and nectar sources flourish. Unfortunately, habitat loss has resulted in colonies becoming isolated, although the best colonies can be very large in good years, where the number of adults emerging is measured by the thousand. Large colonies used to exist on the mosses around Manchester and Liverpool, but these have long since disappeared. This former location was recognised by William Lewin in his *The Papilios of Great Britain* of 1795 where he named this butterfly the 'Manchester Argus' – *"This butterfly was scarcely known in England till lately; when a gentleman found several in a moorish or swampy situation, near Manchester; and, from their local attachment to the same place, he takes them on the wing every year in July"*.

The butterfly is also known for its variability, especially with regard to the eye spots found on its underside. Specimens in the north have almost no spots at all with adults looking like a large Small Heath, while those in the south have very distinctive spots, and this variability has given rise to three named subspecies. Those with the least distinct spots are referred to as the subspecies *scotica*, those with the most distinct spots as the subspecies *davus* (the darkest and most colourful of the three subspecies) and those that are intermediate between the two as the subspecies *polydama*. The differences in appearance certainly confused early entomologists with Adrian H. Haworth, in his *Lepidoptera Britannica* of 1803, considering each a different species, allocating the name 'Scarce Heath' (no doubt due to its similarity with the Small Heath) to what we now call the *scotica* subspecies, 'Small Ringlet' to the *davus* subspecies and 'Marsh Ringlet' to the *polydama* subspecies. In all three subspecies the wingspan is 35 to 40 mm in both sexes.

Many authors believe that this species forms a 'cline' as one subspecies gradually replaces another throughout its distribution and, therefore, there are intermediates between the three subspecies. There have been many studies to explain this variation and the consensus is that it is the result of natural selection based on predation by birds, especially the Meadow Pipit *Anthus pratensis*. The cooler climate in the north results in less active adult butterflies whose plain undersides provide better camouflage and make them difficult to find while at rest on the ground. Adults further south, on the other hand, are much more active and are more likely to attract the attention of birds as a result. In this case, the prominent eye spots deflect the bird's attention away from the body when the butterfly is attacked.

Tim Melling, who studied this butterfly and others for his PhD thesis on butterfly clines, makes the following comments in the Large Heath species description in *The Moths and Butterflies of Great Britain and Ireland*, which provides some 'food for thought' regarding both the subspecies designations as well as the suggestion of any clinal properties: *"The species occurs in numerous local populations which are phenotypically distinct yet have been pigeonholed into one of three subspecies in Britain ... However, the term subspecies implies some degree of geographical isolation as well as taxonomic distinctiveness and by these criteria it would appear that only* scotica *should merit this title ...* polydama *has been used as a convenient depository for all populations showing intermediate characters"*.

Subsp. *scotica* has negligible spotting and looks like a large Small Heath

Subsp. *davus* has the most prominent spots

Subsp. *polydama* is intermediate between *scotica* and *davus*

DISTRIBUTION

The Large Heath lives in isolated colonies from central Wales in the south to Orkney in the north, but it is absent from Shetland. The subspecies *scotica* is found in northern Scotland, north of a line between Glasgow in the west and Aberdeenshire in the east, as well as in most of the Western Isles and also Orkney. The subspecies *davus* is found in north-west England and central England near the border with Wales. Finally, the subspecies *polydama* is found in northern England, central and north-west Wales, southern Scotland and Ireland, where it is concentrated in the north. Of course, these demarcations are somewhat subjective due to the intermediate forms that occur, and few authors agree on the precise boundary between one subspecies and the next. For example, in his wonderful book *Discovering Irish Butterflies and their Habitats*, Jesmond Harding says that both *polydama* and *scotica* subspecies are found in Ireland. On a worldwide basis, the butterfly has a Holarctic distribution and can be found throughout central and northern Europe, through Asia, across the Bering Strait and into Alaska, Canada and the western USA, where it is widespread.

The butterfly is not particularly mobile, with studies showing that individuals rarely move more than 100 m, with an upper limit of 500 m. However, there is evidence of recolonisations after a local extinction and it is thought that the butterfly may well exist within a 'metapopulation' structure, where there is some interchange between sites.

HABITAT

The Large Heath is found in lowland raised bogs (typically referred to as 'mosses'), upland blanket bogs and damp, acidic moorland. These sites, which usually have a base of *Sphagnum* mosses that are able to hold large quantities of water, are often home to the primary larval foodplant of this species, Hare's-tail Cottongrass (although the butterfly has also been recorded on Common Cottongrass and Jointed Rush), as well as favourite nectar sources of Cross-leaved Heath (the primary nectar source), hawkweeds, heathers, Tormentil and White Clover.

Flowering heads of Hare's-tail Cottongrass are impossible to miss during the Large Heath flight period

Habitat management of Large Heath sites is focused on maintaining the required vegetation, including bog mosses, the primary larval foodplant of Hare's-tail Cottongrass, and the primary nectar source of Cross-leaved Heath. Sites soon become overgrown if they dry out and maintenance of the water table is a primary concern at all sites. Removal of competing vegetation, including heathers and coarse grasses, is also managed through appropriate grazing regimes and scrub removal.

Glen Loy in Scotland where subsp. *scotica* flies

Meathop Moss is a raised bog in Cumbria where subsp. *davus* flies

STATUS

The Large Heath has suffered major declines over the last few centuries due to the draining of the boggy habitats it needs, in order to make way for industry and agriculture. As a result, the butterfly is extinct from its former strongholds on the mosses of Greater Manchester, Merseyside and Cheshire. Sites have become unsuitable for several other reasons too, including a natural succession to woodland, afforestation, peat extraction, burning and overgrazing (which reduces the tussocks of foodplant needed by the larvae).

The long-term trend shows a significant decline in distribution and a significant increase in abundance (that may be due to an increased level of recording), with the more recent trend over the last decade showing a stable distribution and a moderate decline in abundance. Using the IUCN criteria, the Large Heath is categorised as Vulnerable (VU) in both the UK and Ireland.

LIFE CYCLE

The Large Heath has one generation each year, with adults emerging from as early as the end of May at some southern sites, and peaking in the second half of June and early July. The butterfly emerges a month later at more northern sites and at high altitudes. In northern Scotland, where the subspecies *scotica* flies, the butterfly emerges in the second half of June, and peaks in the first half of July. In these more northern sites it is thought that a small proportion of larvae undergo a second winter—rather that continuing to feed in the spring, they remain in their third instar and overwinter for another year.

Imago

Given the boggy nature of the sites inhabited by the Large Heath, it can often be quite difficult to get close to the butterfly without risking 'life and limb'. Furthermore, when disturbed, the butterfly will launch itself into the air, often flying some distance before landing again. All in all, this is not the easiest species to get good sightings of. Fortunately, the stewards of several sites are very accommodating—at the Cumbria Wildlife Trust site at Meathop Moss, for example, a significant stretch of boardwalk has been put in place that reaches out into the moss, allowing close-up views of the Large Heath and its habitat.

Fortunately for the keen observer, the adults remain somewhat active even in dull weather, but will remain tucked away in vegetation in strong winds and cool temperatures. The adults always rest with their wings closed, even in good weather, and regulate their temperature by orientating their wings at an appropriate angle to the sun.

Males spend most of their time searching out a mate by flying over the breeding grounds and, when a virgin female is found, she will typically fly into the air with the male following her, before landing on the ground where mating takes place. In his article 'Mating systems of *Coenonympha* butterflies in relation to longevity', published in *Animal Behaviour* in 1992, Per-Olaf Wickman showed that virgin females have an urgency to mate since they are relatively short-lived, and are not only less discriminating towards males, but will even approach passing males in an attempt to solicit courtship. Wickman also found that virgin females more frequently perched higher up in the herb layer with their heads up, no doubt on the lookout for a passing male. Mated females, on the other hand, tend to stay hidden away in grass tussocks unless they are taking nectar or egg-laying.

Ovum

The 0.8-mm high, flat-topped, spherical eggs are laid singly on a leaf blade or dead leaf stem at the base of a mature tussock of the foodplant. The egg has around 50 longitudinal keels running down its sides from top to bottom and is pale yellow when first laid, although brown markings develop after several days, the egg growing even darker as the larva develops within.

The newly laid egg is pale yellow when it is first laid

The egg develops brown markings as it matures

A broken band eventually develops around the egg

Larva

After around two weeks, the 2.5-mm long **first instar** larva emerges from the egg by eating away at the crown, ultimately pushing its way out of the egg with the crown still attached to the remainder of the eggshell, where it forms a 'lid'. Even at this young age, the larva exhibits two points on its anal segment, which is a characteristic that is only found in later instars in some other members of the Satyrinae subfamily.

1st instar: the newly emerged larva is light brown in colour, including the head

1st instar: the larva develops a green ground colour as it ingests food

A distinguishing feature of the first instar is that it has a light brown head, rather than the predominantly green head found in later instars. The ground colour of the first instar is a light brown when the larva first emerges from the egg, but this becomes progressively greener as the larva starts to feed. When fully grown, the first instar is around 3.5 mm long and, in all instars, the larva is very sluggish in its movement. The larva feeds by day on the tenderest tips of the leaves in all instars, retreating out of sight, deep in the grass tussock, when not feeding or when disturbed.

After around two weeks, the larva moults into the **second instar**, and now has a green ground colour that does not change to the same degree as the first instar as food is ingested. The stripes that run along the length of the body are quite varied in colour with a mixture of white, dark green, light green and light brown lines. The most prominent line is a white stripe that runs along each side of the body, just below the spiracles. The head is now a yellow-green that distinguishes it from the first instar, and the anal points take on a pink hue that is present in this and all remaining instars. When fully grown, the second instar is just over 6 mm long.

After between two and three weeks, from the middle of August to early September, depending on site and temperature, the larva moults into the **third instar**, going into hibernation at the end of September at the base of the grass tussock and re-emerging in March when it continues to feed. When fully grown, the third instar is around 7 mm long, which is not much longer than the second instar, although the larva is considerably stouter. Its markings are also similar to the second instar but are much bolder and less varied in colour.

A surprising characteristic of the larva is its ability to survive long periods under water or even being frozen, which are both distinct possibilities in its boggy habitat, although the larva cannot survive indefinitely in such conditions.

2nd instar: the head of the larva is now a yellow-green and the body stripes are quite varied in colour

3rd instar: a pre-hibernation larva— the body stripes are less varied in colour than in the second instar

3rd instar: the fully grown larva is considerably stouter than the second instar

Most larvae moult into the **fourth instar** during March, when their markings are a little more defined than those of the third instar. However, due to its similarity with the third instar, it is unfortunately necessary to resort to the length of the larva to identify the correct instar—when fully grown, the fourth instar is just under 13 mm long rather than the 7 mm of the third instar. Of course, the time of year can also provide a clue since any larva found between October and February is almost certainly the overwintering third instar.

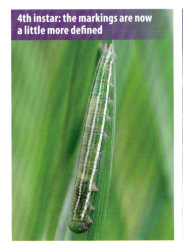

4th instar: the markings are now a little more defined

5th instar: a fully grown larva eating the tip of a grass blade

5th instar: a larva in a pre-pupation J-shape

After between two and three weeks, the larva moults to reveal the final and **fifth instar**. As for the previous instar, the key distinguishing feature of this instar, which in most respects is similar to the fourth instar, is size. When fully grown the fifth instar is just over 25 mm long which is almost twice the size of the fully grown fourth instar.

The final instar larva will wander a little before it finds a suitable site to pupate, which is typically under a grass stem of the foodplant or some other vegetation, where it creates a silk pad from which it suspends itself upside-down, forming a J-shape.

Pupa

After around three weeks, the larva pupates, having spent around 10 months as a larva in total. The 11-mm long pupa hangs head-down, attached by the cremaster to the silk pad created by the larva. The pupa is stunningly beautiful, although rarely encountered, with a ground colour of grass-green and a variable series of dark stripes over the wing cases. The pupa eventually turns brown and ultimately a very dark brown just before the butterfly emerges, having spent three weeks in this stage.

The pupa hangs upside-down on a grass stem or other vegetation

The pupa turns brown as the butterfly develops within

Just prior to the adult emerging, the pupa is a very dark brown

Life Cycles of British & Irish Butterflies

Small Heath
Coenonympha pamphilus

❖ SMALL HEATH IS SUPPORTED BY CAROLINE COLES, GRAEME S DAVIS & JON H-D ❖

	January	February	March	April	May	June	July	August	September	October	November	December
IMAGO												
OVUM												
LARVA												
PUPA												

The diminutive **Small Heath** is one of our most widespread butterflies and will turn up in a variety of habitats and not just heathland as its name implies. In fact, it is often the only species you might find in some of the most remote sites in Britain and Ireland. Indeed, sightings of some of the specialist species found in Cumbria or the highlands of Scotland are often preceded by sightings of this small but robust butterfly—whenever I'm searching out Mountain Ringlet or Large Heath, I almost always find a Small Heath first. This can also be somewhat frustrating, especially when looking for the *scotica* subspecies of Large Heath, which looks like an overgrown Small Heath, and telling the two apart is not as easy as it might sound. The butterfly is also found on the Isle of Rum in the North Ebudes of the Inner Hebrides, where the butterfly has a much greyer overall appearance and is considered to be an endemic subspecies known as *rhoumensis*, although this designation is questioned by some.

The butterfly was given its name by James Petiver in his *Musei Petiveriani* of 1699, despite other authors subsequently suggesting names that include 'Golden Eye', 'Little or Small Gatekeeper', 'Small Argus' and 'Least Meadow Brown', the last of these being, perhaps, the most unfortunate of these alternatives. Petiver's name does at least indicate some affinity with the butterfly's closest relative, the Large Heath, that belongs to the same genus of *Coenonympha*. The main distinguishing feature of this species is that this is the smallest of our 'browns' (the Satyrinae subfamily) and it is closer in size to a skipper, Common Blue or Brown Argus than its relatives, such as the Meadow Brown—the male has a wingspan of between 29 and 33 mm, whereas the female has a slightly larger wingspan of 34 to 37 mm. However, its fluttering flight is quite different from that of the skippers and blues and it is relatively easy to identify in the field.

This charming little butterfly exhibits behaviour that is similar to the Grayling when it lands—it alights with the eye spot on the underside of the forewing visible for a short while, that acts as a decoy to any avian predator, before it eventually lowers its forewing as it settles down to rest. The forewings are also tucked behind the hindwings when roosting and during dull weather, and the butterfly looks quite inconspicuous as the browns and greys on the underside of the hindwings blend in with their surroundings. This butterfly always settles with its wings closed although, in Butterflies of Surrey Revisited, Gay Carr suggests that the female will occasionally open her wings to assess temperature, when she reveals the light brown wing uppersides, which also give the butterfly a much brighter appearance when in flight than the relatively dull undersides might suggest.

♂ and ♀ are similar in appearance [MC]

The butterfly lowers its forewings when roosting and in dull weather

The ♀ will occasionally open her wings to assess temperature [BE]

DISTRIBUTION

This is a widespread butterfly and can be found in grassy areas over most of Britain and Ireland, with the exception of Orkney and Shetland and the most mountainous regions, although the butterfly can be found up to around 700 m above sea level—an altitude that is only exceeded by the Large Heath and Mountain Ringlet. The butterfly lives in small and discrete colonies and adults rarely venture far, although both males and females will occasionally venture further afield and will colonise nearby habitat if it is suitable—adults are sometimes reported several kilometres from known sites. In *The Butterflies of Britain and Ireland*, Jeremy Thomas says that individuals have even been caught on lightships. On a worldwide basis, the butterfly can be found throughout Europe, North Africa and across Asia as far as Mongolia.

HABITAT

Despite its name, the Small Heath can be found in a wide variety of habitats that includes both calcareous and acid grasslands, lowland and upland heathlands, mixed scrub, disused quarries, sand dunes, coastal undercliffs and brownfield sites. The common factors at all suitable sites are that they are dry, well-drained, support low-growing flowering plants (used as nectar sources) and contain fine grasses (used as larval foodplants) that grow in a sparse and relatively short sward when compared with the foodplants used by other members of the Satyrinae subfamily.

Larvae can be found in the relatively short sward at Ballard Down in Dorset

Small Heath is plentiful on the slopes of Ben Lawers NNR in Scotland

STATUS

The most recent analysis of the status of the Small Heath, conducted by a team of conservation scientists, was presented at Butterfly Conservation's Eighth International Symposium in 2018 by Tom Brereton. One fascinating aspect of this analysis is that, over the last 40 years, the population in England has declined by over 60%, has remained stable in Wales, and has increased by just under 90% in Scotland (with insufficient data for Ireland). In other words, the butterfly is declining in the south and east, and increasing in the north and west. Agricultural intensification was not considered a key factor in the decline in England since most Small Heath sites are away from farmland, although loss of habitat due to development was considered a factor, as was higher pollution near urban areas, where elevated levels of CO_2 increases larval development times and the butterfly is potentially sensitive to excessive nitrogen deposition from vehicle emissions and other sources.

The long-term trend shows a significant decline in both distribution and abundance, with the more recent trend over the last decade showing a stable distribution and a moderate increase in abundance. Using the IUCN criteria, the Small Heath is categorised as Near Threatened (NT) in both the UK and Ireland.

LIFE CYCLE

In the southern part of its range the Small Heath typically has two broods, with adults flying seemingly continuously from May until the end of September or early October, with peaks in June and early August. There may even be a third brood in good years. Populations in northern England and Scotland generally have a single brood each year that emerges in the middle of June and flies until the middle of August.

The continuity of sightings of adults from May to early October is not simply the result of the broods overlapping, but also the result of larvae from the previous year reaching adulthood at different times. Specifically, the butterfly can overwinter in one of several instars, and a larva that overwinters in the fourth instar will clearly fly earlier in the year than one that overwinters in the first instar. It has also been found that some larvae from the first brood do not reach adulthood in the same year, but go on to overwinter, whereas their siblings continue to develop and go on to produce another brood.

All in all, the phenology of this species is somewhat blurred as offspring from the current and previous years intermingle as they reach adulthood. Richard South, in his classic work *The Butterflies of the British Isles* of 1906, summarised things quite nicely and his description still stands today: "*May and June butterflies from May and June eggs (twelve months' cycle), July butterflies from August eggs (eleven months' cycle), August and September butterflies (partial second brood) from May and*

June eggs (four months' cycle)". Things would get even more complicated were we to consider the implications of a third brood.

Imago

The Small Heath will spend some time basking, always with wings closed, with its body at an appropriate angle to the sun in order to raise its body temperature. Adults feed on a wide variety of nectar sources, including brambles, buttercups, Common Fleabane, Devil's-bit Scabious, Greater Stitchwort, Kidney Vetch, ragworts, Tormentil and Yarrow, and will roost for the night on plants that provide the best camouflage, such as grass heads, plantain flowers and dead flowerheads.

Much of what we know about the breeding behaviour of the Small Heath is thanks to the studies of Per-Olaf Wickman in Sweden in the late 1980s and early 1990s, and much of the description below is derived from his work.

Males are territorial and will gather in leks that are on the sheltered side of prominent features in the grassland landscape, such as bushes and trees, where they will perch on the ground and await a passing female. Virgin females will also seek out these leks in the hope of finding a mate and will zig-zag repeatedly back-and-forth over the same patch until found by an amorous male. Wickman referred to these leks as 'mating stations' as if to emphasise that they are sought out by both sexes, especially since these leks often have no resources that would otherwise attract either sex, such as nectar sources. Wickman found that both male and female prefer bushes with a narrow base and wide top, a shape that he described as a 'cornet', as opposed to a 'pyramid'.

A mating pair [MC]

As with many other territorial butterflies, a male Small Heath will dart up to inspect any small insect, but an encounter with a virgin female results in the pair landing on the ground, when the male flutters around the female, butts his head up against her, and undoubtedly showers her with pheromones (that act as an aphrodisiac) before they mate. Mating may happen at any time of day and pairings have been observed to last from 10 minutes up to 5 hours.

Females that have already mated avoid male territories and will be found flying over open grassland looking for suitable places to lay their eggs. Suitable plants are inspected before a single egg is laid, normally low down on a grass blade or on nearby dead vegetation. Fine grasses, especially fescues, meadow-grasses and bents, are the preferred foodplants for early instar larvae. In his studies in Sweden, Wickman found that 80% of eggs were laid on Sheep's-fescue (both in captivity and in the wild) and that a female could lay up to 58 eggs in one day. This preference for Sheep's-fescue was also confirmed in English populations by Sarah Meredith, who studied the Small Heath for her MSc in 2009.

Ovum

The egg is 0.7 mm high, has around 50 ribs that run from top to bottom, and has a flattened top. It is usually pale green when first laid, but eventually turns a straw colour and develops a series of blotches as it matures. These blotches may provide a speckled effect or may be joined to create distinct brown zones. As the larva develops the egg turns light brown and becomes increasingly transparent, when the larva can be seen through the eggshell. One interesting characteristic of the egg is that its initial colour will vary, depending on the age of the female that laid it. Wickman discovered that the

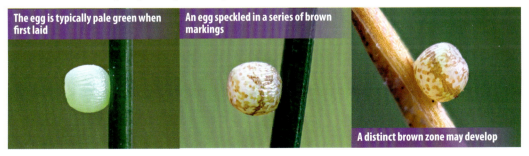

The egg is typically pale green when first laid

An egg speckled in a series of brown markings

A distinct brown zone may develop

first eggs to be laid by any given female were green but that, after 100 eggs or so had been laid, all remaining eggs were yellow in colour. However, no explanation has been given for this change, and both forms appear camouflaged, no matter where they are deposited.

Larva

In his studies in Sweden, Wickman says that the larva can hibernate in any of the first three instars. My own experience of rearing second brood individuals in captivity, and that of F.W. Frohawk as documented in his *Natural History of British Butterflies* of 1924, is that the larva can also hibernate in their fourth instar. This discrepancy may have been answered by Enrique García-Barros who studied the Small Heath in Spain and documented his findings in his paper 'Number of larval instars and sex-specific plasticity in the development of the Small Heath butterfly, *Coenonympha pamphilus* (Lepidoptera: Nymphalidae)', published in *European Journal of Entomology* in 2006, where he reports that larvae that overwintered went through five instars, and those that did not went through four. What we can conclude from these observations is that the Small Heath exhibits a high level of 'plasticity' in terms of both the number of instars, and also the instar in which the larva can overwinter. The description below is of larvae that go through five instars in total, and overwinter while in their fourth instar.

After around two weeks, the 1.7-mm long **first instar** larva emerges from the egg and, even at this young age, possesses two points on the anal segment. Like many of the Satyrinae, the first instar has a light brown ground colour when it first emerges from the egg, with reddish-brown stripes that run the length of the body. It soon turns a light green as it starts to feed, and darker green still as it matures, when it exhibits a prominent white stripe along each side of the body. A characteristic that distinguishes this instar from later instars is that the head of the larva is light brown, rather than green. Larvae spend most of their time tucked away at the base of a tuft of grass, feeding primarily at night on the tender tips of the grass blades. Larvae are very slow-moving, a characteristic that stays with them until pupation. When fully grown, the first instar is 3.5 mm long.

After around a week, the larva moults into the **second instar**. The overall appearance is very similar to the fully grown first instar, with the exception that the head is now light green and a similar colour to the body, and the white stripe along each side is a little more prominent. The fully grown second instar is 5.5 mm long.

1st instar: the newly emerged larva has a light brown ground colour

1st instar: the fully grown larva has a white stripe along each side

2nd instar: the larva now has a light green head

3rd instar: the larva is a brighter green than earlier instars

After between three and four weeks, the larva moults into the **third instar** and is now a brighter green than earlier instars, with a surface that has a rough appearance as a result of numerous short spines that cover the body which, while present in the second instar, are much fewer in number. The larva continues to feed primarily at night but will now rest on a grass blade during the day, rather than disappearing into a tuft of grass. When fully grown, the third instar is 7 mm long.

After around four weeks, the larva moults into the **fourth instar** and a key distinguishing feature of this instar is that the two points on the anal segment have a pink tinge. After feeding for a short while, the larva enters hibernation at the base of a grass tussock where it rests in a straight position. However, the larva never becomes fully torpid and can be found feeding when conditions are mild. Larvae re-emerge in mid-March and start feeding in earnest. The post-hibernation larva is slightly brighter, with several lines that run the length of the body (including a dark line that runs along its back) that give it a striped appearance that is not dissimilar to a Large Heath larva. When fully grown, the fourth instar is 11 mm long.

There is some debate regarding the grasses that certain species of the Satyrinae use as they mature (especially Mountain Ringlet and Scotch Argus) and, as an experiment, I provided a captive-reared fourth instar Small Heath larva with both Sheep's-fescue and Cock's-foot. I was not completely surprised when I found the larva happily munching away on the coarser Cock's-foot since the larvae of related species are known to start their lives feeding on fine grasses, such as fescues, but switch to coarser grasses as they mature.

4th instar: the mature larva has a striped appearance and the points on the anal segment now have a pink tinge (inset)

After another few weeks, in the first half of April, the larva moults into the final and **fifth instar**. The larva is similar in appearance to the fourth instar, but the markings are much more clearly defined. In particular, the dark line that runs along the back of the larva is now bordered by a narrow white line on each side and there is also an additional white stripe on each side of the body that sits halfway between the dark line on the back and the prominent white stripe found along the base of the side. The body is also covered in a sprinkling of white spots and, while these are also present in the fourth instar, they are much more prominent in the fifth instar. After a few weeks, the fully grown and 19-mm long larva prepares for pupation by creating a silk pad on a grass blade from which it hangs upside-down in the shape of a letter 'J'.

The Small Heath often inhabits the same sites as other Satyrines, such as Meadow Brown, Marbled White, Speckled Wood and Wall, and all of these species have similar green larvae. However, the one factor that distinguishes the Small Heath larva from all of these is the lack of hairs on its body and head.

5th instar: the fully grown larva is hairless and around 18 mm long

5th instar: a pre-pupation larva

The pupa hangs upside-down from a grass stem

The pupa darkens as the adult butterfly develops

Pupa

After three or four days, the larva sheds its skin for the last time, revealing a 9-mm long pupa that is attached by its cremaster to the silk pad. The pupa is initially green but, after a few days, develops several streaks along the wing cases and head of the pupa, and the contrast is quite beautiful. The pupa darkens as the adult butterfly develops and the wings can be seen quite clearly through the pupal case just before the butterfly emerges, after between three and four weeks in this stage.

Life Cycles of British & Irish Butterflies

Mountain Ringlet
Erebia epiphron

❖ MOUNTAIN RINGLET IS SUPPORTED BY PAUL DOYLE, ROLF FARRELL & ADAM McVEIGH ❖

Many butterfly enthusiasts set themselves the objective of seeing all of our species and, as they get close to achieving their goal, realise that they have yet to see the Mountain Ringlet, for this is one of our most difficult species to see—the butterfly is only found in discrete colonies in remote locations, it has an extremely short flight period of only a few weeks and the adult only flies when conditions are sufficiently bright and warm. For these same reasons, this is also one of our least-studied butterflies. In very overcast and cool weather the butterfly remains sheltered deep in grass tussocks, but will take to the air in favourable conditions when, in the company of its fellow Mountain Ringlets, it will bring an apparently dormant landscape to life.

The butterfly itself may seem unremarkable—it is, after all, a small and dark butterfly with a number of black-pupiled streaky orange spots and seems to spend most of its time hunkered down in grass tussocks. However, one only has to remember that this is our only montane species, flying high on the mountains of Cumbria and Scotland, to truly appreciate its uniqueness among our butterfly fauna. The butterfly was given its name by Adrian Haworth in his *Lepidoptera Britannica* of 1812.

Subsp. *mnemon* ♂ with a wingspan of 28 to 36 mm

Subsp. *mnemon* ♀ with a wingspan of 28 to 38 mm

Subsp. *scotica* ♂ with a wingspan of 32 to 40 mm

Subsp. *scotica* ♀ with a wingspan of 32 to 42 mm

It is thought that this butterfly was one of the first to recolonise our shores after the last ice age, although it is a relatively recent discovery—it was found at Red Screes, above Ambleside in Cumbria, in 1809. The butterfly was not discovered in Scotland for another 35 years, when it was found at Loch Rannoch in Perthshire in 1844. The Cumbrian population is described as the subspecies *mnemon* and the Scottish population as the subspecies *scotica*, with the latter considered to be slightly larger and with more prominent red spots and black pupils.

The undersides of both sexes and subspecies are similar

DISTRIBUTION

This butterfly is found in two main regions in Britain. It is found in Cumbria in England, and also in west-central Scotland, primarily in Argyllshire, West Inverness-shire and Mid Perthshire, with a few

scattered colonies elsewhere. The butterfly is surprisingly absent from Snowdonia and the Pennines although, on the basis of four specimens in total, all taken in the 1800s, this butterfly is also thought to have previously occurred in Ireland and belonged to a different subspecies than those found in Britain.

As its name suggests, this butterfly is found in mountainous areas accross Europe, and typically between 500 and 700 m above sea level in Cumbria, where colonies are found flying on slopes with various aspects, and between 350 and 900 m in Scotland, where colonies tend to be found on south-facing slopes. The butterfly forms discrete colonies, often on a plateau or in a steep-sided gulley, and the butterfly can be measured by the thousand at the best sites. A study by Keith Porter at Seathwaite Fell in Cumbria in 1974 gave an estimate of 2,870 adults on the peak day during the flight period, with a projection of between 8,500 and 9,000 butterflies flying there that year. Males are more active than females and will make single flights of up to 200 m (average 25 m) over their site when they are quite visible as a small and very dark flapping insect flying clumsily and low over the ground. Females, on the other hand, are much heavier due to their cargo of eggs, and make shorter flights of up to 35 m (average 8 m).

HABITAT

The butterfly prefers moist or boggy ground in sheltered depressions where the primary larval foodplant of Mat-grass and favoured nectar sources grow in abundance. Characteristic plants in preferred areas include sedges and, in drier areas, Bilberry and Wavy Hair-grass. These conditions are often found in small localised areas on a mountainside, which results in colonies that are highly concentrated. In a study described in the article 'The Distribution and Habitat of Mountain Ringlet *Erebia epiphron* (Knoch) in Scotland', published in *Atropos* in 2010, Andrew Masterman demonstrated that areas inhabited by the butterfly had more Mat-grass and more nectar sources

Irton Fell in Cumbria, home to subsp. *mnemon*

Above Lawers Dam at Ben Lawers in Perthshire, Scotland, home to subsp. *scotica*

than uninhabited areas, confirming the theory that the butterfly's distribution is limited by herb-rich *Nardus* grassland, at least in Scotland. In the article 'Changes in Mountain Ringlet *Erebia epiphron* (Knoch) Habitat with Altitude in Scotland, Associations with Topography and Possible Effects of Climate Warming', published in *Atropos* in 2012, Masterman goes on to suggest that the underlying geology and soils, that ensure a high level of soil moisture and mineral enrichment, are essential to the creation of this desired grassland.

STATUS

This is one of our most difficult species to monitor given its remote sites, its short flight period, and differences in the timing of the flight period at different locations and different altitudes. Some surveys have suggested that the butterfly has suffered significant losses, especially at low elevation sites, most of which are found in the Cumbrian population. Ongoing work being undertaken by a team from the University of York, and whose findings were presented at the Butterfly Conservation's Eighth International Symposium in 2018 by Andrew Suggitt, suggests that the average elevation of surviving colonies has risen by 200 m over the last 40–50 years, in a manner consistent with an adverse response to climate change.

Whatever the true picture, the thought that the butterfly is unable to survive at lower elevations due to climate change, and must necessarily move to higher ground, or a different aspect on a mountainside, makes perfect sense. Masterman suggests that the butterfly will, however, be limited as it moves to a higher altitude, with very little suitable habitat found above 900 m, at least in Scotland. Also, where such habitats exist, their increased exposure to wind may render them unsuitable for successful occupation by the butterfly (Andrew Suggitt, pers. comm.).

The long-term trend shows a significant decline in distribution, with the more recent trend over the last decade showing a moderate decline in distribution. No figures are available for relative abundance due to the sheer difficulty in walking regular transects for the species in good weather. Using the IUCN criteria, the Mountain Ringlet is categorised as Near Threatened (NT) in the UK and Regionally Extinct (RE) in Ireland.

LIFE CYCLE

There is one generation each year and, in Cumbria, adults emerge at the start of June (occasionally in late May) and fly for a few weeks only. Emergence is later in Scotland and starts toward the end of June. The peak emergence is difficult to predict each year given the vagaries of the weather and it is best to be flexible in your plans if the Mountain Ringlet is a target species. The butterfly overwinters as a larva and it is thought that the butterfly may spend two winters in this stage if conditions are unfavourable, and this is a phenomenon that has been proven in captivity.

Imago

The butterfly spends a large amount of time at the base of grass tussocks, where their dark undersides provide remarkably good camouflage against avian predators, such as Meadow Pipits *Anthus pratensis*, Pied Wagtails *Motacilla alba*, Skylarks *Alauda arvensis* and Whinchats *Saxicola rubetra*. With an increase in temperature, the butterfly will crawl out of the tussock where, after warming up, it will take to the wing. Most authors say that the butterfly only flies in bright sunshine although, on several occasions, I have also seen the butterfly take to the air even in hazy conditions so long as it is sufficiently bright. Adults will feed on whatever nectar sources are available at their site, including Bilberry, hawkweeds, Heath Bedstraw, Meadow Buttercup, Tormentil and Wild Thyme. Males are also known to feed from damp soil where they take on minerals.

Males are the most often seen of the two sexes as they zigzag low over their habitat, investigating any brown object that might be a virgin female. The female, on the other hand, will typically roost among grass stems until she is located by a male, when she is quickly mated, thereafter flying only to feed or lay eggs.

A mating pair with ♂ above

Eggs are laid singly among vegetation that is between 6 cm and 17 cm tall, usually on a dead grass blade (and not necessarily a foodplant) or on moss that is 1 cm and 6 cm from the ground. Females are able to lay a significant number of eggs per day, some say up to 70 eggs, and such a large number is not surprising given the butterfly's short lifespan. Eggs are most frequently laid on the key larval foodplant of Mat-grass, as well as Sheep's-fescue, which may also be used as a foodplant, given that larvae will readily feed upon it in captivity. Other potential foodplants that have been suggested include Annual Meadow-grass, Early Hair-grass and Tufted Hair-grass.

Ovum

Each 1.2-mm tall egg has between 17 and 20 ribs running down its sides. It is pale cream when first laid, but develops reddish-brown speckled blotches after a few days, ultimately becoming translucent with parts of the larva visible just prior to hatching.

The newly laid egg is pale cream

The egg develops reddish-brown speckled blotches after a few days

The egg becomes translucent just before the larva emerges

Larva

While the adult butterfly is relatively under-studied when compared with other species, the immature stages are even less studied. In completing this book, several larvae were reared in captivity and underwent six instars in total, which are reflected in the descriptions below. Some sources state that the larva goes through five instars, and I believe that in these cases it is the second instar that is overlooked.

After between two and four weeks, depending on temperature, the 1.7-mm long **first instar** larva emerges from the egg by eating away the crown and consumes a good amount of the eggshell as its first meal. The larva has a light brown head with a matching ground colour when first emerged, but its body develops a green hue as it feeds, primarily at night, on the foodplant. An orange-brown line runs along the centre of the back, with three more lines that run along each side, and the body is speckled with tiny dark warts. The larva also has a pale whitish line along the base of each side that becomes more prominent as it matures. An important feature of this instar is that it has no anal points but does have six hairs on the anal segment that point backwards, each of which emanates from a dark base. When fully grown, the first instar is 3 mm long.

After around 12 days, the larva moults into the **second instar** and is similar in appearance to the first instar with a light brown head and no anal points. However, the larva now has a prominent white stripe along the base of each side, just below the spiracles, and a very faint white line that runs along the middle of each side. When fully grown, the second instar is 4.5 mm long.

1st instar: the newly emerged larva has a light brown ground colour

1st instar: the larva develops a green hue as it feeds

2nd instar: the larva has a prominent white stripe along each side

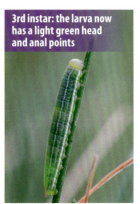

3rd instar: the larva now has a light green head and anal points

After around 10 days, the larva moults into the **third instar** and is similar to the second instar with a few key differences. The head is light green rather than light brown, the larva is now in possession of two points on the anal segment that are tinged with pink, and a dark line runs along the back that is bordered either side with a pale white line. When fully grown, the third instar is just over 5 mm long.

After three weeks, the larva moults into the **fourth instar** and is similar in appearance to the previous instar. The larva enters hibernation while in this instar, sometime in September, remaining at the base of the grass tussock until March, when it recommences feeding. While the length of the larva changes within an instar, the size of the head capsule does not—in my studies, the differences in head capsule size were used to distinguish the last four instars, which are very similar in appearance. The fully grown fourth instar is just over 9 mm long.

At the end of April or start of May, the larva moults into the **fifth instar** and is similar in appearance to the previous instar, apart from size. The fully grown fifth instar is around 16 mm long.

4th instar: this instar grows to 9 mm long

5th instar: this instar grows to 16 mm long

6th instar: this instar grows to between 19 and 21 mm long

After around 20 days, the larva moults into the final and **sixth instar** and, apart from size, is similar in appearance to both the fourth and fifth instars. Although the larva continues to feed primarily at night, it will also feed during the day. After two weeks in this instar, the larva is fully grown and between 19 and 21 mm long. When ready to pupate, the larva moves to the base of the grass tussock and spins a few silk strands between grass stems.

Pupa

The 11-mm long pupa lacks any hooks on the cremaster and simply rests at the base of the tussock, unattached to any surface. It is greenish-brown in colour, although cream examples are also produced, and dark streaks develop on the casing of the wings, antennae proboscis and legs. There are also two white stripes that run in parallel along the back and another on each side of the body, which match the position of the white stripes found in the larva. The pupa darkens as the adult develops and, after three weeks in this stage, the butterfly emerges, when it crawls up a grass stem to expand its wings before taking flight, if conditions are suitable.

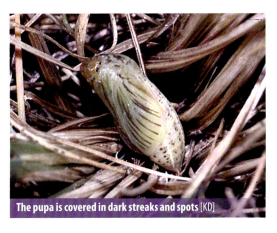

The pupa is covered in dark streaks and spots [KD]

Life Cycles of British & Irish Butterflies

Scotch Argus
Erebia aethiops

❖ SCOTCH ARGUS IS SUPPORTED BY MIKE & GILL BERRY, PAUL KIRKLAND & CALLUM MACGREGOR, IN MEMORY OF GRANDAD ❖

	January	February	March	April	May	June	July	August	September	October	November	December
IMAGO												
OVUM												
LARVA												
PUPA												

Life Cycles of British & Irish Butterflies

A freshly emerged Scotch Argus is a delight to behold, with dark brown velvety uppersides that appear almost black when seen from a distance. When seen close up, and especially when basking with its wings held open, the butterfly is unmistakable with distinctive orange bands on both forewings and hindwings that contain several eye spots. Like the Mountain Ringlet, the ability of this butterfly to survive cool temperatures suggests that it was probably one of the first species to recolonise Britain after the last ice age, over 11,000 years ago.

The butterfly was given its name by Edward Donovan in his *The Natural History of British Insects* of 1807, a name that would imply that this is a butterfly that is only found in Scotland, but this is not the case since there are two significant colonies in England that are the remnants of a much broader former distribution. That being said, the butterfly's strongholds are undoubtedly north of the border, where it is often the commonest butterfly at damp grassland sites where it can fly by the thousand. In *The Lepidoptera of the British Islands* of 1893, Charles Barrett recounts Dr F. Buchanan White: "*Rather open spaces in woods are favourite spots, and these are often 'alive' with the butterflies flitting about in an apparently aimless way. In Glen Tilt one day in August 1877, turn whichever way one would, there was always to be seen a great multitude flitting about over the grass. The monotony of seeing these brown butterflies, zigzagging in every direction, became almost like a dreadful nightmare, and caused a feeling of dizziness which I can still recall to mind*".

Male and female are similar in appearance, although the male is slightly darker. Both sexes exhibit a mix of pale bands on their undersides—the male has a reddish-brown underside whereas that of the female is paler and comes in one of two colour forms—either violet-tinged or yellow-tinged. In his article 'Further Observations on the Scotch Argus *Erebia aethiops* (Esp.)', published in *Atropos* in 2013, Paul Kirkland shows that the two forms appear in more-or-less equal proportions at Smardale Gill in England, but that the yellow form dominated in the two Scottish populations he was studying. In both sexes, the butterfly is well-camouflaged when resting on the ground or within vegetation, where it resembles a withered leaf.

The butterfly has two subspecies on our shores. The nominate subspecies *Erebia aethiops aethiops* is found in two colonies in England and also in north-east Scotland, and the subspecies *caledonia*, which is invariably smaller in size with less prominent markings, is found in western and southern

Subsp. *aethiops* ♂ with a wingspan of 44 to 48 mm [IL]

Subsp. *aethiops* ♀ with a wingspan of 46 to 52 mm [NF]

Subsp. *aethiops* ♂ underside [VM]

Subsp. *aethiops* ♀ (violet form) [IL]

Subsp. *aethiops* ♀ (yellow form) [IL]

Life Cycles of British & Irish Butterflies

Subsp. *caledonia* ♂ with a wingspan of 39 to 45 mm

Subsp. *caledonia* ♀ with a wingspan of 41 to 49 mm

Scotland, although the relative distributions of the two subspecies are not clear cut. The butterfly was also known for its local races, which included a colony at Grass Wood near Grassington in Mid-West Yorkshire that became extinct in 1955 and where adults had reduced orange markings.

DISTRIBUTION

Despite its name, the Scotch Argus is found in two sites south of the border, both in Cumbria—at the National Trust site at Arnside Knott and at Smardale Gill near Kirkby Stephen, that is mostly owned by Cumbria Wildlife Trust. The butterfly was formerly found at other sites in Cumbria; Northumberland; Castle Eden Dene in County Durham; Grassington in Mid-West Yorkshire; and was found as far south as Colne, near Burnley, in East Lancashire. Sites are more numerous in Scotland, where this butterfly can be found in most of northern, western and south-western Scotland. This butterfly is absent from the lowlands of central Scotland, many of the Western

Isles, Orkney, Shetland, Wales, Ireland and the Isle of Man. On a worldwide basis, the butterfly has a Palearctic distribution and can be found from central Europe, through to temperate Asia as far as western Siberia.

This butterfly lives in well-defined colonies that are often very large and extend over several kilometres of habitat, although it is thought that adults are relatively sedentary, flying 500 m at most. Unlike many other species of *Erebia*, such as the Mountain Ringlet, this is not a montane species and is typically found flying below 500 m above sea level, although there are occasional sightings of singletons or small groups at higher altitudes when temperatures are high and when the butterfly is thought to be dispersing, since these sightings are often in habitat that is unsuitable for breeding.

HABITAT

Scottish colonies are found in open but sheltered damp grassland, including moorland, woodland clearings, riverbanks and young plantations—where nectar sources and larval foodplants are to be found. The butterfly is also found in areas where woodland does not form a closed canopy, and this may explain why the English colonies are restricted to limestone grassland sites on nutrient-poor soils that are sheltered by scrub and adjoining woodland. The butterfly favours tall grasses and therefore prefers sites that are ungrazed or lightly grazed. In his 2004 MSc thesis, entitled 'Ecology,

Subsp. *caledonia* flies at Butter Bridge, Cairndow, Scotland

Subsp. *aethiops* flies on the slopes of Arnside Knott, Cumbria [NF]

behaviour and conservation of *Erebia aethiops*, the Scotch Argus butterfly', Paul Kirkland suggested a typical sward height at occupied sites of between 8 and 21 cm.

STATUS

Despite a long-term decline in its distribution, the Scotch Argus has increased in abundance at many monitored sites and is not a priority species for conservation efforts. The long-term trend shows a moderate decline in distribution and a significant increase in abundance, with the more recent trend over the last decade showing a moderate decline in distribution and a moderate increase in abundance. Using the IUCN criteria, the Scotch Argus is categorised as Least Concern (LC) in the UK.

LIFE CYCLE

The butterfly is single-brooded, with a flight period that is relatively short for a butterfly that can appear in significant numbers. Butterflies emerge in the second half of July, peak in early August, and tail off into September. The butterfly is occasionally sighted in the second week of July in England.

Imago

The Scotch Argus is often the first butterfly of the day to be seen, when its dark brown wings allow it to warm up more quickly than other species. Adults also fly only when the sun is shining and tend to retreat among grasses as soon as clouds appear, only to reappear as rapidly when the sun comes out again—it is fascinating to watch an apparently dormant grassland landscape come to life with butterflies as the clouds move away. Of course, there are always exceptions, and males will occasionally fly in overcast conditions, so long as the temperature is sufficiently high. Adults are also known to deliberately retreat into shaded areas on extremely hot days when their usually advantageous dark colouring puts them at a disadvantage.

Adults usually emerge on a warm morning and will delay their emergence until conditions are suitable. Both sexes feed on a variety of nectar sources and usually use whatever is available, including brambles, Devil's-bit Scabious, Field Scabious, hawkweeds, heathers, knapweeds, ragworts, thistles and Wild Marjoram. Females are especially fond of flowers and have been recorded feeding for up to 30 minutes at a time on occasion.

Males spend much of their time searching for a mate and will adopt both perching and patrolling strategies. Perches are usually on grass heads, from which the male will fly up to investigate passing butterflies before returning to his perch. Patrolling flights can be particularly lengthy, when the male will weave slowly among grass stems, searching out any dark brown object that is a potential mate. In an entry in *The Entomologist* in 1895, J.C. Haggart writes: "*The under side of the insect bears a marked resemblance to that of a dead leaf, and I have often watched the males being deceived by withered leaves lying amongst the moss. They would flutter down quite close to the leaf, immediately rise with a disappointed air and fly a little further, only to be deceived again and again*".

Females spend most of their time basking and feeding and are usually mated without any discernible courtship shortly after they have emerged in the morning, with the pair remaining coupled for two hours or so, although those pairing late in the day may remain 'in cop' overnight. An unreceptive female, on the other hand, will avoid a male by fluttering her wings while moving away from him, and flying off if her suitor is persistent.

Subsp. *aethiops* mating pair with ♀ on right

The female selects sheltered sites that are in full sunshine when egg-laying. She will bask for a while before crawling low down into the grass where she lays a single egg, either on a grass blade or stem, or on nearby vegetation or debris. The main foodplant in Scotland is thought to be Purple Moor-grass, with the populations in northern England using Blue Moor-grass, although this is very much a generalisation and other foodplants are used. At Arnside Knott larvae have been found feeding on Blue Moor-grass and Sheep's-fescue. At Smardale Gill, Kirkland found larvae feeding on Blue Moor-grass, Sheep's-fescue, Purple Moor-grass, Cock's-foot and Glaucous Sedge, with Ron Baines from the Cumbria Wildlife Trust also finding eggs on Common Couch. At Blair Atholl in the Highlands of Perthshire, Kirkland found larvae using Purple Moor-grass, Sheep's-fescue, Glaucous Sedge, Tufted Hair-grass, Wavy Hair-grass, Common Bent and Sweet Vernal-grass. Here, Kirkland found that the Scotch Argus, like the Chequered Skipper, will only choose a small proportion of the Purple Moor-grass when available, selecting only those plants that will remain green through the autumn when the larvae are still feeding prior to hibernation.

Ovum

As documented by Enrique García-Barros in 'Egg size in butterflies (Lepidoptera: Papilionoidea and Hesperiidae): a summary of data', published in *The Journal of Research on the Lepidoptera* in 2000, the volume of the Scotch Argus egg relative to its wingspan is 3.1% and is the largest ratio of all of the British and Irish butterflies (the smallest ratio belongs to the Meadow Brown, which is just 0.31%). The 1.3-mm high spherical egg has around 25 shallow ribs that run along the sides from top to bottom.

The egg is pale yellow when first laid

The egg develops reddish-brown blotches after a few days

It is pale yellow when first laid, but turns pale brown with reddish-brown blotches after a few days, becoming a more uniform brown just prior to hatching, when the striped larva can be seen through the eggshell, coiled up and with its head just below the crown.

1st instar: a larva pushing its way out of its egg

Larva

After between two and three weeks, the **first instar** larva nibbles around the top of the egg to create a hinged 'lid', before it pushes its way out, subsequently eating a good proportion of the eggshell as its first meal.

Given the large size of the egg relative to the adult butterfly, it is not surprising that the larva is equally large, being 2.5 mm long when newly emerged. The larva has a pale brown ground colour with a rust-coloured line along the middle of its back, and three more on each side of the body. It is covered in several rows of small black warts that each bear a short hair. In this instar, the larva has an anal flap that has six

1st instar: the newly emerged larva is pale brown

1st instar: the mature larva develops a greenish hue

long hairs that project backward away from the body. As it feeds, the larva develops a greenish hue, together with a pale line along each side between two of the rust-coloured lines, and another pale line that is in line with the spiracles at the base of each side. When fully grown, the first instar is 5.5 mm long.

Early instar larvae feed by both day and night but develop slowly and it is two to three weeks before the larva moults into the **second instar**. While superficially similar to the first instar, the larva now has two stubby anal points and is covered in short backward-pointing curved hairs all over the surface of the body, with each emerging from a conical base. The ground colour is now a creamy brown and the reddish-brown stripes and pale lines along the back and sides are much more prominent. The larva overwinters in this instar, occasionally in the first instar, and enters hibernation in leaf litter at the base of the foodplant at the end of October or in November, when it is just over 7 mm long. The larva recommences feeding at the end of March and early April and, when fully grown, the second instar is 8.5 mm long. The larva now feeds primarily at night, remaining at the base of grass stems during the day

2nd instar: a fully grown post-hibernation larva

2nd instar: a pre-hibernation larva with two stubby anal points

3rd instar: the larva has a pale grey-green ground colour

Toward the end of April, the larva moults into the **third instar** and now has a pale grey-green ground colour with a dense sprinkling of hairs that give the body a rough and granular appearance. The dark lines along the body now have a darker purplish-brown hue and the divisions between segments are light brown. In all other respects the larva is similar to the previous instar and, when fully grown, the third instar is around 14 mm long.

The larva moults into the final and **fourth instar** around the middle of May and now has a light yellowish-brown ground colour. While the brown line along the back remains, all other dark lines are less distinct than in previous instars, with the whitish lines along the middle and base of each side the most prominent features. The larva remains hidden at the base of grass stems during the day and slowly crawls up a grass stem at dusk to feed. When fully grown, the larva is around 25 mm long and prepares for pupation by creating a slight hollow at the base of grass stems, typically in mosses or some other soft material, where it spins a few silk threads while remaining somewhat upright.

4th instar: the larva has a yellowish-brown ground colour

Pupa

The 13-mm long pupa is formed from mid-June and is a uniform and semi-transparent light brown when first formed, becoming a deeper brown and more opaque as it matures. As the adult develops, the eyes darken and the wings turn cream, darkening further just before the adult emerges. After between two and three weeks, depending on temperature, this northern grassland butterfly emerges.

The pupa is formed at the base of grass stems

Life Cycles of British & Irish Butterflies

Ringlet
Aphantopus hyperantus

❖ RINGLET IS SUPPORTED BY TUC AHMAD, TIMME KOSTER, THE NETHERLANDS, NL BUTTERFLIES & JANET PYCROFT ❖

	January	February	March	April	May	June	July	August	September	October	November	December
IMAGO												
OVUM												
LARVA												
PUPA												

The **Ringlet** is a relatively common butterfly and, when in flight, can be confused with the Meadow Brown with which it often shares its habitat. When at rest, however, there is no mistaking this butterfly—the uniform chocolate brown ground colour and the rings found on its wings, which are especially prominent on the undersides, are unlike any other butterfly found on our shores. A newly emerged adult has velvety wings that provide a striking contrast with the delicate white fringes of the wing edges and makes for a surprisingly beautiful insect. The dark ground colour also allows this butterfly to warm up quickly, and it can be found bobbing around its habitat on overcast days and even in light rain so long as the temperature is sufficiently high. Male and female are similar in appearance, although the male is slightly darker, almost black, and it is sometimes possible to make out the faintest of sex brands on his forewings that contains specialised scent scales used during courtship.

♂ and ♀ undersides are similar

The ♂ has a wingspan of 42 to 48 mm [MC]

The ♀ has a wingspan of 46 to 52 mm [IL]

The butterfly was first named 'The brown ey'd Butterfly with yellow circles' by James Petiver in his *Musei Petiveriani* of 1699. In his *Papilionum Britanniae* of 1717, Petiver gave the butterfly the names 'Brown 8 Eyes' and 'Brown 7 Eyes', presumably thinking that they may be different species, rather than variation within the same species. The butterfly is subject to a significant amount of variation, especially with regard to the number and significance of the spots. The *lanceolata* aberration is particularly striking, where the rings are elongated to form teardrops. The name we use for the butterfly today was provided by Moses Harris in his *The Aurelian* of 1766.

DISTRIBUTION

This butterfly can be found throughout most of Britain and Ireland, but is absent from north and north-west Scotland, the Outer Hebrides, Orkney, Shetland, the Isle of Man and the Channel Islands. The butterfly's range is also expanding and the gaps in its distribution in parts of north-east Wales and north-west England are gradually being filled. Worldwide, the butterfly is found across much of the Palearctic, from northern Spain to southern Scandinavia and eastwards across temperate Asia as far as Japan.

This butterfly forms discrete colonies although numbers vary from a few dozen in small patches of grassland to several thousand at the largest sites. This damp-loving species also flourishes in the year

Life Cycles of British & Irish Butterflies

following a relatively rainy season and, conversely, suffers after a hot and dry summer. Adults are fairly sedentary and will typically remain within the confines of their site but, if their site is large, will fly freely through it. That being said, the butterfly may be more mobile than is thought, given the expansion of its range in recent decades. In a study of a population in Switzerland that is described in 'Distribution and dispersal patterns of the ringlet butterfly (*Aphantopus hyperantus*) in an agricultural landscape', published in *Bulletin of the Geobotanical Institute ETH* in 2003, Regula Billeter, Isabella Sedivy and Tim Diekötter suggest that the butterfly may reside within a metapopulation structure—a network of interconnected colonies.

HABITAT

Preferred sites are characterised as being sheltered and damp, with plentiful nectar sources, and includes woodland clearings, rides and glades; mixed scrub; hedgerows; meadows and pasture; arable field margins and set-aside; river banks; road verges; and disused railway lines—wherever the full heat from the summer sun can be avoided and where grasses are lush and tall, although dense woodland is avoided. The butterfly is not typically found in open areas, such as grassland or heathland, and prefers areas with partial shade, at least in the south of its range where conditions are warmer.

Bell Coppice Meadow, one of the many damp meadows within the Wyre Forest on the Worcestershire/Shropshire border [JBi]

STATUS

This is one of the few species that is doing well, with evidence of increases in both distribution and abundance. It is not, therefore, a priority species for conservation efforts. The long-term trend shows a significant increase in both distribution and abundance, with the more recent trend over the last decade showing a moderate increase in distribution and a significant increase in abundance. Using the IUCN criteria, the Ringlet is categorised as Least Concern (LC) in both the UK and Ireland.

While the butterfly has been lost in some areas due to the draining of the damp habitats on which it depends, with the subsequent loss of the tall and lush grasses that it favours, the general outlook is positive. In *The Millennium Atlas of Butterflies in Britain and Ireland*, Jim Asher and colleagues suggest that the Ringlet may have been adversely affected by atmospheric pollution, which explains losses from industrial areas during the 19th century, and subsequent expansion as a result of cleaner air: "*Although the theory has not been thoroughly investigated, the distribution of the Ringlet coincides with the distribution of lichens, which are adversely affected by sulphur dioxide pollution*". Other beneficial factors that have been suggested includes a warming climate and an increase in nitrogen deposition that, in essence, acts as a fertilizer and encourages the growth of coarse grasses.

LIFE CYCLE

There is one generation each year, with adults emerging in the second half of June, peaking in mid-July, with some individuals flying into August. Males emerge before females and the flight period is relatively short when compared with its close relatives.

Imago

The Ringlet can be a difficult butterfly to observe, since it is often restless as it makes short flights in and out of shade. Early morning is a good time to see this species, when it sits with wings outstretched as it warms up on a suitable platform, such as a Bracken frond or bramble leaf. The butterfly will always settle with its wings closed in hot weather, when the beautiful eye spots on its undersides are on full display. When roosting and during particularly inclement weather, the butterfly will hide deep within vegetation. Both sexes are particularly fond of bramble flowers as a nectar source, but will also use Common Fleabane, Hemp-agrimony, Kidney Vetch, ragworts, thistles, Wild Marjoram and Wild Privet.

A mating pair [BE]

Males adopt an exclusive strategy of patrolling for mates and are often seen in ones and twos fluttering among the tall grasses that typify their habitat. Most females are mated in late morning or early afternoon and pairs remain coupled with wings closed for around half an hour.

Like the Marbled White, the female lays her eggs in a somewhat-chaotic fashion—she alights at the top of a tall grass stem and ejects a single egg, which simply falls to the ground. In *The Butterflies of Britain and Ireland*, Jeremy Thomas suggests that the female's choice of location is not entirely random: "*I suspect that there is more to this than meets the eye, and that they are responding to the scents of specific – and possibly fungus-infected – coarse grasses that waft upwards on the cool, humid air*". Various coarse grasses are selected, including Cock's-foot, Common Couch, False Brome, meadow-grasses and Tufted Hair-grass.

The egg is not attached to any surface

Ovum

The 0.8-mm high dome-shaped egg is pale yellow when first laid, but gradually darkens in colour, with the larva visible through the shell just before it emerges. The egg is shiny and is covered in a very fine but regular network pattern that gives the illusion of a number of delicate ribs that run the length of the egg from top to bottom.

Larva

After between two and three weeks, the 1.8-mm long **first instar** larva emerges from the egg. The ground colour is pale cream when newly emerged, and a greenish hue develops as the larva starts to feed. A brownish line runs along the middle of the back, with another along the middle of each side, and a pale line also runs along the base of each side. The larva has several rows of hairs that run the length of the body, and the longest hairs are found on the back, which all curve backward except for those on the first three segments which curve forward. The brown head also bears a number of relatively long and dark hairs. The anal points are mere stumps and each bears two hairs that project away from the body. The larva is nocturnal and hides by day at the base of a grass stem, emerging at night to feed on the tenderest parts of the foodplant. When fully grown, the first instar is around 4 mm long.

1st instar: the larva is covered in several rows of hairs

Life Cycles of British & Irish Butterflies

2nd instar: the larva now has a granular appearance

3rd instar: a mix of browns run down the head capsule

After two weeks, the larva moults into the **second instar**. The cream ground colour develops a greenish hue as the larva recommences feeding, and a series of brown lines run the length of the body, the most prominent of which run along the centre of the back, along the middle of each side, and below the spiracles. There is also a noticeable pale band that runs along the base of each side. The larva is now covered in numerous hairs of different lengths and the body has a granular appearance. When fully grown, the second instar is just under 6.5 mm long.

The larva is slow growing and, after around six weeks, moults into the **third instar** and resembles the previous instar in most respects, although the head is now a mix of browns that run down the head capsule. The larva overwinters while in this instar, tucked away at the base of a grass tussock. The larva does not become fully torpid, however, and will feed on particularly warm evenings during the winter. Regular feeding resumes in the spring when larvae can be found by torchlight feeding on grass stems, although they will fall to the ground with the slightest disturbance. When fully grown, the third instar is 8 mm long.

In 'The Seasonal Development Cycle of *Aphantopus hyperantus* (L.) (Lepidoptera, Nymphalidae: Satyrinae) in Leningrad Province', published in *Entomologicheskoe Obozrenie* in 2016, M.V. Ryzhkova and E.B. Lopatina show that both temperature and photoperiod (day length) work in combination to accelerate or decelerate the development of the larva, thereby regulating the

4th instar: the ground colour is now a light reddish-brown

4th instar: the larva has an unbroken chocolate-brown stripe along its back timing of the larva going into hibernation and maintaining its single-brooded nature. The study also showed that the trigger for resuming feeding in the spring is day length, rather than temperature.

Towards the end of March, the larva moults into the **fourth instar**. While the ground colour is now a light reddish-brown and the white stripe along each side is more prominent, the most obvious difference with the previous instar is that the larva now has a more prominent and continuous chocolate-brown stripe along the length of its back, although this is rather faint on the first two segments. The fully grown fourth instar is 14 mm long.

After around six weeks, this slow-growing larva moults for the last time into the **fifth instar**. As in the fourth instar, the key distinguishing feature is the line that runs along the back, which is now largely absent on the first three segments but is made up of a series of chocolate-brown marks that are found at the division between each pair of segments and that grow in size toward the anal segment. In all other respects it is similar to the previous instar. After four or five weeks, when fully grown, the larva is just over 21 mm long, when it retires to the base of the plant where it spins a few silk threads, but does not attach itself to any surface.

5th instar: the larva is superficially similar to the previous instar

5th instar: a chocolate-brown mark is found at the division between segments along the back

Pupa

The stout pupa is 11 mm long in the male and just under 13 mm long in the female. Its most striking feature, especially in those individuals that are the most heavily marked, are a number of streaks found on the wing cases, with an especially wide mark found on the central 'cell' of the forewing. The pupa darkens as the adult develops and becomes a dull black when the adult is fully formed. After two weeks in this stage, the butterfly emerges and crawls up a grass stem to inflate its wings, before ultimately taking flight in its damp grassland home.

The pupa is covered in brown markings

Meadow Brown
Maniola jurtina

❖ MEADOW BROWN IS SUPPORTED BY TIM BERNHARD, SUE & MICK OLDHAM & ROB PARTRIDGE ❖

	January	February	March	April	May	June	July	August	September	October	November	December
IMAGO												
OVUM												
LARVA												
PUPA												

The Meadow Brown is one of our most widespread butterflies, probably the commonest, and is a familiar sight throughout the summer months in any grassy area across Britain and Ireland. Male and female differ in the amount of orange markings found on the forewing uppersides, which are more significant in the female. The male also possesses sex brands that appear as a dark patch in the central area of each forewing. These differences led James Petiver to believe that the two sexes were different species, naming the male the 'Brown Meadow Ey'd Butterfly' and the female the 'Golden Meadow Ey'd Butterfly' in his *Musei Petiveriani* of 1695. The name we use today was provided by Eleazar Albin in his *A Natural History of English Insects* of 1720.

The ♂ has a dark sex brand on each forewing [IL]

The ♀ has orange patches on her forewings [IL]

The ♂ has less contrast on his underside than the ♀ [IL]

The ♀ has a larger eyespot than the ♂

The markings on the adults vary geographically and this has led to the identification of four subspecies within Britain and Ireland, although the differences between them are subtle. The most widespread is the subspecies *insularis* that is found throughout England, Wales and south-east Scotland. The subspecies *cassiteridum* is found in the Isles of Scilly and has a greater contrast in underside markings. These subspecies have wingspans of 49 mm in the male and 53 mm in the female. The third subspecies is *iernes* that represents the population found in Ireland and, it is thought, the Isle of Man, where the orange markings on the forewing uppersides are more extensive, the butterfly appearing brighter as a result. The fourth subspecies is *splendida* that is found in north-west Scotland, including the Hebrides, where the orange markings on the forewing uppersides are also more extensive, but where the underside is somewhat darker in appearance. These last two subspecies are also the largest, with wingspans of 52 mm in the male and 56 mm in the female.

A brightly coloured ♀ found in Cornwall [NF]

Life Cycles of British & Irish Butterflies

This variation and other attributes of the Meadow Brown, such as its abundance and ease of identification, made it the perfect subject when studying ecological genetics, according to W.H. Dowdeswell, E.B. Ford and their colleagues at Oxford University who studied the species on the Isles of Scilly, with Dowdeswell documenting their findings in *The Life of the Meadow Brown*, published in 1981.

DISTRIBUTION

This species can be found in all parts of Britain and Ireland, with the exception of the most mountainous regions and Shetland. On a worldwide basis, the butterfly is found across Europe and temperate Asia, and in parts of North Africa. The butterfly is considered to be relatively sedentary but will regularly fly up to 150 m within its breeding site, only moving further afield if its site becomes unsuitable, when the adults will fly over roads, arable crops and open fields in pursuit of a new home. At the best sites, adults can be measured in their thousands.

HABITAT

Given the broad distribution of this species, it should come as no surprise that it can be found in a variety of habitats, wherever the grass sward grows to a medium height and is not heavily grazed. Aside from the obvious habitats of both acid and calcareous grasslands, the butterfly can be found in large, open areas of woodlands; mixed scrub; hedgerows; disused quarries; sand dunes; arable field margins and set-aside; brownfield sites; gardens, parks and churchyards; and road verges. The butterfly tends to avoid windswept habitats such as heathland and moorland.

The butterfly is found in the meadows at Ashford Hill NNR in Hampshire

STATUS

As one of our commonest butterflies, this is not a species of conservation concern. Both long-term and short-term trends show a stable distribution, and the long-term trend shows a stable abundance, with the more recent trend over the last decade showing a moderate decline in abundance. Using the IUCN criteria, the Meadow Brown is categorised as Least Concern (LC) in both the UK and Ireland.

LIFE CYCLE

There is one generation each year and the flight period can be extremely protracted, especially on the chalk downs of southern England, with adults found from the beginning of June until early October in most years and, at all locations, males always emerge before females. The butterfly overwinters as a larva, remaining in this stage for between eight and nine months.

Imago

The adult butterfly is a familiar sight across most of Britain and Ireland, so much so that it is often overlooked, even though it will take to the wing in dull weather, including light drizzle, so long as the temperature is sufficiently high. Both sexes roost low down within tall grass clumps or other vegetation, such as bushes and hedges, although they will also roost higher up in trees. Early morning is a good time to see this species, when it sits with its wings outstretched as it warms itself up in the sun. That being said, the butterfly is easily alarmed and, despite its familiar presence, is one of the most difficult species to photograph. Both sexes are avid nectar feeders and use a variety of flowers, especially those of knapweeds and thistles. Other nectar sources include bramble, buttercups, Common Fleabane, Devil's-bit Scabious, Hemp-agrimony, ragworts, Selfheal, Wild Marjoram, Wild Privet, Wild Teasel and Yarrow.

The male is the more active of the two sexes and adopts both perching and patrolling strategies when attempting to find a mate, when any butterfly that is encountered is investigated. If an already mated female is approached then she will flutter her wings rapidly, whereas a receptive female will participate in a brief courtship, where the male flies around her while showering her with pheromones from the scent scales on his sex brands. The pheromones act as an aphrodisiac that seduces the female and mating quickly follows, the pair remaining coupled for between one and two hours. In his classic work entitled, simply, *Butterflies* (volume 1 of the 'New Naturalist' series, published in 1945), E.B. Ford compares the smell from the pheromones to an "*old cigar-box*". If the pair is disturbed then they will fly to a new spot, with the female always carrying the male. The Meadow Brown also has difficulty in finding the right partner on occasion since, for some unexplained reason, both sexes occasionally pair with other species, including Gatekeeper, Ringlet, Marbled White, Small Tortoiseshell and even Silver-washed Fritillary, although no viable offspring have ever been recorded.

A mating pair with the ♀ at the top [NKi]

An egg-laying female is quite easy to spot as she flutters slowly over patches of short to medium turf. Once landed, she will crawl through tufts of grass before depositing a single egg either on the foodplant or on nearby vegetation. Eggs are occasionally dropped into vegetation near the foodplant without being attached to any surface, and a bout of egg-laying may result in several eggs being laid in the same vicinity. Early instar larvae feed on grasses with fine leaves, such as fescues, bents and meadow-grasses, although more mature larvae feed on coarse grasses such as Cock's-foot, Downy Oat-grass, and False Brome.

The female is able to lay over 200 eggs and this partially explains why the ratio of the volume of its egg relative to its wingspan is just 0.31% and is the smallest of any butterfly found in Britain and Ireland, as documented by Enrique García-Barros in 'Egg size in butterflies (Lepidoptera: Papilionoidea and Hesperiidae): a summary of data', published in *The Journal of Research on the Lepidoptera* in 2000 (the largest ratio is that of the Scotch Argus, which is an incredible 3.1%).

Ovum

The 0.5-mm high egg is conical with a flattened top and has between 20 and 24 ribs that run down its sides. Eggs are a pale yellow when first laid, but soon turn light brown with a mottling of dark brown blotches, turning grey just prior to hatching.

Larva

After two to four weeks, depending on temperature, the 1.4-mm long **first instar** larva emerges by eating a hole in the side of the egg,

The egg develops dark brown blotches after a few days

The newly laid egg is pale yellow

before proceeding to eat the majority of the shell as its first meal. The larva has a brown ground colour when newly emerged, but soon turns green as it starts to feed on grasses. The body has a brownish line running along the middle of the back, along with three broken and irregular lines just below this line along each side, and the last segment is adorned with two stubby anal points. The larva also has three rows of long hairs above the spiracles along each side, together with shorter hairs found below the spiracles. Its head is greenish-brown, and it is also covered in a good number of hairs. When fully grown, the first instar is 2.5 mm long, its ground colour is a brighter green, and a faint white line has developed along each side.

The relatively long hairs found on the first instar larva (most of which curve backwards) are a feature that carries through to the final instar and is one of the diagnostic features that help distinguish Meadow Brown larvae from those of other species of the Satyrinae subfamily. Young larvae feed by day, although more mature larvae tend to feed at night, resting head-down on a grass stem during the day, deep in vegetation. In all instars, the larvae are sensitive to any vibration and will fall to the ground if disturbed.

After two weeks, the larva moults into the **second instar** but does not eat its cast skin, and now feeds mainly at night. The head is now green rather than greenish-brown and the line running along the back is now dark green rather than brown, although the main distinguishing feature is that the larva now has a prominent white line running along each side. The larva is still covered in long and dark hairs and, when fully grown, the second instar is 4.2 mm long.

1st instar: the head and body are covered in long hairs

2nd instar: the larva now has a prominent white line along each side

3rd instar: the larva has more subdued colours and whitish hairs

4th instar: the anal points now have a pinkish tinge

After between two and three weeks, the larva moults into the **third instar**. The larva is similar to the previous instar, but all of the colours are more subdued, the spiracles are orange and the hairs on the body are now more translucent and appear whitish as a result. When fully grown the third instar is around 6.5 mm long and, when disturbed, will curl the front part of its body or simply drop to the ground.

After around four weeks, the larva moults into the **fourth instar**. The larva is of a similar colour to the previous instar, although there are several subtle differences in appearance. While the third instar has a number of whitish warts all over its body that each bear a hair, these are more numerous in the fourth instar, and the smaller warts bear correspondingly short hairs that gives the skin a rough appearance. The white line found along each side is also finer, yet more prominent, and the anal points are also much more conspicuous and have a pinkish tinge. When fully grown, the fourth instar is 10 mm long.

With the onset of a cooler September and October, development of the larva slows down considerably, but the larva is only torpid in the coldest weather when it is hidden away at the base of a grass clump—it will continue to feed if conditions are warm enough and it therefore only partially hibernates. While there are many accounts of larvae overwintering while in their fourth instar (and this is my own experience), some authors suggest that larvae can overwinter while in their second or third instars also. It is therefore reasonable to conclude that the Meadow Brown is able to successfully overwinter as a larva in one of several instars and that this also explains, to some extent, the protracted emergence of the adult butterfly.

The moult into the **fifth instar** takes place the following spring. The larva is now covered in white warts, the largest of which bear a grey or black hair, and the larva now has a significant dark green

5th instar: the larva now has a significant dark green line along its back

stripe that runs along its back. The body, which is now a paler green that is more attuned to the grasses on which the larva feeds, continues to exhibit a white line along each side, and shows a clear tapering towards the anal segment, with the anal points continuing to exhibit a pink tinge. The larva continues to feed at night, resting deep down in the tussock during the day. When fully grown, the fifth instar is around 15 mm long.

Given the protracted flight period of this butterfly, the moult into the final and **sixth instar** may happen any time between March and July. The larva is now a whitish-green, thanks to the profusion of white warts and hairs that cover its surface, and the relatively hairless head now appears slightly darker than the body. The larva retains the dark green line along its back, the thin white line along each side and pink-tinged anal points. When fully grown, the larva is around 25 mm long and suspends itself upside-down from a silk pad attached to a grass blade, low down in vegetation, pupating after two days.

The fully grown larva is particularly easy to find by torchlight in May, when it can be found feeding about halfway up a lush stem of a variety of coarse grasses, such as Cock's-foot. When undertaking such a search it is important not to disturb the larva, otherwise it will simply drop into the grass tussock where it will curl up and take some time to reascend a grass stem.

6th instar: a fully grown larva

The pupa is invariably attached to the old larval skin

Some pupae are less heavily marked than others

Pupa

The 16-mm long pupa is attached to the old larval skin which itself remains attached to the silk pad. The retention of the old larval skin is deliberate, since the cremaster at the tail end of the pupa has no hooks that would allow it to attach directly to the silk pad. This may also explain why the pupa is formed low down in a grass tussock where it will avoid being damaged should the precarious hold on the old larval skin fail.

The pupa is quite beautiful, with a ground colour that is a mix of greens and yellows, splashed with a variable number of dark markings on the wing cases and body. Studies have shown that the colour of the pupa is influenced by its environment, including the amount of light, temperature and relative humidity, and that pupae formed low down on a plant tend to be darker than those formed higher up. The pupa grows progressively darker as the butterfly develops, with the wing cases becoming almost black just prior to emergence, and the adult butterfly emerges after three or four weeks.

Gatekeeper
Pyronia tithonus

❖ GATEKEEPER IS SUPPORTED BY CHRIS FREEMAN, PETER KING & ANDY LODGE ❖

The Gatekeeper is the epitome of high summer when its golden wings brighten up our countryside throughout its range. The two sexes spend much of their time basking with wings open, when they are easy to tell apart since the male has a distinctive dark sex brand on each forewing that is absent in the female. This species is occasionally confused with its close relative, the Meadow Brown, and James Petiver, in his *Musei Petiveriani* of 1695, attempted to distinguish between them by naming the Gatekeeper the 'Lesser Double-eyed Butterfly', recognising that the eye spots found on the forewings of the Gatekeeper normally have two pupils, whereas those on the Meadow Brown normally have one. A subtler distinction is that the tiny spots on the underside of the Gatekeeper are white, whereas those on the Meadow Brown, if they are present at all, are black.

The similarity to the Meadow Brown was also recognised in the name of 'Small Meadow Brown' given by George Samouelle in his *The Entomologist's Useful Compendium* of 1819. The name we use today was provided by Moses Harris in his *The Aurelian* of 1766 and, fortunately, the rather confusing name of 'Large Heath' given to this butterfly by Adrian Haworth in his *Lepidoptera Britannica* of 1803, which is now used for a completely different species, never caught on. The butterfly is also referred to as the 'Hedge Brown' by many authors and has been one of the names in common use from the late 1800s.

For such a common butterfly, it is surprising that this is one of our least studied species. The sentiments expressed by B. Anderson in 'Preliminary studies on the Gatekeeper butterfly *Pyronia tithonus* at an urban site in Bedford', published in the *Bedfordshire Naturalist* in 1995, would still seem to hold true: *"The amount of research devoted to any particular British butterfly species often seems inversely proportional to its abundance; probably this is due to the prospect of a species imminent extinction generating a sense of urgency"*.

♂ and ♀ have similar undersides

The ♂ has distinctive sex brands and a wingspan of 37 to 43 mm

The slightly larger ♀ has a wingspan of 42 to 48 mm

DISTRIBUTION

This is a common and widespread species that is found across England and Wales, south of a line between Cumbria in the west and County Durham in the east, although its range is expanding at its northern edge. In Ireland the butterfly has a more local distribution and is found within 50 km of the coast of southern counties, with its range extending northward along the east coast towards Dublin. The butterfly is also found in the Channel Islands, but is absent from Scotland and the Isle of Man. Since the butterfly has a southern distribution in both Britain and Ireland and, given that suitable habitat is found outside of its current range, we can assume that its range is governed

Life Cycles of British & Irish Butterflies

primarily by climate and that increasing temperatures are allowing the butterfly to move further north. On a worldwide basis, the butterfly is found throughout most of Europe and western Asia and is found locally in Morocco.

Colonies vary greatly in size, ranging from a hundred individuals during the course of a season to several thousand, with numbers proportional to the extent of suitable habitat. In her PhD thesis 'Dispersal of Satyrid Butterflies', published in 1988, Susan Clarke conducted a mark-release-recapture exercise at Bernwood Forest in Buckinghamshire, and found that adults dispersed around 140 m on average, confirming that this butterfly is rather sedentary. However, this butterfly's expansion in the northern part of its range would indicate that it must fly greater distances on occasion.

HABITAT

This species can be found anywhere within its range where shrubs grow close to rough grassland, including mixed scrub; hedgerows; arable field margins and set-aside; road verges; woodland clearings, rides and glades; and scrubby areas of heathland, sand dunes and undercliffs.

Gatekeepers fly in good numbers at Holtspur Bottom in Buckinghamshire [JA]

STATUS

Both long-term and short-term trends show a moderate increase in distribution and a moderate decline in abundance, indicating that the northwards expansion of this butterfly has not been accompanied by a corresponding increase in numbers. Using the IUCN criteria, the Gatekeeper is categorised as Least Concern (LC) in the UK and Near Threatened (NT) in Ireland. Factors that negatively affect this butterfly are a loss of unimproved grassland to agriculture and development, the removal of hedgerows and a decrease in arable field margins.

LIFE CYCLE

There is one generation each year, with adults emerging at the end of June in favourable years, peaking at the end of July and early August, with only a few adults remaining until the end of the month. In contrast with its close relative, the Meadow Brown, this butterfly has a later and relatively short flight period that is better synchronised across its distribution. In *The Butterflies of Sussex*, Neil Hulme writes that the Gatekeeper persists "*just long enough to mercilessly tease and frustrate the hopeful hunters of the Brown Hairstreak*" and I can personally vouch for this—my initial 'Brown Hairstreak' sightings at Noar Hill in Hampshire were, indeed, of Gatekeepers that were dancing around the Blackthorn hedgerows.

Imago

Both sexes feed from whatever nectar sources are available, especially bramble, Common Fleabane and ragworts. Other sources include Devil's-bit Scabious, Hemp-agrimony, Red Clover, thistles, Water Mint, Wild Privet and Wild Thyme and, at woodland sites, the butterfly will also feed on aphid honeydew. The butterfly roosts in hedges or scrub at night, where it also shelters during dull or wet weather.

Males set up small territories, where they will perch on a particular shrub or bush, flying up to investigate passing butterflies in the hope of finding a mate. In his unpublished 1979 PhD thesis, 'An experimental study of the maintenance of variation in spot pattern in *Maniola jurtina*', Paul Brakefield makes some observations regarding the Gatekeeper and showed that males may remain in the same area, and even the same bush, for several days before eventually moving up to 150 m to take up a new perch in a new territory.

A mating pair with ♀ on the left [VM]

Pairing occurs without any discernible courtship and, once mated, a female will typically lay between 100 and 200 eggs. She lays her eggs on finer grasses, such as bents, fescues and meadow-grasses, selecting those that are shaded and that are growing beneath shrubs and hedgerows—at Greenham Common in Berkshire, I occasionally find females laying on fescues growing beneath bramble stands.

Ovum

The 0.65-mm high eggs are laid either on the foodplant or nearby vegetation, or are randomly dropped when the female is settled over a suitable patch of foodplant. Eggs are conical with a flat top and base, and have 16 or 17 ribs running along their sides. The egg is pale yellow when first laid but soon develops brown blotches on its surface, becoming a more uniform dark brown as the larva develops within the egg, when the blotches are less distinct.

The newly laid egg is pale yellow

The egg develops brown blotches after a few days

Larva

After between two and three weeks, depending on temperature, the 1.6-mm long **first instar** larva emerges from the egg by eating away the crown, consuming the remainder of the shell as its first meal, before proceeding to feed on the tips of grass blades by day. The larva is initially brown, but develops a green tinge as it starts to feed, growing increasingly green as it matures, although its head always retains its brown colouring. Its resemblance to a Meadow Brown larva is uncanny, since it also possesses a brownish line along its back, with three broken and irregular lines along each side. The larva is also covered in long hairs, most of which curve backwards over the body, although a first instar Gatekeeper larva does not possess any points on its anal segment, which are present in the Meadow Brown. When mature, the first instar has a whitish line along each side and is around 3.5 mm long.

1st instar: the newly emerged larva is brown

1st instar: the larva grows greener as it feeds

Larvae feed slowly and do not moult into the **second instar** for several weeks, at the end of September or in early October, when the weather is already turning cooler. Shortly after moulting, the larva enters hibernation, usually within a withered grass blade, at the base of a grass clump. The larva is now bright green, has a darker green line running along the middle of its back, a green rather than brown head, a white line along each side of the body and two whitish points on the anal segment. The body is now covered in relatively short hairs, many of which emerge from a white bulbous base, giving

2nd instar: the larva is now a brighter green and has a green head

2nd instar: the larva will fall from the foodplant at the slightest disturbance

the impression of white spots covering the body when seen close up. Unlike larvae of the Meadow Brown, overwintering Gatekeeper larvae become fully torpid and do not feed intermittently on warm days. The larva recommences feeding with the onset of warmer weather, but feeds only at night, when it moves very slowly, spending the day hidden away at the base of grass stems. The second instar is fully grown by the end of March when it is around 6.5 mm long.

After two or three weeks of feeding in the spring, the larva moults into the **third instar** and continues to feed at night, resting head-down at the base of a grass blade during the day. The larva is similar in appearance to the previous instar, although the body is more densely covered in hairs with a corresponding increase in the density of white spots found on the body. The larva now starts to exhibit two colour forms—one has a more subdued green ground colour than that of the second instar, and the other has a grey-green ground colour. In *The Moths and Butterflies of Great Britain and Ireland*, the authors suggest that these forms are environmentally rather than genetically controlled. When fully grown, the third instar is 10.5 mm long.

3rd instar: the green form of the larva

4th instar: the green form of the larva

3rd instar: the grey-green form of the larva

4th instar: the grey-green form of the larva

After around three weeks, the larva moults into the **fourth instar**. The larva continues to exhibit two colour forms and the green form has a prominent dark green stripe running along its back, whereas that of the grey-green form is dark brown. Both forms are covered in pale spots and still possess a white stripe along each side of the body. The two points on the anal segment remain pale, but now have a pink tinge. The larva continues to feed solely at night, remaining at the base of grass stems during the day and, when fully grown, the fourth instar is 16 mm long.

After another three weeks, the larva moults for the last time into the **fifth instar**. It no longer continues to exhibit two clear-cut forms and all larvae have a pale brown ground colour with a variable green tinge and a brown head. The larva also has a dark brown line along its back, a white stripe that runs along the base of each side, and a less distinct pale stripe that runs between the two. The hairs are now relatively short and, in combination with numerous pale spots, give the body a rough appearance. The anal points are still pale, but now have a stronger pink tinge than those of the fourth instar. Larvae continue to feed only at night, resting at the base of the plant during the day.

5th instar: a fully grown larva feeding at night

5th instar: a larva preparing to pupate

The fully grown larva is 22 mm long and can be found by carefully searching taller grasses growing beneath shrubs and hedgerows on warm May nights. When ready to pupate, the larva suspends itself upside-down from a silk pad that is attached to vegetation at the base of the plant. After three or four days in this position the larva pupates, having spent around four weeks in the final instar and around 240 days in the larval stage.

Pupa

The stout 12-mm long pupa is attached to the old larval skin using hairs on its abdomen, but it is the old skin and not the pupa that is attached to the silk pad. The pupa has a very pale ground colour, almost white, and is splashed with a number of dark streaks on the wing cases and dark blotches on the body. These markings vary in intensity and some pupae appear much paler than others. After three weeks in this stage, and just prior to emergence, the wings of this summer butterfly are clearly visible through the pupal case.

The pupa is always attached to the old larval skin

Some pupae have very few dark markings and appear relatively pale

A ♂ pupa just prior to emergence

Life Cycles of British & Irish Butterflies

Marbled White
Melanargia galathea

❖ MARBLED WHITE IS SUPPORTED BY JANET ASHWELL, PENNY BRIGGS & GARY WELSBY ❖

	January	February	March	April	May	June	July	August	September	October	November	December
IMAGO												
OVUM												
LARVA												
PUPA												

Despite its name, the Marbled White belongs to the subfamily commonly known as the 'browns' (the Satyrinae) rather than the family of whites (the Pieridae), although the butterfly's colouring would suggest otherwise. The two-tone and striking appearance is acknowledged in every name that has been given to this species, from 'Our half Mourner' given by James Petiver in his *Musei Petiveriani* of 1695, to the name we use today given by Moses Harris in *The Aurelian* of 1766. Even the genus, *Melanargia*, is a combination of '*melas*' (old Greek for 'black') and '*arges*' (old Greek for 'brightness'). Male and female are similar in appearance, although the female has a brown hue to her undersides and a creamy brown leading edge to the forewings.

The ♂ has a wingspan of 53 mm [JA]

The ♀ has a wingspan of 58 mm and a creamy brown leading edge to the forewings [PC]

The ♂ underside is a mix of white, greys and blacks [JA]

The ♀ has a brown hue to her underside [MC]

It is thought that the butterfly exhibits aposematic colouring that advertises its unpalatability to potential predators such as birds. In all known cases of such colouring, including the bold colours of the male Orange-tip, this distastefulness comes from the food ingested by the larva. Studies by Angela Wilson in the 1980s showed that the cream markings on the butterfly's wings contain flavonoids that are derived from Red Fescue. However, these chemicals are only mildly toxic and, in *The Butterflies of Britain and Ireland*, Jeremy Thomas suggests that the main benefits derived from these flavonoids are that they contribute to the butterfly's camouflage when it is roosting on grasses. These flavonoids also make the butterfly more visible to potential mates thanks to the ultra-violet reflection from the flavonoid-rich white areas that contrast with the blacks on the wings.

In the article 'The Marbled White (*Melanargia galathea*), a Toxic Butterfly', published in *Antenna* in 2001, Miriam Rothschild acknowledges Robert Nash's discovery that the ingestion of fescues also results in the consumption of an *Acremonium* fungus and that this, in turn, leads to a build-up of loline (a different chemical to flavonoids) that makes the adult (and pupa) obnoxious to birds. All this being said, there is no evidence that the female Marbled White deliberately selects areas where fescues are growing and this subject is, therefore, worthy of more study.

DISTRIBUTION

This butterfly is found in distinct colonies, primarily south of a line between south-west Wales and north-east Yorkshire, although it is not found in much of eastern England and is absent from Scotland, Ireland and the Isle of Man. The strongest colonies are found in southern England with scattered colonies further north. On a worldwide basis, the butterfly is found throughout most of Europe, eastwards as far as the Caucasus, and in parts of North Africa.

Colonies vary in size, usually in proportion to the amount of suitable habitat available—some are therefore fairly small but some, such as those found on extensive patches of calcareous grassland, can be huge, where the number of adults is measured by the thousand.

While most adults spend their entire lives within their colony, the butterfly will disperse on occasion, and there are regular sightings of individuals flying across unsuitable habitat, such as heathland. Exchange between colonies has also been proven, both directly through mark-release-recapture studies and also indirectly through genetic analyses.

HABITAT

The butterfly is found in unimproved grassland where grasses grow tall. The largest colonies are undoubtedly on calcareous grasslands where the butterfly is often found in the company of the Chalk Hill Blue, although the butterfly will persist in a variety of other habitats where patches of tall grasses occur, including woodland clearings, rides and glades; mixed scrub; disused quarries; arable field margins and set-aside; road verges; and disused railway lines. In *The Butterflies of Britain and Ireland*, Jeremy Thomas writes: "*In Dorset, I have found high densities flying and breeding along*

Marbled Whites nectaring on Greater Knapweed at Stockbridge Down in Hampshire [TB]

the 2–5 m-wide strip of the central reservation of a busy dual carriageway, with the adults apparently oblivious to the noise and fumes of the traffic streaming past on either side".

STATUS

The butterfly does particularly well in years that follow a warm and dry summer, especially a dry August when newly emerged larvae are entering hibernation, and it has increased in terms of both distribution and abundance in recent years, moving northwards and eastwards. The long-term trend shows a moderate increase in distribution and a significant increase in abundance, with the more recent trend over the last decade showing a moderate increase in both distribution and abundance. Using the IUCN criteria, the Marbled White is categorised as Least Concern (LC) in the UK.

LIFE CYCLE

This single-brooded species emerges from the middle of June, reaches a peak in early July, and flies until the end of the month, with a few stragglers occasionally seen in August.

Imago

Both sexes are avid nectar feeders and can be found on every flowering knapweed and thistle at some sites, sometimes with several adults jostling for position on the same flowerhead. Other nectar sources include clovers, Wild Marjoram, Wild Thyme and Yarrow. This species is also fairly conspicuous, even when seen from a distance, since adults may be the only black-and-white butterflies inhabiting their grassland home. They are particularly easy to locate in weak sunlight in early morning and late afternoon, or in overcast conditions, when the adults will open their wings to warm up while sat atop tall flowering plants. The adults will also roost on flowerheads and grass heads, where they position themselves head-down.

The ♀ on the left was mated before her first flight

An adult carrying several *Trombidium breei* parasitic red mite larvae

Males adopt a patrolling strategy when finding a mate, when they can be found flying, seemingly randomly, over their habitat. Mating occurs quickly without any discernible courtship and, on several occasions, I have found a male mating with a newly emerged female that has yet to dry her wings, the pair remaining 'in cop' for between 30 minutes and two hours.

I am always surprised at the apparent lack of care when a female is egg-laying—she will land on a grass stem or some other vegetation before releasing a single egg that simply drops to the ground, partly due to the lack of a glutinous covering that would allow it to stick to any surface it touches. Given the larval requirements, eggs are laid in the vicinity of both fine-leaved grasses such as Red Fescue and Sheep's-fescue that are used by young larvae, and coarse-leaved grasses such as Cock's-foot, Timothy, Tor-grass, Yorkshire-fog and undoubtedly others too, that are used by more mature larvae.

The adult butterfly is often found with the larvae of parasitic red mites, especially *Trombidium breei*, attached to the soft membrane between the thorax and abdomen, or between the head and thorax. The mite larvae pierce the membrane to feed on the bodily fluids, although this parasitism does not appear to affect the butterfly in any way. Mite larvae are found on species from several families, including the Small Skipper (Hesperiidae), Meadow Brown (Nymphalidae) and Common Blue (Lycaenidae).

The white spherical egg is dropped to the ground

Ovum

The almost-spherical egg is 1 mm in diameter and is pale green when first laid, but almost immediately turns an opaque white when exposed to air, turning slightly darker as the larva develops.

Larva

After three weeks, the 3-mm long **first instar** larva emerges from the egg by eating around the crown and pushing its way out, with the crown still attached to the remainder of the eggshell, where it forms a 'lid'. The larva eats the remaining egg shell and immediately enters hibernation, tucked away deep down in vegetation at the base of a grass clump. The larva becomes active at the end of January and early February, feeding by day and resting along the length of a withered and brown fescue stem or blade where it is remarkably well camouflaged. The head is pale brown and is covered in white tubercles that each bear a pale hair. The body is also pale brown and has several red-brown stripes and four rows of long hairs that run its length, together with two points on the anal segment. The fully grown larva is just over 4 mm long and may be tinged with green.

1st instar: a newly emerged larva

1st instar: a fully grown larva with a greenish tinge

2nd instar: the brown form of the larva

2nd instar: the green form of the larva

The larva moults into the **second instar** in early April and will be one of two colour forms—light brown or light green. In both cases the stripes along the body are correspondingly brown or green, although the head is light brown in both forms. The body remains covered in hairs (although these are relatively shorter than in the first instar), a dark stripe now runs along the middle of the back and, in both forms, the anal points have a pink tinge. Larvae continue to feed by day and rest along the length of a grass blade and, when fully grown, the second instar is 8.5 mm long.

3rd instar: the brown form of the larva

3rd instar: the green form of the larva

After two weeks, the larva moults into the **third instar** and continues to exhibit the two colour forms. The markings are now clearer in both forms and the body is covered in numerous white hairs. The larva starts to feed primarily at night and, when fully grown, the third instar is 15 mm long.

After a few weeks, the larva moults for the last time into the **fourth instar** and still comes in two colour forms. The body remains covered in

4th instar: the brown form of the larva

4th instar: the green form of the larva

4th instar: the larva has a dark line along its back

numerous white hairs and the two anal points now have a reddish tinge. The dark line that runs along the back is especially prominent and the spiracles along each side of the body are now black and stand out from the pale ground colour. Larvae are extremely easy to find at night in late May when they are feeding high up on grass blades, and this is when their light brown heads are a great help in distinguishing them from Meadow Brown larvae, whose heads are green. In mid-June, having spent around 340 days in this stage, the larva is fully grown and 28 mm long. Before pupating, the larva moves deep into the base of the grass tussock, but does no more than conceal itself under moss or other vegetation.

The author photographing larvae at dusk [MC]

Pupa

The larva pupates after a few days and the resulting 12- to 15-mm long and plump pupa is yellow-brown and somewhat translucent, with a dark spot at the base of each wing case near the eye, and a chestnut-brown cremaster. The adult butterfly is quite visible within the pupal case just prior to emerging and, after two or three weeks in this stage, this unmistakable and beautiful two-tone butterfly takes to the wing.

A recently formed pupa

A fully coloured-up pupa

Life Cycles of British & Irish Butterflies

Grayling
Hipparchia semele

❖ GRAYLING IS SUPPORTED BY APITHANNY BOURNE, JERRY DENNIS, HOWARD & MARIE-CHRISTINE FROST & ROBERT PENNINGTON ❖

	January	February	March	April	May	June	July	August	September	October	November	December
IMAGO												
OVUM												
LARVA												
PUPA												

The Grayling, the largest of the 'Satyrines' or 'browns' with a wingspan of 51 to 56 mm in the male and 54 to 62 mm in the female, is a master of disguise and was once known as the 'Rock Underwing', a name that pays homage to the butterfly's ability to mysteriously disappear as soon as it lands, when it is perfectly camouflaged against a background of bare earth and stones, always settling with its wings closed. One of my most memorable experiences of this butterfly was at Godshill in the New Forest in Hampshire where Graylings were in such good numbers that I would disturb one with every other footstep while walking along a track through the heathland. Each butterfly would take to the air with a characteristic series of hops and glides, fly several yards in front of me and, despite being conspicuous while in flight, disappear as soon as it landed.

When the butterfly first lands, or when disturbed, it follows a three-step movement. The butterfly first lands with its forewings raised, revealing two dark eye spots that stand out against a beautiful spectrum of browns. After a second or so, the forewings are partially lowered before, after another second, they are fully lowered. This movement is an obvious defence against a bird attack, where the eye spots distract the bird away from the body. This is almost certainly the reason why the butterfly raises its forewings when feeding at flowers, when it is less well camouflaged, and this is also one of the best times to obtain a photo of the beautiful forewing underside.

A ♂ camouflaged against its background

A ♀ raising her forewings while nectaring on Buddleja

A ♂ subsp. *thyone*

A ♂ subsp. *clarensis*

This butterfly is also known for its variation, with six named subspecies within Britain and Ireland, the most of all our species. The nominate subspecies, *Hipparchia semele semele*, is found in England, Wales, the Isle of Man and the Channel Islands; subspecies *scota* is generally distributed around the coast of Scotland with the exception of the Western Isles and adjacent mainland where subspecies *atlantica* is found; subspecies *thyone* is confined to the Great Orme near Llandudno in North Wales; and subspecies *hibernica* is found throughout Ireland, primarily in coastal areas, with the exception of the Burren in County Clare and south-east Galway, where subspecies *clarensis* is found.

I am fortunate to have seen all six subspecies and my conclusion is that, based on the size and undersides of the butterfly (which is all you would realistically see in the field), there are two subspecies that exhibit significant differences from all others. The first is subspecies *thyone* that is substantially smaller in size than other subspecies (the average wingspan of the male is 48 mm and the female 52 mm), flies earlier in the year and has the longest flight period. The second is subspecies *clarensis* which, even when newly emerged, is relatively pale with a washed-out appearance.

Irrespective of subspecies, the two sexes are most easily distinguished by their uppersides—the male has a distinctive sex brand on each forewing—and it is a shame that this key diagnostic feature belongs to a butterfly that always settles with its wings closed. However, as a general rule, the male has a more prominent white band on the underside of each hindwing than the female, although this is an extremely variable characteristic both within and between subspecies.

DISTRIBUTION

The butterfly is found throughout Britain and Ireland, except for Orkney and Shetland, and its distribution would suggest that it is primarily a coastal species, although there are significant inland colonies, such as those found on the heathlands of the New Forest. Such large expanses of land have colonies that are measured in the thousand, although there is considerable variation in colony size and the smallest colonies reach only a couple of dozen adults at the peak of the flight period. On a worldwide basis, the butterfly is found throughout Europe and in northern and western Asia.

HABITAT

This butterfly is found on sheltered, sunny, dry and well-drained sites with an abundance of fine-leaved grasses and where vegetation is sparse, providing the bare ground that this butterfly requires. Coastal sites include sand dunes and undercliffs, and inland sites include lowland heathland, limestone pavement, disused quarries and brownfield sites. Calcareous grassland sites, where colonies rely on a short sward and patches of bare ground maintained by grazing rabbits, were severely impacted from the 1950s to 1980s when the rabbit population was decimated by myxomatosis and sites became unsuitably overgrown, although a few Grayling populations remain. This includes a strong population at Arnside Knott in Cumbria, where Grayling can be found in good numbers flying around the limestone scree slopes.

Lowland heathland near Fordingbridge in the New Forest

Ballyteigue Burrow Nature Reserve, County Wexford, Ireland is home to subsp. *hibernica*

Limestone scree at Arnside Knott

Habitat management regimes depend on the site, but usually involve scrub control and grazing by cattle and ponies which results in a varied sward and also disturbed ground, providing the bare patches that the butterfly favours. Heathland sites are also managed through rotational cutting and burning.

STATUS

The long-term trend shows significant declines in both distribution and abundance. Developments such as links golf courses and their associated facilities on the coast, and housing and road development on and through heathland, have resulted in large amounts of suitable habitat being lost. Agricultural intensification and ploughing have also pushed the butterfly literally to the edge in some areas, where the butterfly survives only because coastal cliffs cannot be worked. These factors, along with changes in habitat management, have led to increasingly fragmented and isolated sites. The more recent trend over the last decade is less severe, with a moderate decline in distribution and a moderate increase in abundance. Using the IUCN criteria, the Grayling is categorised as Vulnerable (VU) in the UK and Near Threatened (NT) in Ireland.

LIFE CYCLE

This single-brooded species generally emerges at the start of July, or late June in early years, peaking at the end of July, with numbers dropping off throughout August with a few stragglers flying into September. The population of subspecies *thyone* found on the Great Orme emerges earliest, often at the beginning of June, and, curiously, also has the longest flight period of all subspecies.

Imago

This butterfly is often found perching on the bare ground that typifies its habitat or, occasionally, on a tree trunk or boulder. Regulation of body temperature is especially important to this butterfly, with adults orientating their bodies toward the sun to catch any early morning rays and, conversely, turning their bodies away from the sun in the heat of the day which, in their sun-drenched habitats, could easily cause overheating. Both sexes nectar on a variety of plants, including bird's-foot-trefoils, brambles, heathers, Red Clover, thistles, Wild Marjoram and Wild Thyme, usually in early morning or late afternoon. They will also feed on Buddleja if found in the vicinity and will also gorge themselves on sap runs.

The male is territorial and the more active of the two sexes, and is quite conspicuous when flying up to investigate passing objects, in the hope of encountering a virgin female. When the male is successful in finding a potential mate, he undergoes an elaborate courtship that was first spelled out by the Nobel prize-winning Dutchman, Nikolaas (Niko) Tinbergen, who dedicated his life to the study of animal behaviour and found the Grayling an interesting enough subject to include in his work. I can still remember the first time I encountered (and filmed) this behaviour at Ballyteigue Burrow Nature Reserve in County Wexford, Ireland, and was simply amazed at just how intricate the courtship is.

The courtship is normally preceded by the male flying upwards toward and overhead a passing female. The pair land with the male usually behind the female, before he moves around her until they are facing each other. An unreceptive female will flutter her wings to deter the male, while a virgin female will remain still, encouraging the male to perform an elaborate courtship that comprises four main stages–quivering, fanning with antenna spinning, bowing and clasping. This may give the impression that the courtship always rigidly follows this sequence, but this is not the case—the sequence may be repeated, steps may be performed out of sequence and some steps may be omitted altogether. The stages are, however, an excellent rule of thumb.

Wing quivering: the male starts the courtship by slowly raising his forewings (so much so that the orange patches found on the underside of the forewings are revealed) with the wings held slightly apart, before he rapidly closes and lowers them.

The ♂ (on the left) draws the ♀'s antennae across his sex brands when 'bowing'

Life Cycles of British & Irish Butterflies

Fanning with antenna spinning: the male then vibrates his wings rapidly in front of the female, with forewings held up and the wings held slightly apart. During this fanning, the male spreads his antennae horizontally and rapidly rotates them.

Bowing: the male then undertakes the most elaborate aspect of the courtship—with his wings steadied and held open, he bows forward, drawing the female's antennae across the sex brands on his forewings, while gently closing his wings with the female's antennae held between them. The androconia on the sex brands, to all intents and purpose, emit a scent that seduces the female. This stage of the courtship is one of the few occasions when it is possible to catch a glimpse of the beautiful Grayling upperside.

Clasping: with a willing subject, the male then moves to the side and slightly behind the female while facing the same direction. He then bends his abdomen round to mate with her, before moving to adopt a typical mating position with male and female facing away from each other. The pair remain coupled for between 45 minutes and two hours.

The female is more secretive than the male and is most often encountered when she is egg-laying, an activity that typically peaks either side of midday when the temperature is at its highest. The primary larval foodplants are Early Hair-grass, Red Fescue and Sheep's-fescue on chalk and limestone sites, Bristle Bent on heathland sites and the much coarser Marram grass at coastal sites. Larvae have also been recorded on Tufted Hair-grass. The plants selected are usually those in full sun and that form tussocks surrounded by bare ground. However, in the heat of the summer of 2018, almost every female I watched lay in the New Forest heathlands chose areas that were in full or partial shade. When egg-laying the female will fly between these tussocks, land and then crawl through the grass before curling her abdomen forward to test for a suitable location before depositing a single white egg. Another egg may be laid soon after, either on the same plant or a nearby plant. Eggs are almost always laid on dead material, typically a dead leaf of the larval foodplant or nearby plant debris, and Tinbergen suggests that this might be to protect against grazing animals such as rabbits.

A mating pair with ♀ on the left [NH]

Ovum

The 1-mm high egg is almost spherical with a flattened base and has a series of approximately 30 fine ribs running down its sides that can just be made out with a powerful hand lens. Eggs are white when first laid, but gradually turn pale yellow before darkening slightly as the larva develops within. Unlike the eggs of closely related species, the egg of the Grayling retains a uniform colour and does not develop any brown blotches.

A newly laid Grayling egg

Larva

After two to three weeks, the 1.5-mm long and cream-coloured **first instar** larva cuts a hole toward the top of the egg, from which it emerges by pushing its way out, ultimately feeding on the tender tips of the foodplant and from where it will fall into the grass tussock at the slightest disturbance. The larva has the faintest of stripes that run the length of its body and two stumpy points on the anal segment. Early instar larvae can occasionally be found feeding during the day, but the increasingly nocturnal larvae are most easily found at night, when the pale larva contrasts nicely against the green of its foodplant in torchlight. Of course, knowing where an egg has been laid is a considerable advantage. After around three weeks, and just prior to its first

1st instar: a larva feeding on its eggshell

1st instar: a fully grown larva

2nd instar: the larva now has a series of dark brown marks along the back

3rd instar: the larva has more prominent points on the anal segment

moult, the larva is just over 4 mm long and has developed a green hue, along with a brown stripe at the base of each side from the fourth to the 11th segment, although this does vary in intensity between individuals.

After another two or three weeks, the larva moults into the **second instar** and is now straw brown in colour. It retains the brown band along each side from the 4th to the 11th segment, although these are now bordered above by a prominent white stripe that runs the length of the body, and these same segments have a broken series of dark brown marks along the back. The larva now has a more striped appearance, and this extends to the head, where there are six faint bands running down the face that are each made up from a series of minute black dots. When fully grown, the second instar is 7 mm long.

The larva is slow growing and it is only after several more weeks of intermittent feeding that it moults into the **third instar**. It is similar in appearance to the second instar, but now has a yellowish tinge, the markings are better defined, and the anal points are more prominent. The larva overwinters at the base of a grass tussock or just under the soil surface while in this instar, although it does not become fully torpid and will continue to feed intermittently if the weather is mild. When fully grown, the third instar is 10 mm long.

Feeding resumes in earnest in the spring and, around the end of March, the larva moults into the **fourth instar**. The larva is now noticeably different from earlier instars since there is a series of distinct stripes that run the length of the body. These more mature larvae are now truly nocturnal, crawling up the grass stems to feed only at night and retreating deep within the base of a grass tussock during the day, so deep that it takes considerable effort to prise the base of a tussock apart to locate its occupant.

At the end of April, and when 13 mm long, the larva moults for the last time into the **fifth instar**. The larva is similar in appearance to the previous instar but now has much bolder stripes and a continuous dark brown stripe along its back. After six weeks, when fully grown, it is between 28 and 31 mm long. When preparing to pupate, the Grayling larva exhibits a moth-like tendency of burrowing a little way into the soil where it creates a cell and will also deposit a small amount of silk to bind grains of earth together. The resulting construction has surprisingly smooth walls.

4th instar: the stripes are much more distinct

5th instar: the larva has bolder stripes and a continuous dark brown stripe along its back

Pupa

Around the start of June, depending on location and temperature, the larva pupates. The pupa is 16 mm long and is completely unattached to the cell in which it resides. After approximately four weeks, the adult emerges and makes its way out of its earthen cell before crawling up vegetation to expand its wings.

A recently formed Grayling pupa

A pupa in the process of colouring up

Life Cycles of British & Irish Butterflies

Pearl-bordered Fritillary
Boloria euphrosyne

❖ PEARL-BORDERED FRITILLARY IS SUPPORTED BY ANTHONY McCLUSKEY, THERESA TURNER & JOHN WILLIAMS ❖

	January	February	March	April	May	June	July	August	September	October	November	December
IMAGO												
OVUM												
LARVA												
PUPA												

The **Pearl-bordered Fritillary** gets its name from the series of 'pearls' that run along the outside edge of the underside of the hindwing. It was originally named 'April Fritillary' by James Petiver in his *Musei Petiveriani* of 1695 in recognition of the month in which its flight period started. In 1717 Petiver changed its name to 'April Fritillary with few spots' in order to distinguish it from the Small Pearl-bordered Fritillary, which has more silver patches on its underside. The adoption of the Gregorian calendar in 1752 advanced the calendar by 11 days and this meant that the 'April Fritillary' became a May fritillary although, as a result of a changing climate, the butterfly has advanced its flight period so that it is now an April fritillary once again. The name we use today was provided by Moses Harris in his classic work *The Aurelian* of 1766, where he gave this butterfly the name 'Pearl Border Fritillaria'.

♂ and ♀ have similar undersides

F.W. Frohawk writes in his *Natural History of British Butterflies* of 1924: "*This is one of the commonest of our woodland butterflies in the spring and early summer months; it occurs in almost all the larger woods and forests throughout England and Wales*". It is therefore a sad fact that one of the most notable aspects of the Pearl-bordered Fritillary since those halcyon days is its decline and entomologists of yesteryear would be horrified to learn that this once common butterfly is now one of our most threatened. Its decline within living memory is shocking—against a 1976 baseline, the butterfly has suffered a 95% decline in distribution and a 71% decline in abundance.

The ♂ has a wingspan of 38 to 46 mm

The slightly larger ♀ has a wingspan of 43 to 47 mm

DISTRIBUTION

In England and Wales this butterfly can be found in scattered and isolated colonies south-west of a line running between Denbighshire in North Wales to Sussex in south-east England (it was lost from Kent in 2002), although there are also colonies in south Cumbria and north Lancashire. It is also widespread in central Scotland, but very local or absent in the north and south of this central concentration. In Ireland it is found in the Burren in County Clare and south-east County Galway. The butterfly is not found in the Isle of Man or the Channel Islands. On a worldwide basis, the butterfly is found throughout most of Europe and through temperate Asia.

Based largely on the availability of suitable habitat, nectar sources and larval foodplants, colonies vary in numbers from only a few dozen adults at the peak of the flight period, to over a thousand at the best sites and when conditions are optimal. Matthew Oates describes the mobility of this butterfly in his 2004 *British Wildlife* article

'The ecology of the Pearl-bordered Fritillary in woodland': "*When there were strong colonies in Cirencester Park Woods, the butterfly readily colonised new habitat patches up to 600m distant, often despite the absence of interconnecting open rides through dense woodland. However, only a small minority of individuals moved, and then only during sunny anticyclonic weather. One marked female travelled 4.5km, across arable land and a deep wooded valley*". In *The Butterflies of Sussex*, Neil Hulme describes an instance of the butterfly apparently flying up to 12 km to colonise a new patch, with no interconnecting habitat, although, given the lack of similar sightings, this must be considered an exception.

HABITAT

In Britain and Ireland, the Pearl-bordered Fritillary occupies three types of habitat and surveys in the late 1990s showed that 50% of colonies are found in woodland clearings, 30% in bracken/scrub/grassland habitats and 20% in open woodland. In all habitat types, the butterfly requires nectar sources and an abundance of violets, the larval foodplant, that grows within short and sparse vegetation and where there is a good amount of leaf litter that creates the required microclimatic conditions for the successful development of the immature stages. The adult butterfly also requires sunny and sheltered spots that are usually warmer than surrounding areas. Because of these requirements, the butterfly may colonise only a small part of the available habitat at any given site, such as particular clearings in more extensive woodland.

In the south of its range, the butterfly is found in the first habitat type of woodland clearings where there is a regular cycle of coppicing, or clear-felling before replanting. Studies by Martin Warren and Jeremy Thomas in Hampshire in 1992 showed that populations of the Pearl-bordered Fritillary peak under the most open conditions provided by the coppice cycle but decline rapidly thereafter. Oates highlights the significance of Bracken at most woodland sites, which provides the breeding habitat required by the butterfly by suppressing the growth of grasses while allowing violets to flourish. Changes in woodland management, and particularly the decline of coppicing, sadly mean that suitable open habitats in woodland are increasingly fragmented, even within heavily wooded areas and, as a result, most woodland populations are now relatively isolated, with limited movement between them (Dan Hoare, pers. comm.).

The second habitat type is represented by a mosaic of dense stands of Bracken, light scrub and grassland. These sites are found in the south-west in Cornwall, Devon and Dorset, as well as on the limestone hills of south Cumbria and north Lancashire, and in parts of Scotland, Wales and northern England. In this more extensive habitat, it is thought that the species is more mobile and forms metapopulations that represent a network of breeding areas.

In Bracken-dominated sites, such as those on Dartmoor in Devon, the butterfly lays its eggs wherever violets grow under Bracken stands (which, to all intents and purposes, act as a surrogate woodland canopy in this often treeless habitat), preferring areas where dead Bracken is also present, which allows larvae to bask and raise their body temperature, thereby aiding their development (especially in the spring, post-hibernation). Bracken stands soon become unsuitable for fritillaries if

The Pearl-bordered Fritillary flies in the New Copse inclosure in the New Forest

neglected, since dense shade will prevent violets and other plants from growing. Sites of this habitat type are often maintained with an appropriate grazing regime involving cattle and ponies.

In limestone-dominated sites, such as those in the Burren and the area around Morecambe Bay, the butterfly uses violets growing beneath regenerating scrub. Although these sites can remain suitable for several years without any intervention, they will eventually scrub over due to natural succession if abandoned.

The butterfly is found in a third habitat type (which occurs only in Scotland) of open deciduous wood pasture (grassland with an open cover of mature trees, usually birch or oak) where patches of Bracken are present and, again, the colonies here exist within a metapopulation structure and are more mobile than those colonies found in woodland.

STATUS

Once considered common and widespread, the Pearl-bordered Fritillary is now one of our most threatened species. In a review of key sites in central-southern England, Martin Warren found that the rate of loss was 39% per decade in the 1980s. The butterfly has undoubtedly declined as a result of woodland management changes (such as the cessation of coppicing), general habitat loss due to development and agriculture, and fragmentation of sites that prevents natural colonisation (and recolonisation following a local extinction). Conservation efforts have focused on habitat management and the more recent trend shows that the situation seems to be stabilising.

The long-term trend shows a significant decline in both distribution and abundance, with the more recent trend over the last decade showing a stable distribution and a moderate increase in abundance. Using the IUCN criteria, the Pearl-bordered Fritillary is categorised as Endangered (EN) in both the UK and Ireland.

LIFE CYCLE

This is the earliest of our fritillaries to emerge and, in early years, can be seen from the middle of April or even earlier (there are some records from late March). Most adults emerge at the start of May or towards the end of May at more northerly sites. The butterfly normally has a single brood but, in exceptional years, there may be a partial second brood at some southern sites (which is the norm on the continent), with adults emerging in August.

Imago

The butterfly exhibits a flit-and-glide motion when in flight, when it can only be mistaken for a Small Pearl-bordered Fritillary that has a similar flight pattern. Both male and female are avid nectar feeders and can often be found in sunny and open areas feeding from the flowers of Bugle, their favourite nectar source. Less widely used nectar sources include Bluebell, buttercups, Common Bird's-foot-trefoil, dandelions, hawkweeds, Ragged-Robin, Selfheal, thistles and Tormentil. A good time to observe this species is early in the morning when the butterfly spends a good amount of time refuelling. At night and in dull weather, the adults will rest on flowerheads and the leaves of various bushes.

Both sexes are extremely difficult to track when in flight since the colour of their wings provides excellent camouflage against the dead Bracken that is often found at their sites. Males are the most active of the two sexes and they can often be seen patrolling swiftly and low across the breeding site, investigating any reddish-brown object that might be a newly emerged female waiting in the undergrowth for a passing male to find her. When a male encounters another male, the briefest of skirmishes occurs before the two go their separate ways. If a male encounters an unreceptive female, then she will rapidly flutter her wings until the male loses interest and flies off. When a receptive female is found then she will fly to a suitable platform, sometimes at some height, where the pair mate and stay together for

A mating pair [NH]

A ♀ egg-laying on the underside of a leaf of Common Dog-violet

between 30 and 60 minutes. Mated females spend a couple of days waiting for their eggs to mature before starting to lay and, when basking in the sun, will snap their wings shut when a male flies nearby, in order to avoid unwanted attention.

Egg-laying females are relatively easy to follow in flight as they flutter slowly and deliberately, low down over vegetation, searching out suitable patches of foodplant. They have a preference for young plants found in warm areas, such as those growing over bare ground or through dry and heat-absorbing leaf litter, with eggs rarely laid in grassy areas. The most widely used larval foodplant is Common Dog-violet, although other violets are also used, such as Heath Dog-violet and, in the north of its range, Marsh Violet. Eggs are laid singly, occasionally in pairs, either on the foodplant (on the underside of a leaf or on a stem) or on nearby dead vegetation.

Ovum

The 0.8-mm high conical egg is greenish-yellow when first laid, but gradually turns light yellow and finally a leaden grey prior to the larva emerging. The egg has around 25 ribs that run irregularly down its sides, although there is no set pattern for where each rib starts and ends.

A newly laid egg is greenish-yellow

The egg turns pale yellow after a day or so

Larva

The 1.4-mm long **first instar** larva emerges from the egg after around two weeks, depending on temperature, and, if the egg has been laid on dead vegetation, crawls to a nearby violet plant where it feeds by day on the tenderest leaves. The larva has a light yellow-brown ground colour, with brown blotches on the 4th, 6th, 8th and 10th segments that give it a banded appearance. The larva has a black head and is covered in warts, each of which bears a long hair. When fully grown, the first instar is just over 3.5 mm long.

After around 10 days, the larva moults into the **second instar**. The larva now has an olive-brown ground colour and is in possession of six rows of black spines (rather than the hairs of the first instar) that run the length of the body. While still relatively young, the second instar continues to exhibit slightly darker banding on the 4th, 6th, 8th and 10th segments. A dull, grey stripe also runs along the base of each side and, when fully grown, the second instar is around 6.5 mm long. When not feeding or basking on dead Bracken or other leaf litter, the larva will rest under a leaf or on a stem of the foodplant.

After between one and two weeks, depending on temperature, the larva moults into the **third instar**. The larva now has a much darker overall appearance with an almost black ground colour. It also has two grey lines that run in parallel along the length of its back and the subdued grey stripe along each side is much more prominent than in the previous instar.

1st instar: the larva has a banded appearance

2nd instar: the larva is now covered in spines rather than hairs

3rd instar: the larva has a darker appearance than earlier instars

Life Cycles of British & Irish Butterflies

After another week or so, the fully grown 8.5-mm third instar moults into the **fourth instar**—the instar in which it will overwinter. The ground colour is a purplish-brown, speckled with grey markings that coalesce somewhat to form a dull stripe along each side, and the larva continues to show two grey lines that run in parallel along the length of its back. A couple of weeks after moulting into this instar, and when just over 10 mm long, the larva enters hibernation among

4th instar: a larva hibernating under a Bracken frond

4th instar: a pre-hibernation larva

dead vegetation, often on the underside of a crumpled leaf. It shrinks somewhat over the winter and, when it awakens from its slumber from early March, is only 6.5 mm long, when it can be found basking on dead Bracken and other leaf litter. Larvae start feeding by eating the lobes at the base of violet leaves, leaving characteristic feeding damage that can give away their presence, although whole leaves may also be eaten, with only the stem left intact. Developing flower buds are also eaten. The fully grown fourth instar is around 13 mm long.

5th instar: the larva is now covered in yellow spines

After two or three weeks of post-hibernation feeding, the larva moults into the final and **fifth instar**. The larva is now largely black, with a light band that runs along each side of the body and a light brown underside. The most distinctive characteristic, however, is two rows of prominent black-tipped yellow spines that run from the first to the 11th segments. The colour of these spines is somewhat variable, however, and larvae with completely black spines are found on occasion. Larvae continue to feed by day, typically move off the foodplant at night, and, in early April, can be found warming themselves on a dead Bracken frond or dead leaf, especially following a cold spell when they are looking to raise their body temperature. Final instar larvae can also move very rapidly and cover large distances when necessary. After two to four weeks in this instar, the larva prepares for pupation by spinning a silk pad low down in vegetation on a leaf or stem, from which it suspends itself upside-down. When fully grown, the larva is around 25 mm long.

Pupa

After a few days, the larva moults for the last time, revealing a 14-mm long pupa that is attached to the silk pad by the hooks on its cremaster, resembling a dead leaf. After between nine days and three weeks, depending on temperature, the adult butterfly emerges.

5th instar: a larva preparing to pupate

A pupa suspended from a Bracken frond

Life Cycles of British & Irish Butterflies

Small Pearl-bordered Fritillary
Boloria selene

❖ SMALL PEARL-BORDERED FRITILLARY IS SUPPORTED BY ALAN KERR, RORY MARLAND & GARY NORMAN ❖

	January	February	March	April	May	June	July	August	September	October	November	December
IMAGO												
OVUM												
LARVA												
PUPA												

The **Small Pearl-bordered Fritillary** is a delightful butterfly that is similar in both appearance and behaviour to the closely related Pearl-bordered Fritillary and, like that species, gets its name from the series of 'pearls' that run along the outside edge of the underside of the hindwing. The two species often fly together and, no matter where I have come across both species on the same day, whether this is in southern England or in northern Scotland, the Pearl-bordered always emerges around two to three weeks before the Small Pearl-bordered and generally appears very faded in comparison with the bright orange of its newly emerged cousin. Earlier taxonomic listings assigned the Scottish populations to the subspecies *insularum* on account of their lighter appearance, although this designation is not recognised in the most current taxonomic checklist.

The underside reveals a series of 'pearls' along the outside edge of the hindwing

The naming of the Small Pearl-bordered and Pearl-bordered Fritillaries was mired in confusion in the early 1700s. The Pearl-bordered Fritillary was originally named the 'April Fritillary' by James Petiver in his *Musei Petiveriani* in 1695, only for John Ray to confuse matters by using the same name to describe the Small Pearl-bordered Fritillary in his *Historia Insectorum* in 1710. Recognising that there were two species to describe, James Petiver made an attempt to rectify matters in his *Papilionum Britanniae* of 1717, where he applied the name 'April Fritillary' to the Small Pearl-bordered Fritillary and 'April Fritillary with few spots' to the Pearl-bordered Fritillary. The adoption of the Gregorian calendar in 1752 advanced the calendar by 11 days (when 2 September was followed by 14 September) and confused matters further, with both species now becoming May species, despite their names. The impending confusion was addressed by Moses Harris who, no doubt, felt that naming the two fritillaries based on the month in which they appeared was singularly unhelpful. In his classic work *The Aurelian* of 1766, Harris introduced the basis of the names that are used today (albeit with the names 'Pearl Border Fritillaria' and 'Small Pearl Border Fritillaria') although, it has to be said, identifying an individual based on whether it is 'small' or not is a bit of a stretch in the field when the wingspans of the two species are so similar—the wear and tear of a given individual is a much better initial guide.

The ♂ has a wingspan of between 35 and 41 mm

The ♀ has a wingspan of between 38 and 44 mm

DISTRIBUTION

This butterfly is found in discrete colonies that can contain anything from a couple of dozen adults to over 100 at the peak of the flight period, the latter figure usually recorded from one of the strongholds of this species which are undoubtedly Scotland, Wales, and the north-western and south-western counties of England. A case in point is the area just south of the River Strontian near Ariundle NNR where this is the commonest butterfly to be found in early June, where it flies in the company of the equally scarce Chequered Skipper. The butterfly is absent from Ireland, the Outer Hebrides, Orkney,

Shetland, the Isle of Man and the Channel Islands. On a worldwide basis, the Small Pearl-bordered Fritillary is found throughout the Holarctic, where its range extends from central and northern Europe through to Asia as far as Korea, and it is also found in North America, where it is widespread.

In the coppiced woodlands of southern England, the butterfly is extremely sedentary, with only a limited capacity for colonising new areas and most success is where there is a continuous supply of new and accessible clearings nearby. In *The Butterflies of Britain and Ireland*, Jeremy Thomas writes: "*Just 100m of tall shrubs may represent an insurmountable obstacle to the colonisation of a new clearing in a wood*". In more northern areas, where colonies exist in open woodland, grassland and moorland, the butterfly is provided with a more continuous habitat, and it is thought that the butterfly here is more mobile, possibly living within 'metapopulations' that are spread over a wide area. Indeed, studies by Liz Davidson at Warton Crag in Morecambe Bay have shown that the Small Pearl-bordered Fritillary has a more dispersed population structure and is much more mobile than the Pearl-bordered Fritillary, with the latter forming discrete colonies with little exchange.

One of my most vivid memories of this species was when I accidentally stumbled across a colony in a favoured clearing at Pamber Forest in Hampshire, where numerous adults jostled for position while nectaring on Bugle. Unfortunately, like so many sites in southern England, this colony has now disappeared, although the possibility of a reintroduction at some point in the future is hopefully on the horizon, once the ecology and practical conservation of this species is better understood. However, if the root cause of the butterfly's demise turns out to be a loss of its preferred wet grassland habitat due to climate change, then this may not be a realistic prospect.

HABITAT

The butterfly exists in a wider variety of habitats than any other fritillary and has a preference for damp areas where the larval foodplants grow particularly vigorously. Habitats include woodland clearings and glades (especially in southern England where other habitat types have been largely destroyed as a result of agricultural intensification); woodland edges and adjoining open pasture (in Scotland); moorland, marshland and damp grassland (in the West Country, Wales and northern England); and Bracken-covered and lightly grazed grassland with patches of scrub (in the Mendips and the Morecambe Bay area). The butterfly is also widespread along the cliff systems of western Britain.

In *The Butterflies of Sussex*, Neil Hulme provides a checklist of desirable features of Small Pearl-bordered Fritillary woodland habitat: "*Widespread and common violets, growing in mats; dappled*

Woodland edge habitat at Ariundle NNR in the Scottish Highlands where two rarities are found side by side (inset)—a ♂ Chequered Skipper with a ♂ Small Pearl-bordered Fritillary

sunlight/shade adjacent to fully sunny areas; drier refuge areas (slopes and banks); moisture-loving graminoids (rushes and sedges); Bugle; Marsh Thistle; Ragged-Robin; Bluebell, and horsetails". Ragged-Robin is a good indicator of the wet grasslands this butterfly likes—the plant itself is becoming less common, showing how many areas are drying out (Dan Hoare, pers. comm.).

STATUS

Despite some stability in the west and north, this species has suffered a long-term decline in both distribution and abundance with extinctions across most of central, eastern and south-eastern England. The cessation of coppicing in woodlands, which creates the right habitat for the adult several years after a clearance and encourages vigorous growth of the foodplant, is believed to be a primary cause of the decline here as sites become too shaded for the butterfly to survive. It is also thought that climate change has exacerbated the situation across the drier parts of England, with some former sites now lacking the damper areas favoured by this species (Neil Hulme, pers. comm.). Based on experiences from captive breeding programmes, larvae are becoming increasingly active in mid-winter (something unheard of in older texts) due to mild winters and, in the wild, this may also be a factor in the butterfly's decline, with larvae potentially starving due to a lack of suitable foodplant (Dan Hoare, pers. comm.).

The long-term trend shows a significant decline in both distribution and abundance, with the more recent trend over the last decade showing a moderate decline in distribution and a stable abundance. Using the IUCN criteria, the Small Pearl-bordered Fritillary is categorised as Near Threatened (NT) in the UK.

LIFE CYCLE

This butterfly first emerges in south-west England, where it may be seen from the beginning of May, and this early emergence results in a second brood there, which appears in late July or August. A second brood is occasionally produced in other areas too, such as South Wales, and is a regular occurrence on the continent. However, most colonies emerge in the second half of May, peak in early June and fly to the end of the month. The flight period is a little later in Scotland, especially in the north, where the butterfly emerges in the second half of June and flies until the end of July.

Imago

The male is the most conspicuous of the two sexes when it is patrolling just a couple of feet off the ground, alternating between a burst of rapid wing beats and a short glide as it searches out newly emerged females in surrounding scrub. The wing pattern, however, makes the adult difficult to follow in flight and it is much easier to observe this butterfly when it is basking or feeding, and early morning is often a good time to observe this species as adults warm themselves in the sun before becoming active.

A mating pair with ♀ on the left [TP]

With luck, adults may also be found roosting on Bracken fronds and on the flowers of rushes or sedges. Both sexes are avid nectar feeders and can be seen at a variety of flowers, although Bugle is a particular favourite. Other flowers visited include those of Common Bird's-foot-trefoil, Bluebell, buttercups, dandelions, hawkweeds, heathers, Ragged-Robin, thistles, Wild Thyme and Wood Spurge.

The male can be seen dipping down from its patrolling flight to examine any orange object that might be an emerging female and, on more than one occasion, I have watched a male locate a potential mate that was just a few feet from where I was standing, but that was quite invisible to me. When a virgin female is found, the pair quickly mate and generally remain hidden low down in vegetation, remaining coupled for between 30 minutes and two hours. Conversely, an unreceptive female that has already mated will rapidly flutter her wings in an attempt to dissuade the male from attempting to mate.

Egg-laying females are easy to follow as they flutter slowly and deliberately low over vegetation, searching out suitable areas of foodplant. Eggs are laid in areas that have an abundant growth of violets growing in lush vegetation. Preferred areas are also slightly warmer than their surrounds, presumably because the immature stages benefit from the local microclimate. The most widely used foodplants are Common Dog-violet and Marsh Violet where the latter is used at particularly damp sites, especially in Scotland and Wales.

The straw-coloured eggs are often laid on vegetation near the foodplant

Ovum

The 0.6-mm high eggs are laid singly and, while some are laid on the underside of a leaf of the foodplant, many are laid on nearby vegetation. The female will also simply drop her eggs once she has alighted on the foodplant or nearby, and some authors say that she will even lay while in flight, although such indiscriminate egg-laying would seem to be the exception. Eggs are strongly ribbed and are greenish when first laid, but soon turn a straw colour and ultimately grey as the dark head of the larva shows through the eggshell.

Larva

After just under two weeks, the 1.4-mm long **first instar** larva emerges from the egg and proceeds to eat most of its shell. The larva has a yellow-brown ground colour when first emerged, although this changes to a light green once it has started feeding on violets, when dark bands on the 4th, 6th, 8th and 10th segments also become more apparent. The body is also covered in numerous and relatively long hairs, each of which emerges from a black tubercle. The young larvae feed on the foodplant by day and will rest on a shaded part of the plant, such as the underside of a leaf, or in nearby leaf litter, since, unlike the Pearl-bordered Fritillary, larvae of the Small Pearl-bordered Fritillary tend to avoid direct sunlight. This makes the larvae particularly difficult to locate since they are only visible when feeding and are never found basking on dead Bracken or leaf litter. The larvae do, however, leave distinct crescents where they have fed on the edges of the heart-shaped leaves of the foodplant. When fully grown, the first instar is just over 3 mm long.

After a couple of weeks, depending on temperature, the larva moults into the **second instar**. The larva now has several rows of spines that run the length of the body, rather than the hairs found in the first instar. The second instar initially appears quite dark, when the black spines are relatively close together. As the larva grows, the spines are spread further apart, and the olive-green ground colour of the body becomes more visible. The fully grown second instar is 4.5 mm long

1st instar: the larva has distinct bands on the 4th, 6th, 8th and 10th segments

2nd instar: the larva is now covered in spines

Aside from its size, the **third instar** larva is similar in appearance to the second instar, although there are two distinguishing features. The first is that the third instar has a prominent dorsal stripe that runs the length of its body, and the second is that it now has a series of dull yellow-orange spots on each side of the body. The fully grown third instar is just over 6 mm long.

While most authors agree that the larva hibernates in dead leaves or other debris, which provide both shelter and camouflage, they disagree on the instar in which the larva overwinters—some say third instar and others say fourth. In order to shed some light on this discrepancy, I conducted my

3rd instar: the larva has a prominent dorsal stripe and yellow-orange spots on its sides

3rd and 4th instar: larvae hibernating together under a Bracken frond

4th instar: a post-hibernation larva

own analysis by rearing larvae in captivity and proved that larvae will overwinter in either their third or fourth instar, although there was no discernible bias to one instar or the other, and all larvae successfully overwintered. Whatever the overwintering instar, the larvae will feed intermittently when the weather is warm enough.

If the third instar does not overwinter then, after a couple of weeks, it moults into the **fourth instar**, otherwise it delays moulting until early March of the following year. The third and fourth instars are extremely similar in appearance, so much so that the only real distinguishing factor, unfortunately, is size—a fully grown fourth instar is around 10 mm long and almost twice the length of the third instar.

After two to three weeks, the larva moults for the final time, revealing a **fifth instar** larva that is very different in appearance from earlier instars. The larva now has a relatively uniform ground colour (although this does vary from light brown to light grey), is covered in colourful spines (which may be dark yellow, orange or pink), and has two very prominent tubercles that emanate from the first thoracic segment and point forward over the head. The larva feeds primarily by day, always rests in a straight position, and, when preparing to pupate, will suspend itself head-down from a silk pad low down in leaf litter or on low-growing plants. When fully grown, the larva is around 21 mm long.

5th instar: a larva with brown ground colour and orange spines

5th instar: a larva with grey ground colour and dark yellow spines

5th instar: a pre-pupation larva suspended head-down

5th instar: a larva with grey ground colour and pink spines—the inset shows that the spines on the first segment protrude forward over the head

Pupa

The pupa is around 15 mm long and is suspended head-down, attached by the cremaster to a silk pad. Like the pupae of several other Nymphalids, the pupa is adorned with several metallic conical projections on its back, which are thought to mimic water droplets. After two to three weeks, this characteristic butterfly of damp habitats emerges.

The pupa hangs upside-down from a suitable surface

The same pupa just prior to the adult emerging

Life Cycles of British & Irish Butterflies

Silver-washed Fritillary
Argynnis paphia

❖ SILVER-WASHED FRITILLARY IS SUPPORTED BY JIM BALDWIN, TREVOR GOODFELLOW & TONY ROGERS ❖

	January	February	March	April	May	June	July	August	September	October	November	December
IMAGO												
OVUM												
LARVA												
PUPA												

The **Silver-washed Fritillary** is the largest of our fritillaries and gets its name from the beautiful silver streaks found on the underside of its hindwings, which distinguish it from every other fritillary found in Britain and Ireland. It is also a quintessential butterfly of woodland and is unmistakable when seen flying along a woodland ride, or when feeding from a favourite nectar source, such as bramble flowers. The bright orange male is quite distinctive as it flies powerfully along, pausing only briefly to feed or investigate anything with an orange hue that could be a potential mate. The male has four prominent black sex brands on each forewing that contain the 'androconial' scent scales used during courtship. The female is slightly larger than the male, has more prominent spots and, of course, lacks the sex brands that are only found in the male butterfly.

The ♂ has a wingspan of 69 to 76 mm and prominent sex brands on its forewings

The ♀ is slightly larger than the ♂, with a wingspan of 73 to 80 mm [PB]

An adult feeding on a damp patch of earth, showing its silver streaks [MC]

The *valesina* form of the ♀ [NF]

Whenever I think of the Silver-washed Fritillary, two things stand out in my mind—the first is the enchanting courtship flight, described below, and the other is an incredible form of the female known as '*valesina*', where the orange-brown ground colour is replaced with a deep bronze-green, the butterfly appearing at a distance to be an oversized Ringlet, a species that flies at the same time of year. This form occurs in a small percentage of females, primarily in the larger colonies in the south of England. The legendary lepidopterist F.W. Frohawk was so taken with this form that he named his youngest daughter after it, albeit with the spelling 'Valezina' rather than the spelling of the original definition of this form that was provided by E.J.C. Esper in 1800, which uses an 's' rather than a 'z'. The *valesina* form is genetically controlled and, while the male carries the relevant form of the gene, its effect is only physically manifest in the female. What is most surprising about this sex-linked form is that it is, in genetic terms, 'dominant'. Without getting into the details of genetic inheritance, this should mean that we should see more *valesina* females than we actually do—so why don't we?

Life Cycles of British & Irish Butterflies

Thanks to the studies of Dietrich Magnus, whose findings were published in 1958, we now know that the *valesina* form of the female is at a disadvantage in its environment because it takes longer for it to be found by a male since certain visual clues are missing. Specifically, the optical stimulus provided by the female to attract a mate (a distant fluttering brown object) is missing in a *valesina* female. As a result, the female may not find a mate at all and, if she does, there is potentially a longer time between the butterfly emerging and her egg-laying, leading to a greater possibility of predation during this critical period. Conversely, it could be argued that the butterfly is at an advantage in dense woodland where it can warm up more quickly than a normal female due to the darker colouring of the wings and can become active earlier in the day and be the first to grab the attention of amorous males. Others suggest that *valesina* eggs are less viable. Suffice to say, the *valesina* form of the female Silver-washed Fritillary remains one of our most enigmatic forms and would make an excellent research subject.

DISTRIBUTION

The butterfly is found in woodlands primarily in the southern half of England and Wales, with a few scattered colonies elsewhere, such as in south Cumbria and west Lancashire. This species is also widely distributed in Ireland and is expanding throughout its range, but is absent from Scotland, the Isle of Man and the Channel Islands. On a worldwide basis, the butterfly has a Palearctic distribution and can be found throughout most of Europe, eastwards as far as China and Japan, and is also found in some parts of North Africa.

Wherever it is found, the butterfly forms somewhat-discrete colonies within specific woodlands, where it flies rapidly between clearings, along rides and over the tree canopy as it moves from clearing to clearing. Individuals have been recorded some distance from known sites, allowing it to colonise new areas on occasion and, in the hot summer of 2018, I was astonished to find two females laying on various surfaces inside my suburban garage that happened to have a good growth of violets, the larval foodplant, just outside the door.

HABITAT

The Silver-washed Fritillary is found primarily in deciduous woodland, although it will occasionally use coniferous woodland or a mix of the two, where its larval foodplant, Common Dog-violet, grows in shady or semi-shady conditions on the woodland floor. Oak woodland is preferred, with Ash and Beech woodlands appearing unsuitable. The butterfly may also be encountered in hedgerows if they are sufficiently populated with mature trees, especially in south-west England and in Ireland. The butterfly will also wander into any adjacent habitat—I have a fond memory of finding the *valesina* form of the female feeding on thistle on the chalk downland at Stockbridge Down in Hampshire where it was flying in the company of Chalk Hill Blues, approximately 200 m from the nearest wood.

Woodland ride habitat at Pamber Forest in Hampshire

The butterfly spends some of its life in the canopy where it rests, basks and feeds on aphid honeydew and, while it can be found flying deep within a wood (especially the female when she is looking to find suitable places to lay her eggs), it is most often observed in woodland rides, glades and clearings. Here it can be found gorging itself on a favourite nectar source of brambles or thistles, although adults will also feed on Common Fleabane, knapweeds, ragworts, Water Mint and Wild Privet. Habitat management involves thinning of woodland to maintain an open canopy, along with rotational cutting of clearings and rides in autumn and winter, that will allow a good amount of sunlight to enter and nectar sources to flourish.

STATUS

The status of the Silver-washed Fritillary in Britain and Ireland is relatively stable when compared with other species and could even be said to be doing well given the increases in both distribution and abundance with regard to both short- and long-term trends. The long-term trend shows a significant increase in both distribution and abundance, with the more recent trend over the last decade showing a significant increase in distribution and a stable abundance. Using the IUCN criteria, the Silver-washed Fritillary is categorised as Least Concern (LC) in both the UK and Ireland.

LIFE CYCLE

The Silver-washed Fritillary has a single brood each year and is on the wing from the middle of June until late August, with the butterfly overwintering as a young larva. As is the case with most species, the butterfly may emerge slightly earlier in good years, and be found flying later in poorer years and in the most northern part of its range.

Imago

Adults spend much of their time in the woodland canopy and out of sight when feeding on aphid honeydew, or when roosting at night and during inclement weather, but are frequently seen at ground level when taking nectar. F.W. Frohawk describes their arboreal habits in his *Natural History of British Butterflies* of 1924: "*In unsettled weather, with sunny intervals, it was an interesting sight to observe the* paphia *dropping from the trees into the rides as soon as the sun appeared, when their numbers were so great that they resembled an autumnal shower of falling leaves, and directly the sun became obscured they all fluttered up again to settle among the foliage, where this butterfly rests during dull weather and also passes the night. On calm, fine evenings they frequently fly to the topmost branches of the tallest trees and retire for the night among the leaves*".

When in search of a mate, males will fly along woodland rides, clearings and woodland edges, pausing to examine any golden-brown object, especially one that is moving and is more likely to be a female. If a virgin female is encountered by a male, then she will also emit pheromones from the end of her abdomen to encourage the male to begin the spectacular courtship flight that is unlike any other seen in the butterflies of Britain and Ireland and is simply enchanting—the female flies in a straight line while the male repeatedly loops under, in front of, and then over the top of the female, showering her with pheromones that emanate from his androconial scales as he goes. This flight is usually undertaken at a pace that allows the observer quite clear views of the courtship and, when it is over, the pair land on a convenient platform where the male draws the female's antennae over his sex brands before mating with her, the pair remaining coupled for around two hours. Research has shown that the male not only transfers his sperm to the female, but also chemical compounds that coagulate and that, to all intents and purposes, serve as a repellent to other males that might attempt to mate with her.

When egg-laying, the female will fly low over the floor of semi-shaded woodland in search of violets and may land amongst them, before flying onto the shaded and often moss-covered north-facing trunk of a nearby tree. She then probes the

A mating pair, with ♂ on the right

surface with her abdomen before laying a single egg in a crevice in the tree bark or on moss, and several eggs may be laid on the same tree. Observers have seen eggs laid on other surfaces too, such as dead Gorse, and it is thought that such locations provide a suitable microclimate for the overwintering larva.

Most eggs are laid between 1 and 2 m off the ground, in a crevice of bark of both deciduous and coniferous trees, although trees with smooth bark are avoided. There are always exceptions, of course, and I have seen females at Pamber Forest in Hampshire laying on moss at almost ground

A ♀ egg-laying on Yew [PB]

level at the base of a young oak, as well as several metres up on the main trunk of a mature oak. It is surely a minor miracle that the larvae emerging from eggs laid so high up are ultimately able to make their way down to the ground and find violets on which to feed.

The egg is a light yellow-brown

The egg darkens as the larva develops

Ovum

The 1-mm high eggs, which are relatively small compared to the size of the butterfly, are conical in shape and have around 25 ribs that run longitudinally from top to bottom. The egg is a light yellow-brown but darkens as the larva develops.

Larva

1st instar: the larva overwinters on a silk pad

1st instar: a fully grown larva

After around two weeks, the 2-mm long **first instar** larva emerges from the egg by eating away at the crown, which forms its only meal before it immediately moves into a crevice in the bark, within moss or some other nook and cranny, where it spins a silk pad on which to pass the winter. The larva descends to the woodland floor the following spring (although some think it may simply drop to the ground), when it searches out the violets on which it feeds intermittently during the day, eating the most tender leaves and shoots. In all instars, the larva enjoys basking in sunlight, typically on dead vegetation near the foodplant, such as leaf litter, or on bare ground, and in later instars this may be a little distance from the foodplant. This makes finding larvae particularly difficult since there is no guarantee that they will be located close to their food source. The first instar is covered in long hairs and several rows of prominent tubercles, the largest of which run the length of each side of the body and, unlike later instars, the first instar does not possess any spines. Of course, the larva is tiny, and these features are only be visible when using a very strong hand lens.

In early April, the fully grown 3-mm long first instar moults into the **second instar**. The larva is now covered in six rows of black spines that run the length of the body and distinguish the second instar from the first instar. Each spine has an orange base and sports a number of bristles. The larva now has a pale-bordered dark line that runs along the middle of its back, which is found in all

2nd instar: the larva is now covered in spines

3rd instar: the spines take on a golden appearance

4th instar: the first pair of spines now appear significantly larger

remaining instars also. These markings become more prominent as the larva grows, when the spines are spread further apart. When fully grown, the second instar is 5 mm long.

After around a week, the larva moults into the **third instar**. The change from early-instar to late-instar can be quite dramatic, and this is exemplified no better than in the Silver-washed Fritillary, since the third instar now exhibits two rows of prominent golden spines along its back that look quite majestic, along with two rows of shorter spines along each side. The fully grown third instar is 11 mm long.

After between two and three weeks, the larva moults into the **fourth instar**. On first inspection, the newly moulted larva is similar in appearance to a newly moulted third instar, since it also possesses two rows of golden spines along its back. However, the first pair of spines in the fourth instar, which are not as bright as the other spines, are significantly longer than the others and project over the head. The ground colour is also somewhat darker than the previous instar and the body is adorned with a beautiful pattern of markings along each side. When fully grown, the fourth instar is around 20 mm long.

5th instar: the first pair of spines are particularly prominent (inset), but the brownish larva is difficult to find even when basking

5th instar: a larva preparing to pupate

After around 10 days, the larva moults for the last time into the **fifth instar**. The first pair of spines are now particularly long, all other spines are light brown rather than 'golden', and the ground colour is now lighter than in the fourth instar, the larva having a browner appearance overall—even the dark line along the back is bordered by a light brown, rather than the white found in earlier instars. A key difference with other instars is size and the fully grown 40-mm long larva has to be one of the most spectacular larvae of all our species. In his *Natural History of British Butterflies* of 1924, Frohawk commented on the voracious appetite of the final instar larva: *"The larva feeds at frequent intervals during the day and with remarkable rapidity, often devouring the whole of a large violet leaf in less than two minutes"*.

When preparing to pupate, the larva creates a silk pad beneath a leaf or twig of a tree or shrub, 1 or 2 m off the ground, from which it suspends itself, hanging head-down in a J-shape. After a day or two, and 11 days in the final instar, it pupates, having spent 10 months as a larva in total.

Pupa

The 22-mm long pupa is formed head-down attached by the cremaster. It resembles a shrivelled leaf covered in water droplets, thanks to the mirror-like spots that are found on its back. The adult emerges after between two and three weeks, taking wing in its arboreal home.

The pupa resembles a shrivelled leaf

Dark Green Fritillary
Argynnis aglaja

❖ DARK GREEN FRITILLARY IS SUPPORTED BY FRANK FOWLER, THE MAKIN FAMILY & ROB SANDERCOCK ❖

	January	February	March	April	May	June	July	August	September	October	November	December
IMAGO												
OVUM												
LARVA												
PUPA												

The Dark Green Fritillary is our most widespread fritillary and is found throughout much of Britain and Ireland, where it can be seen flying powerfully over its grassland habitats, stopping frequently to nectar at knapweeds and thistles. Male and female are similar in appearance, although the female is the more heavily marked and, while the male possesses a patch of specialised scent scales on certain veins on each forewing, these are barely discernible, unlike those of closely related fritillary species.

The butterfly's name in use today, which was provided by Moses Harris in *The Aurelian* of 1766 (albeit with a spelling of 'Fritillaria'), recognises the green hue found on the underside of the hindwings. The undersides are also peppered with large silver spots that inspired the botanist Reverend Adam Buddle (who is immortalised in the 'butterfly-bush' genus of *Buddleja*) to name this butterfly the 'Great Sylver Spotted Fritillary' in c.1700.

♂ and ♀ undersides are similar with a green hue and large silver spots

The ♂ has a wingspan of 64 mm [IL]

The slightly larger and darker ♀ has a wingspan of 70 mm

DISTRIBUTION

This butterfly can be found in all four corners of Britain and Ireland and is the only fritillary found in the Outer Hebrides and Orkney, where it is at the northern limit of its distribution. It can be found in most parts of southern England, Wales and central and western Scotland, but is primarily a coastal species elsewhere, especially in Ireland, where it can be found flying over sand dunes, undercliffs and river valleys, wherever violets, the larval foodplant, grow. In England, the butterfly's strongholds are in the west, including (from north to south) the Lake District, the Peak District, the Cotswolds, Salisbury Plain, Exmoor and Dartmoor. The butterfly is less common in northern, central and eastern England, where it has suffered substantial declines over the last century. On a worldwide basis, the butterfly is found throughout Europe, where its northern distribution reaches the Arctic Circle. It is also found across Asia as far as China and Japan, and in the Atlas Mountains in Morocco.

Where the butterfly does occur, it is relatively mobile and is usually found in small numbers over a large area, although concentrations do build up in suitable areas, especially where nectar sources are growing in full sun. In *The Butterflies of Britain and Ireland*, Jeremy Thomas recalls one special moment: *"On one memorable day in the 1990s, at Porton Down, Wiltshire, the adults were so abundant that they 'were swarming like flies'"*. Despite its mobility, the butterfly will typically stay within its breeding grounds, flying several hundred metres, although the butterfly will turn up several kilometres away from any known population, suggesting that it is able to colonise suitable areas nearby.

HABITAT

The characteristic habitat of this butterfly is open, windswept calcareous grassland, although the butterfly is also found in bracken-dominated grassland; upland heathland; sand dunes and undercliffs in coastal areas; and woodland clearings and wide rides following clearance work. In all cases, suitable habitats are rich in nectar sources contain suitable violets used by egg-laying females, which are those growing through a medium sward of between 8 and 20 cm in scrubby areas.

Habitat management regimes include light, rotational grazing with cattle or sheep that maintains this medium sward. Sites that become overgrown with coarse grasses and scrub benefit from rotational scrub clearance and, on Bracken-dominated grassland, trampling by grazing cattle and ponies can break up dead Bracken trash, which encourages growth of violets.

The butterfly is attracted to abundant nectar sources at Cressbrook Dale in Derbyshire [PC]

Bockerley Dyke provides plenty of shelter at Martin Down on the Hampshire-Dorset border [PB]

STATUS

Although this species has declined in northern, central and eastern England over the last century, especially since the 1970s, the butterfly remains our most widespread fritillary and is not considered a priority species for conservation efforts. Thankfully, this butterfly has not suffered to the same degree as its close cousin, the High Brown Fritillary, which may be due to its use of a wider range of habitats, including those that are wetter and cooler. The long-term trend shows a moderate decline in distribution and a significant increase in abundance, with the more recent trend over the last decade showing a moderate increase in both distribution and abundance. Using the IUCN criteria, the Dark Green Fritillary is categorised as Least Concern (LC) in the UK and Vulnerable (VU) in Ireland.

LIFE CYCLE

The butterfly has a single brood each year, with adults flying from early June, reaching a peak in early July, with the last adults flying until the end of August and even into September on occasion. The emergence is slightly later in more northern parts of the butterfly's range.

Imago

In early morning and in cooler conditions, adults can be found basking with wings open on bare earth, such as a rabbit scrape or a bare track, in order to raise their body temperature. Adults are also avid nectar feeders and typically feed in early morning and early evening on the purple flowers of thistles and knapweeds, as well as brambles, clovers, and even Buddlejas (if they are present) at woodland sites. Whatever the location, adults will constantly fly from flower to flower, staying at each for only a few seconds, making them very difficult to observe—it is sometimes easier to wait near a favourite nectar source for the butterfly to come to you.

The males are the more conspicuous of the two sexes and can be found patrolling over large areas of habitat with a 'flit-and-glide' motion, where they will drop down every now and again to investigate any brown object in the hope of finding a virgin female low down in vegetation. Once found, mating takes place immediately without any discernible courtship.

The female will hide away in vegetation while waiting for her eggs to ripen, and is more conspicuous when she is either feeding or egg-laying, when she can be found fluttering over vegetation in an attempt to detect the scent of violets. When a suitable area is found she will land

Mating pair with the ♀ on the left

and then crawl around the vegetation and, once she has detected violets growing through the grass sward, will lay a single egg, occasionally two or three in the same area, before flying off. Eggs are usually laid on vegetation near the foodplant, such as a dead leaf, twig or grass stem, rather than the foodplant itself.

The butterfly uses a variety of violet species, including Hairy Violet on calcareous grasslands, Marsh Violet in wetter habitats in the north and west, and Common Dog-violet in woodland and other habitats. This species also breeds in more open areas than the other violet-feeding fritillaries and, at Stockbridge Down in Hampshire, I have seen females laying within isolated patches of regenerating scrub growing within a medium sward within an otherwise uniform and open grassland landscape.

Ovum

The 1-mm high conical egg is pale yellow when first laid and, after five days or so, develops a purple band in its upper half with purple mottling at its base, and eventually turns dark grey just before the larva emerges. Each egg has between 19 and 22 ribs that run along its sides, and each pair of ribs has around 20 finer ribs running transversely between them.

The newly laid egg is pale yellow

The egg develops purple markings as it matures

Larva

After between two and three weeks, the 2-mm long **first instar** larva emerges from the egg, eats most of its eggshell, and immediately enters hibernation in a curled-up leaf or other piece of debris. The larva has a shiny black head, a narrow black collar on the top of the first segment and is covered in eight rows of shiny brown warts that each bear a long hair. Around the end of March, having spent the best part of 230 days in hibernation, the larva starts feeding on young and often unfurled violet leaves when it nibbles away at the leaf edges. The larva remains hidden away in warm conditions, but will bask on top of dead vegetation when it is cooler. When fully grown, the first instar is 3 mm long.

1st instar: newly emerged larvae

1st instar: a fully grown larva in the spring

2nd instar: a fully grown larva

After two weeks of feeding, the larva spins a silk pad underneath a leaf, where it subsequently moults into the **second instar**. The larva is now covered in six rows of black spines that run the length of the body. Each spine is encircled by a number of smaller spines that each bear a stiff hair, and those that run along the bottom of each side, from the 4th to 11th segment, also have an amber base. The larva also has a striped appearance due to a series of alternating white and brown markings along its body, the most noticeable of which run along its back, where a broad pale line is divided by a relatively thin dark line. When fully grown, the second instar is 5 mm long. As a general rule, the larva grows darker as it matures through its remaining instars, and the amber markings on its side become more vibrant, eventually turning deep orange-red.

After two weeks, the larva moults into the **third instar**. Its colouring now has more contrast, with its pale markings becoming whiter, and its darker markings become blacker, resulting in a slightly darker appearance overall. The spines running along the bottom of each side, from the 4th to 11th segment, continue to possess an amber base. When fully grown, the third instar is around 7.5 mm long.

3rd instar: the larva has prominent pale stripes along its back

4th instar: a fully grown larva

After a week, the larva moults into the **fourth instar** and retains the stripes that run the length of the body, although the amber markings are a deeper orange than previous instars and much more prominent as a result. When fully grown, the fourth instar is 13 mm long and almost twice the length of the third instar. The black body is designed to absorb heat to aid digestion and larvae can occasionally be found basking on the foodplant or a dead leaf as they absorb the heat of the sun. Once warmed up, the larvae are particularly mobile, and have no hesitation in rapidly crawling away into the undergrowth when disturbed. They may also be seen wandering across bare ground or short turf in search of their foodplant.

After another week, the larva moults into the **fifth instar** and, apart from its size, is similar in appearance to the fourth instar. It does, however, have a slightly darker appearance and also pale orange markings on its back from the 2nd to 10th segments, in addition to the vibrant amber markings on its sides. When fully grown, the fifth instar is around 20 mm long.

After another 10 days, the larva moults for the final and **sixth instar** and is now a magnificent beast with a velvety appearance. It is now completely black, with pale orange markings along its back, and deep orange-red markings along each side from the 4th to 11th segments. Larvae feed during the day and are extremely mobile, moving over vegetation and bare ground at some pace. After two weeks, around the end of May, the larva is fully grown and around 40 mm long and prepares for pupation by creating a tent by drawing together various pieces of vegetation, under which it builds a silk pad from which it hangs head-down.

5th instar: the larva has a darker appearance

6th instar: a fully grown larva scrambling over moss to find fresh foodplant

Pupa

The 19-mm long pupa is formed head-down, with its cremaster attached to the silk pad. The pupa is a mix of browns and shiny, which some have suggested gives the pupa a scorched or burnt appearance, with the wing cases particularly heavily marked. It also has a very curious pose, with the last few abdominal segments bent at such an acute angle to the body that the cremaster and the ends of the wing cases almost touch. After three or four weeks, depending on temperature, the adult butterfly emerges, and it is always a delight to watch a newly emerged bright orange adult soaring over its habitat.

The pupa is suspended upside-down underneath vegetation

High Brown Fritillary
Argynnis adippe

❖ HIGH BROWN FRITILLARY IS SUPPORTED BY RICHARD BANCE, PHILIP BROMLEY & DR MAX WHITBY, NATUREGUIDES LTD ❖

	January	February	March	April	May	June	July	August	September	October	November	December
IMAGO												
OVUM												
LARVA												
PUPA												

The **High Brown Fritillary** is one of our most threatened butterflies and has suffered a 96% contraction in its range and a 67% decline in abundance in the last 40 years. The butterfly is now confined to just four landscapes, which is in marked contrast with its former distribution—in *The Lepidoptera of the British Islands* of 1893, Charles Barrett writes: *"Apparently found in most of the larger woods of the southern counties, from Kent, Essex, Suffolk, and Norfolk, on the east, to Devonshire, Glamorganshire, and Merionethshire* [a historic county in north-west Wales] *on the west; also in similar situations through the north-western counties and the more sheltered woods of the Midlands to Herefordshire, Shropshire, Derbyshire, and Lincolnshire. Found in several localities in Yorkshire, in the favoured Grange and Silverdale districts of Lancashire, and near Lake Windermere in Westmoreland; its extreme northern boundary being reached in Cumberland"*. The cessation of coppicing, coniferisation, agricultural abandonment (including the cessation of grazing), a reduction in traditional forms of Bracken management and climate change are all thought to have contributed to the butterfly's long-term demise.

Male and female are similar in appearance although the male is less yellowish than the female, has less rounded wings, and has sex brands on the forewings on the thickened second and third veins up. On the underside, the male is less richly marked than the female and has smaller silver spots, especially on the forewing. James Petiver gave this butterfly the name 'The greater Silver-Spotted fritillary' in his *Musei Petiveriani* of 1699, with Benjamin Wilkes providing the name we use today in his *Twelve New Designs of English Butterflies* of 1742.

This butterfly is easily mistaken for the Dark Green Fritillary and the two often fly together, making a positive identification almost impossible unless the butterfly is at rest. The two species are most easily distinguished by their undersides—the High Brown Fritillary has a row of spots between the outer margin and the silver spangles, which are missing in the Dark Green Fritillary. A less reliable identification guide is that, as its name suggests, the High Brown Fritillary has a predominately brown hue to its underside, rather than green. It is much more difficult to distinguish the Dark Green Fritillary from the High Brown Fritillary based on their uppersides. However, the first row of dots from the outside edge of the forewing upperside do provide a clue—the third dot

The ♂ has sex brands on its forewings and a wingspan of 55 to 62 mm [IL]

The ♀ has a wingspan of 62 to 69 mm [IL]

♂ underside

♀ underside

Life Cycles of British & Irish Butterflies

from the apex of the forewing is in line with the other dots in the Dark Green Fritillary but indented toward the body in the High Brown Fritillary. Also, if the butterfly is male, then the sex brands on the High Brown Fritillary are much more conspicuous than those on the Dark Green Fritillary, which are barely discernible.

DISTRIBUTION
The butterfly is now confined to four landscapes—the Bracken-covered southern edge of Dartmoor in South Devon; the Heddon Valley on Exmoor in North Devon; the Alun Valley on the western side of the Vale of Glamorgan in Wales; and the limestone outcrops in the Morecambe Bay area of North Lancashire and South Cumbria. The latter also extends to sites in the South Cumbria Low Fells. On a worldwide basis, the butterfly is found throughout Europe and across temperate Asia as far as China and Japan and is in decline in several countries.

The butterfly forms discrete colonies, although the majority are small and contain fewer than 50 adults at the peak of the flight period and cover a flight area of less than 5 hectares (220 by 220 m) (Sam Ellis, pers. comm.). Both sexes will also wander up to 2 km from their breeding sites to feed, especially when nectar is in short supply, when they can be found feeding together on suitable flowerheads. Adults have also been sighted up to 5 km from their breeding areas and it is now known that the butterfly resides within a metapopulation structure, with some interchange between colonies spread over a wide area.

HABITAT
The butterfly uses two main habitat types in Britain—those dominated by Bracken (used by the Dartmoor, Exmoor, Vale of Glamorgan and South Cumbria Low Fells populations) and limestone rock outcrops that support a mosaic of habitats, including grassland, pavement, Bracken, scrub and woodland (used by the Morecambe Bay population). The butterfly formerly used woodland clearings extensively, where regular coppicing opened up new areas of suitable habitat that the butterfly was able to colonise once existing sites had become overgrown.

Suitable Bracken-dominated habitats are typically found on south-facing slopes, where Bracken litter provides the particularly warm microclimate that is essential for the successful development of the larvae. These sites can become unsuitable if the Bracken litter is left unchecked and becomes too dense for ground flora to penetrate, and they are often managed using grazing livestock, such as cattle and ponies, that break up the litter and allow violets, the larval foodplant, to flourish. On limestone rock outcrops, the butterfly usually breeds in sheltered clearings where scrub or woodland has been cleared or coppiced.

Bracken appears to be particularly important for this species, with the butterfly found predominantly in areas where Bracken litter is present. This finding confirms a suggestion made by

Bracken-dominated habitat in the Heddon Valley on Exmoor [SE]

Martin Warren in 1995 that Bracken was probably present in woodland clearings where the butterfly has since become extinct. Outside of woodland, Bracken acts as a surrogate woodland canopy and suppresses the growth of grasses that would otherwise make the habitat unsuitably cool for the development of larvae.

STATUS

Although there has been some recovery at sites that are specifically managed for this butterfly, the High Brown Fritillary remains our most threatened butterfly and has been lost from six landscapes since 1994—the Herefordshire

Arnside Knott in Cumbria, overlooking Morecambe Bay

Commons and the Malvern Hills in the West Midlands, the Culm Measures in Devon and, in Wales, the Mid Glamorgan Coalfield, the Gwaun Valley in Pembrokeshire and the Montgomery Hills.

Given the butterfly's former abundance in woodland, it is clear that this species is particularly sensitive to changing woodland management practices. In *The Butterflies of Britain and Ireland*, Jeremy Thomas writes: "*The High Brown Fritillary was usually the first of the five violet-feeding fritillaries to disappear from a district following the abandonment of coppicing*". Thomas goes on to speculate that the butterfly's abrupt decline in woodland may have been accelerated by the loss of rabbits following the introduction of myxomatosis in the 1950s, when the reduction in grazing allowed vegetation to swamp the violets and warm patches needed by the larvae.

A recent article that focuses on this butterfly, 'Are habitat changes driving the decline of the UK's most threatened butterfly: the High Brown Fritillary *Argynnis adippe* (Lepidoptera: Nymphalidae)?', published in *Journal of Insect Conservation* in 2019, is authored by a 'who's who' of those involved in the conservation of this species—namely Sam Ellis, Dave Wainwright, Emily Dennis, Nigel Bourn, Caroline Bulman, Russel Hobson, Rachel Jones, Ian Middlebrook, Jenny Plackett, Richard Smith, Martin Wain and Martin Warren (all are affiliated with Butterfly Conservation)—although many others have also been involved in the conservation of this species—not least the well-known naturalist, Matthew Oates. This article concludes that climate change and nitrogen deposition (or a combination of the two) may be overwhelming any benefit to be gained from 'best practice' habitat management, and that these practices must therefore be kept under constant review. For example, while it could be assumed that a changing climate would benefit this warmth-loving butterfly, higher temperatures may also encourage the growth of grasses earlier in the year, with a net effect of a lower overall temperature at ground level that negatively impacts the butterfly.

Various landscape-scale projects targeting this species have required the collaboration of an impressive list of organisations, landowners and volunteers. The butterfly is undoubtedly benefiting, with the butterfly not only stabilising but increasing at some sites. The UK Butterfly Monitoring Scheme report for 2017 shows that the butterfly has experienced a 225% increase in abundance from 2008 to 2017, driven primarily by successes at some Bracken-dominated sites, although the population on the Morecambe Bay Limestones continues to decline.

The long-term trend shows a significant decline in both distribution and abundance, with the more recent trend over the last decade showing a moderate decline in distribution and a stable abundance. Using the IUCN criteria, the High Brown Fritillary is categorised as Critically Endangered (CR) in the UK.

LIFE CYCLE

The butterfly has a single brood each year and, at southern sites, adults emerge throughout June, peak at the end of the month and the first week of July and fly until the end of July. Further north, the butterfly emerges from mid-June and peaks in mid-July, with the butterfly flying into August.

Imago

Both sexes have a powerful flight and are difficult to approach when basking on Bracken fronds or when resting on dead vegetation in a sheltered depression on the ground, and are much more amenable when feeding from flowers. Brambles and thistles are favourite nectar sources, although the butterfly will also feed from Betony, knapweeds, ragworts and Wild Thyme. When roosting, and during periods of dull weather, the adults will settle among the foliage of trees and, at Bracken-dominated sites, under Bracken fronds, where they are remarkably well-camouflaged.

Males will patrol large areas of habitat when looking for a mate and will investigate any brown object that might be a virgin female. When a receptive female is found, often in late morning on the day she has emerged, then the pair mate and remain coupled for around an hour.

Females are most often seen when egg-laying, when they flutter low over the ground in search of the larval foodplant. Sunny and sheltered sites are preferred, and the female can be seen crawling in the dappled sunlight beneath vegetation,

A mating pair with ♀ on left [NF]

before laying a single egg on a dead leaf, dead Bracken frond, twig or some other substrate that will not rot down over the winter, and several eggs may be laid in the same area. Eggs may also be laid on moss growing on limestone outcrops. At both Bracken-dominated sites and limestone rock outcrops, the butterfly uses Common Dog-violet as the larval foodplant, although Hairy Violet is also used at limestone sites. Heath Dog-violet and Pale Dog-violet may also be used at some sites.

Ovum

The 0.8-mm high and conical egg has between 13 and 15 ribs that run from top to bottom, with finer 'struts' that run between them. The egg is greenish-brown when first laid, but gradually develops a pinkish-grey hue as it matures, with the fully formed larva overwintering while inside the egg. The egg turns dark grey before the larva emerges in the spring, when the head of the larva shows as a dark patch through the eggshell.

Eggs are often laid on dead vegetation near the foodplant

Larva

From early March, depending on temperature, the 2-mm long **first instar** eats an untidy hole, usually toward the top of the egg, before making its exit, although it does not feed further on the shell. The larva has a black head, a short dark collar on the first segment, and is covered in several rows of black warts—those on the back each bear two hairs that curve forward, while those on the sides bear four bristles that project away from the body. The ground colour is a dull purplish-brown, with both darker and paler blotches that form longitudinal bands that run the length of the body. The larva rests in a straight line (usually on a leaf stem or dead vegetation near the foodplant) and feeds by day, when it nibbles on the edges of a violet seedling, stalks and stems. The fully grown first instar is 3 mm long.

After around three weeks, the larva moults into the **second instar** and now has a more chequered appearance, with an off-white ground colour that is covered in several rows of wavy dark lines along the back and down each side. The most noticeable difference from the first instar, however, is the presence of six rows of spines, each covered in bristles, that run the length of the body. The spines just below the back have a white base with a faint amber hue, those along the sides have a black base, and those just below the spiracles have a white base. When fully grown, the second instar is 5 mm long.

1st instar: the larva is covered in several rows of long hairs

2nd instar: the larva is now in possession of six rows of spines

3rd instar: the lower half of the spines along the back are dull yellow

After just over a week, the larva moults into the **third instar**. The general colour and pattern are similar to the previous instar, although the lower half of the spines along the back are now dull yellow, with the upper half black. When fully grown, the third instar is just under 8 mm long.

After a week, the larva moults into the **fourth instar** and has a similar pattern and colouring to the previous two instars, although all of the spines are now completely amber with those on the back the most vivid, and those at the base of each side the least vivid. The two lobes of the head also show some light brown blotching. When fully grown, the fourth instar is just over 12 mm long.

4th instar: all spines are now amber

5th instar: the lobes of the head are now light brown

After just over a week, the larva moults into the **fifth instar** and, while similar in colour to the previous instar, has a much richer colouring, especially the spines that are all now a vivid amber. The two lobes of the head are now a noticeable light brown, with darker mouth parts. When fully grown, the fifth instar is around 20 mm long.

After around 10 days, the larva moults for the last time into the **sixth instar** and is a very different beast from earlier instars. The larva now has a light reddish-brown ground colour, with various black flecks that run along the length of the body. A prominent white stripe runs along the back, and the front half of each segment, either side of this stripe, is a velvety black. The head is now brick red and there are both light and dark forms of the larva—the light form tends to have light-red spines and a light brown ground colour, whereas the dark form has pinkish spines and a greyish-brown ground colour. The highly mobile final instar is extremely well-camouflaged when basking on a dead Bracken frond, other dead plant debris or bare soil, where the local microclimate (whose temperature can be 15 to 20°C higher than surrounding grassy vegetation) allows the larva to develop quickly, although the larva will retreat under vegetation during particularly hot spells. After two weeks, the fully grown larva is between 38 and 42 mm long and it prepares for pupation by spinning silk between a few dead leaves, close to the ground, where it suspends itself upside-down from a silk pad.

6th instar: a relatively light form of larva

6th instar: the dark form of larva

Pupa

The 20-mm long pupa, which is suspended by the cremaster that is affixed to the silk pad created by the larva, resembles a shrivelled leaf and its back is covered in several rows of raised silver points that give the illusion of water droplets. Like the final instar larva, the pupa comes in both a light and dark form. After between two and three weeks, this most threatened of our butterflies emerges.

Dark form of pupa

Light form of pupa, formed beneath a 'tent' of leaves

Life Cycles of British & Irish Butterflies

White Admiral
Limenitis camilla

❖ WHITE ADMIRAL IS SUPPORTED BY PATRICK BARKHAM, PETER CLARKE & COLIN, ANDREA, ROSIE & JAMES GIBBS ❖

	January	February	March	April	May	June	July	August	September	October	November	December
IMAGO												
OVUM												
LARVA												
PUPA												

If someone were to ask me "As an aficionado of immature stages, is there one species in Britain and Ireland that really stands out?" then my answer would have to be the White Admiral. My reasoning is quite simple—the White Admiral is a spectacular insect in all of its stages and the two mobile stages, the adult and the larva, also exhibit unique behaviours. Along with the Chequered Skipper, the White Admiral also happens to be a species that I've studied more than any other. All in all, these were sufficient reasons for placing this iconic woodland species on the cover of this book.

The White Admiral is a delight to behold as it glides along woodland rides, flying from tree to woodland floor and back up with only a few effortless wing beats. For this reason, some of its closest relatives on the continent are known as 'gliders' which is a wonderfully descriptive and wholly applicable term for this species also. When basking or feeding from a favourite nectar source, such as bramble, the adults are unmistakable, with black uppersides intersected by prominent white bars. The undersides of this butterfly are much more colourful and really belong to a species that you would expect to find in the tropics, they are that stunning. This is surely one of the most beautiful species of butterfly found on our shores.

James Petiver originally named this butterfly the White Legorn Admiral in his *Gazophylacium naturae et artis* of 1703, with the scientific name of *Papilio livornicus*. Both vernacular and scientific names refer to the Italian port city of Livorno in Italy, which is traditionally known as 'Leghorn' in English, where Petiver's first specimen originated and before the butterfly was found in London. Petiver later dropped the location from the vernacular name, leaving us with the name we use today, although I do have a fondness for the name 'The Honeysuckle' that was given to the butterfly by James Rennie in 1832, which recognises the primary larval foodplant. Petiver's scientific name was replaced 50 years later with the adoption of the binomial Linnaean system, when Linnaeus himself assigned this butterfly the scientific name of *Papilio Camilla*.

♂ and ♀ are similar in appearance with a wingspan of 56 to 66 mm

The undersides are quite exquisite

DISTRIBUTION

The distribution of this species hit an all-time low in the early 1900s when it was restricted to southern England but, thankfully, there has been a reversal of fortunes, with the butterfly now reaching its former distribution that extends as far north as Lincolnshire. The butterfly is now found in central and southern England, with a few scattered colonies in the eastern counties of Wales, although it is not found in Scotland, Ireland or the Isle of Man. The White Admiral forms discrete colonies within a woodland, suggesting that it is not a great wanderer, although its relatively recent expansion would seem to indicate otherwise, and it has been shown that the butterfly is capable of travelling several kilometres on occasion. On a worldwide basis, the butterfly is found throughout most of central Europe, across Asia to Japan.

Life Cycles of British & Irish Butterflies

HABITAT

The White Admiral is found in woods throughout its distribution. Most sites are deciduous woodland dominated by oak, whereas those woodlands dominated by Ash and Beech are devoid of White Admiral since they do not tend to support the suitable Honeysuckle growth required by the larvae. The butterfly can also be found in conifer plantations, so long as Honeysuckle is available. Enthusiasts are often surprised to find that the type of Honeysuckle that the butterfly needs is not the lush and dense growth found in full sun, but the rather wispy and grubby growth that is found in partial or full shade, often near a woodland ride or clearing. Another requirement is that the habitat has abundant nectar sources, especially bramble, that the adults can be found feeding upon throughout the day and especially in the morning when temperatures are relatively cool.

Pamber Forest in Hampshire is one of our best sites for White Admiral, and is dominated by oak

STATUS

The White Admiral has declined along with many other butterfly species, but it may have a more promising future. One reason for this positive outlook is that climate change is allowing the species to thrive at sites that had become too cool. Another is that the cessation of coppicing, which has been detrimental to so many woodland butterflies, has actually benefited this species given that it requires Honeysuckle growing in shade for the successful development of its larvae. The long-term trend shows a moderate decline in distribution and a significant decline in abundance, with the more recent trend over the last decade looking slightly more favourable with a moderate decline in both distribution and abundance. Using the IUCN criteria, the White Admiral is categorised as Vulnerable (VU) in the UK.

LIFE CYCLE

Adults start to emerge in the second week of June in most years, peak in the first part of July and fly to the beginning of August, with a slightly later emergence at the most northern sites. There is usually one brood each year although there are reports of second brood individuals in September in exceptional years, such as 2018. Studies by Ernie Pollard at Monks Wood in Cambridgeshire in the late 1970s also showed that numbers of adults increases when June temperatures are warm, when the late-instar larvae and pupae develop relatively quickly, resulting in less predation by birds.

Imago

Adults feed on aphid honeydew but also take nectar from a variety of plants such as Betony, thistles, umbellifers, Wild Privet and especially brambles. That being said, my first encounter with this species was a male that was nectaring on the flowers of Alder Buckthorn—an experience I shall never forget as I glimpsed its exotic underside for the first time. On good sites, it is not uncommon to find several White Admirals all feeding from the same bramble patch and their fondness for this nectar source is undoubtedly the reason that their wings get tatty very quickly as they move around bramble blossoms, probing for nectar. Adults will also feed on salts and minerals from moist earth and animal droppings, especially in hot weather.

The White Admiral is best sought out while it is feeding and knowing the location of a favourite patch of flowers is a distinct advantage—I have visited the same bramble-rich area in a local wood for over a decade on mid-June mornings and have never failed to connect with this denizen of the woods. William Lewin described this best in his *The Papilios of Great Britain* of 1795: "*It frequents the south sides of woods and lanes near them; and may be readily taken as it is feeding on the various flowers then in bloom, before nine o'clock in the morning; after which time, as the sun grows hot, it sports and flies about with great swiftness, frequently settling on the tops and sides of high trees*".

The male and female of many species of the Nymphalidae are very difficult to tell apart, and the White Admiral is no exception, although females are slightly browner and larger than the males and have more-rounded wings. Very few observers have witnessed any courtship (or found a mating pair) and, like many related species, this probably occurs out of sight and almost certainly in the tree canopy.

An egg-laying female is easy to spot as she flies slowly and deliberately through woodland in search of non-flowering Honeysuckle that is in full or partial shade, that is usually within a few feet of a woodland ride or clearing. Suitable plants are often isolated and hang down from a supporting tree or bush and, if selected, the female sits across the top of a leaf before curving her abdomen toward the opposite edge, where she lays a single egg before immediately flying off. Early entomologists were aware that the larvae would feed on snowberries as well as Honeysuckle in captivity and, as recently as 2018, Julia Huggins (White Admiral species champion for the Upper Thames branch of Butterfly Conservation) found snowberry being used exclusively in one south Oxfordshire wood that had very little Honeysuckle growth present.

Most eggs are laid between 1 and 2 m above ground level, although I often find eggs both lower down and higher up—I once saw an egg being laid on Honeysuckle growing over the woodland floor at Chiddingfold Forest in Surrey and, conversely, I once found an egg that was three metres up a growth of Honeysuckle at Pamber Forest in Hampshire. The latter was found while observing a first instar larva on an adjoining Honeysuckle spray, which I was able to bring down to eye level thanks to a walking stick that my youngest son had given me as a 50th birthday present (it was intended as a joke for his now-ancient dad, but has turned out to be surprisingly useful).

Two eggs were found on this discoloured Honeysuckle spray

Two eggs were laid on the upperside of this leaf

Eggs are typically laid on isolated Honeysuckle plants growing in full or partial shade

Life Cycles of British & Irish Butterflies

Ovum

It's hard to wax lyrical over an egg but, if push came to shove, I would have to say that the egg of the White Admiral, with the appearance of a golf ball, is the most unusual of all of the butterflies of Britain and Ireland. When seen close up, the 1-mm wide egg reveals a delicate honeycomb pattern of hexagons and pentagons that are neatly interwoven, with a hair protruding from the angle of each cell. The egg is a light olive green when first laid but turns darker as the larva develops and when its head is visible through the eggshell.

The egg looks like a golf ball, and is a light olive green when first laid

The egg darkens as the larva develops

Larva

After about a week, the light brown and 2-mm long **first instar** larva emerges and proceeds to eat all of the eggshell, leaving a 'halo' that is quite easy to locate once a larva has been found since the first instar always feeds on the same leaf as the egg. The newly emerged larva then moves around the edge of the leaf until it reaches the leaf tip (although it may occasionally crawl more directly over the leaf surface) where it starts to feed either side of the midrib.

1st instar: a newly emerged larva eating its eggshell

What happens next is absolutely fascinating and was spelled out by Barry Fox who studied the larvae of both the White Admiral and Broad-bordered Bee Hawk-moth *Hemaris fuciformis* for his 1996 PhD thesis entitled 'Alternative foraging strategies of the White Admiral butterfly and the Broad-bordered Bee Hawk-moth on Honeysuckle'. Based on observations of dozens of newly emerged larvae, Fox found that first and second instar White Admiral larvae employed two defence mechanisms.

The first defence mechanism is a 'pier' (as Fox called it) that the larva constructs from silk and frass and that extends beyond the midrib of the leaf. In an article published in the *Entomologist's Gazette* in 2005, Fox says: "*The larva utilised its own faecal pellets or frass as building blocks, using its head to manipulate the pellets into the appropriate position. The pellets were bonded into place with silk and the occasional swiped hair that the larva carried from the leaf midrib. The larva was using the same technique plasterers used years ago when adding horsehair to plaster to improve bonding*". Fox considered the pier to act as a defence refuge since early-instar larvae always moulted on the pier, and always returned to the pier when alarmed. He also provides an indication of the strength of the pier, and the larva, in the following anecdote when he was watching a video recording of the larva: "*On one occasion the VCR replay revealed a bluebottle fly trying to suck the tiny larva off its pier. The 2-mm long larva (less than half the size of the fly) anchored itself to its pier by gripping layers of silk with the crochets of its anal prolegs. There was a tug of war for a few seconds before the fly gave up. This larva survived but was obviously upset as it refused to feed for many hours afterwards and remained anchored to its pier*".

1st instar: the 'pier' combines silk and frass and eventually extends beyond the midrib

1st instar: the piers are easy to find in favoured spots and give away the presence of the larva

The second defence mechanism is an 'aerial latrine' that is made out of leaf cuttings that are silked together and attached to the midrib. The larva uses this latrine to deposit frass, both directly by crawling into the bundle, and by firing frass into it. Fox recalls: *"Positioned on the midrib 1–2 cm away from the latrine, the larva lifted its anus and with great accuracy and speed, fired a pellet of frass into the latrine. Not once did I observe a miss"*. The latrine is used by the larva until it enters hibernation in the third instar.

Fox gives two possible reasons for the creation of a latrine. The first is that the silk that surrounds the latrine neutralises any chemical scent given off by the frass (which contains an amount of undigested food) that might otherwise attract predators such as parasitoid flies. The second is that the larva may use the latrine as a store of frass that it uses to maintain the pier when needed although, despite the frequency with which a larva managed to cover itself in its own frass after a visit to its latrine, it was not known if any movement of frass was deliberate or accidental.

1st instar: faecal pellets may stick to the larva following a visit to its latrine

1st instar: a leaf showing characteristic feeding damage

1st instar: the head of the larva has no significant protrusions

Two or more eggs may be found on the same leaf, either as a result of the same female revisiting the leaf, or two different females laying on the same leaf. This does beg the question of what the larvae do when there is only one midrib on which to build a pier. In these cases, only the first larva to emerge uses the midrib, and any other larvae will create a pier on one side of the leaf, although these are typically less robust than those formed at the end of the midrib. When not resting on the pier, the larva feeds on the leaf each side of the midrib, but leaves the midrib intact, resulting in characteristic feeding damage that is quite easy to spot in the right light. When fully grown, the first instar is a little over 3 mm long.

After nine days or so, the larva moults into the **second instar**, eating its cast skin, which it does after each moult. Whereas the first instar is covered in numerous bumps and a few hairs, with no significant protrusions on its head, the second instar is covered in short spines and has several prominent protrusions on its head. It also sports a pale white stripe down each side of the body that is most visible when it is fully grown and around 6 mm long.

2nd instar: the larva now sports a pale white stripe along each side

3rd instar: the larva has more prominent spines than earlier instars and a brown head

After another nine days or so, the larva moults into the **third instar**. The larva now has more prominent spines on its body and it retains the pale white stripes on its sides, although its head is now predominantly brown rather than the black of earlier instars.

The larva overwinters in this instar and, towards the end of August, proceeds to create a hibernaculum from a Honeysuckle leaf in which it will overwinter, often the leaf on which the egg was laid. Honeysuckle is deciduous and loses its leaves over the winter, and so the larva secures the leaf stalk to the stem by using a

Life Cycles of British & Irish Butterflies

number of silk strands that ensures that the leaf does not detach from the plant, even in strong winds.

As the leaf dies and the leaf stalk breaks away from the stem, it is only these silk strands that stop the leaf from falling or being blown away. I am grateful to Matthew Oates for showing me a technique for finding hibernacula once the leaves have died—gently move or blow on the plant to see if any of the leaves sway from side to side, thereby indicating a potential hibernaculum that is held on to the plant only with silk. Any other leaves that are present do not tend to move in this manner and are usually those that were either the last to die and their stalk is still attached to the stem, or they have yet to feel the full force of a winter storm, after which they are normally blown from the plant.

3rd instar: a hibernaculum attached to the plant stem with silk

3rd instar: a close-up of the silk bindings between the leaf stalk and the plant stem

Given my fascination with immature stages, I have spent a considerable amount of time studying the formation of the White Admiral hibernaculum and realised that much of what has been written is very simplistic. In the course of my studies I was able to identify four different formations of hibernacula.

The first formation is what I call 'cut and fold'. In this type of hibernaculum, the larva deliberately cuts the leaf so that the half furthest from the stem falls away. The larva then lays down silk strands that span each side of what remains of the leaf, causing the sides to come together as the silk dries and contracts. The larva remains in the resulting abode and eventually seals the open ends with silk and occasionally creates an overlapping flap. This is one of the commonest types of hibernacula that I have come across, accounting for 29.2% of those recorded, from a sample of 48.

3rd instar: the 'cut and fold' hibernaculum under construction [WW]

3rd instar: the 'cut and fold' hibernaculum in its final state, with a flap sealing the open end

3rd instar: a 'partial cut and fold' hibernaculum under construction

3rd instar: a newly formed 'symmetric fold' hibernaculum

The second type of hibernaculum is 'partial cut and fold' and is so-named because it is the start of the 'cut and fold' construction, but only half of the leaf is cut, and this half is then folded over with silk, creating a chamber within which the larva overwinters. Also, the fold may not be at the base of the leaf and may be formed at the tip of the leaf, or even in the middle of the leaf where two separate cuts are made, and the leaf folded over. Overall, this type was the most common and accounted for 45.8% of those recorded.

The third type of hibernaculum, which I call 'symmetric fold', is where the larva neatly silks the edges of the leaf together, leaving only the ends free of silk. The leaf is not cut, but simply allowed to wither. This is the least common type of hibernaculum, accounting for only 8.3% of those recorded.

The fourth type of hibernaculum, 'asymmetric fold', is the simplest of all, where the leaf is simply folded on itself, with the leaf edges silked together. The end result is often quite messy and certainly not as symmetric as the previous type. This type was not very common, accounting for just 16.7% of hibernacula recorded.

3rd instar: a larva in spring on the outside of its 'asymmetric fold' hibernaculum

The larva emerges from the hibernaculum in early April and may wander up to a metre from its hibernaculum before spinning a pad of silk on a leaf stalk or on a nearby stem that it treats as its base. From there is moves off to feed on the edge of a fresh Honeysuckle leaf before returning to the same spot to rest, only moving home when its current leaf has been eaten. When fully grown, the third instar is around 8 mm long.

Around the middle of April, the larva moults into the **fourth instar**. The larva now looks quite exotic, with large red-brown branching spines on the second, third and fifth segments, and relatively smaller spines on the remaining segments. The larva also has a greenish tinge when first moulted that becomes more obvious as it grows. It also has a white stripe along each side and a greater number of spiny protrusions on its head than earlier instars. The head also darkens as the larva matures so that

4th instar: a newly moulted larva

4th instar: the larva in a characteristic pose

4th instar: the head is darker than the third instar, and with more spiny protrusions

Life Cycles of British & Irish Butterflies

5th instar: a fully grown larva in its classic pose

5th instar: a larva hanging upside-down, ready to pupate

5th instar: the protrusions on the head are most developed in the final instar

it appears quite different from the mainly brown head of the third instar. The larva crawls with a very jerky motion and, when at rest, takes on a characteristic pose, forming an S-shaped curve with its head down and its tail up, a pose that it also exhibits to some degree in the third instar. When fully grown, the fourth instar is around 12 mm long.

After around 12 days in the fourth instar, the larva moults for the final time into the **fifth instar** and is now a spectacular beast. It is now bright green although the white stripes along the sides are still present. The most obvious characteristic, however, is the set of beautiful red branching spines that, as in the fourth instar larva, are most prominent on the second, third and fifth segments. Around the end of May and beginning of June, after two weeks or so in this instar, the larva is fully grown and around 28 mm long. When preparing to pupate, the larva constructs a silk pad, usually under the midrib of a Honeysuckle leaf and occasionally on a stem, from which it hangs head-down.

Pupa

The 22-mm long green and brown pupa is formed upside-down, attached to the silk pad by its cremaster. Like all previous stages, the pupa has a very interesting shape, with two prominent horns on the head, and a curious protrusion at

A view from the rear shows silver spots on the back and on the two horns on the head of the pupa

The pupa hangs head-down from a Honeysuckle leaf or stem

The wings of the adult are clearly visible through the pupal case just before emergence

the back that sits above two series of silver spots. The overall appearance is one of a partly dead and shrivelled Honeysuckle leaf covered in beads of water. For those readers that are old enough to remember 'Punch and Judy' seaside shows, F.W. Frohawk, in his *Natural History of British Butterflies* of 1924, wrote: "*The pupa bears a close resemblance to a profile portrait of Punch*".

Very few observers have found the pupa in the wild although H.T. Stainton, editor of William Buckler's classic volumes on *The Larvae of the British Butterflies and Moths*, wrote the following footnote in 1886: "*Mr. C. G. Barrett visited Woolmer Forest … and in the course of several hours' search found four pupae and two suspended larvae … Of these, five were spun up to leaves of honeysuckle, and one to the leaf of* Rhamnus frangula *[Alder Buckthorn] growing contiguously, and in every case were firmly suspended to a button of silk on the underside of the midrib of the leaf. Not a single specimen was found attached to a stem or branch*".

After two to three weeks the adult butterfly emerges, exhibiting its beautiful underside while drying its wings.

The adult butterfly emerges after two to three weeks

Life Cycles of British & Irish Butterflies 223

Purple Emperor
Apatura iris

❖ PURPLE EMPEROR IS SUPPORTED BY MARK COLVIN, GREENWINGS WILDLIFE HOLIDAYS (greenwings.co.uk) & NATUREGUIDES ❖

	January	February	March	April	May	June	July	August	September	October	November	December
IMAGO												
OVUM												
LARVA												
PUPA												

The Purple Emperor is, without doubt, the most sought after of our butterflies and its appeal is not surprising—this butterfly is an elusive and hard-to-come-by creature, the male is of breathtaking beauty and both sexes exhibit fascinating behaviours. The butterfly is so admired that the male is sometimes referred to as 'His Imperial Majesty' or simply 'HIM' by observers (his 'subjects'). Originally given its vernacular name by Benjamin Wilkes in his *Twelve New Designs of English Butterflies* of 1742 (albeit with the spelling 'Emperour'), this is one of our most widely written about butterflies and the classic work *Notes and Views of the Purple Emperor* by Ian Heslop, George Hyde and Roy Stockley, published in 1964, is dedicated solely to this species. That being said, there are surprisingly few scientific studies on this species, no doubt due to its elusive behaviour.

The adult butterfly spends most of its time in the woodland canopy, offering the occasional close encounter when the male comes down, usually mid-morning and in late afternoon / early evening, to feed on animal droppings, carrion or moist ground that provide much-needed salts and minerals, some of which are passed onto the female during mating, especially sodium that helps with the development of her eggs. Both sexes also have a particular fondness for sap bleeds where they benefit from the minerals and sugars obtained. This butterfly clearly has different tastes to most, and those that make pilgrimages to see this spectacular creature will often try and lure the males down from the canopy using all manner of temptations, including foul-smelling shrimp paste.

The ♂ has a wingspan of between 70 and 78 mm and displays a blue-purple sheen when at certain angles to the light

The larger ♀ has a wingspan of between 76 and 92 mm

A ♂ feeding from a sap bleed [MC]

Two ♂s feeding on animal dung

Life Cycles of British & Irish Butterflies

The male Purple Emperor is one of the most beautiful of all our butterflies. From certain angles it appears to have black wings intersected with white bands. However, when the wings are held at a certain angle to the light, the most beautiful blue-purple sheen is displayed, a result of light being refracted from the structures of the wing scales. In Greek mythology, the goddess '*iris*' (the specific name of the Purple Emperor) was the personification of a rainbow, a name that applies so well to the changing colour of the male's wings. The slightly larger female, on the other hand, is a deep brown, lacks the multi-coloured display of the male, and has broader white bands on her wings.

♂ and ♀ (shown) have similar undersides [MC]

DISTRIBUTION

This butterfly's historic strongholds are deciduous woodlands of central-southern England, although the species has been expanding its range in the last decade or so. Core sites are to be found in the counties of Wiltshire and Hampshire in the west, to Surrey and Sussex in the east, and to Oxfordshire, Buckinghamshire, Northamptonshire, Bedfordshire and Hertfordshire in the north, with scattered colonies elsewhere. The butterfly is not found in the north of England, Scotland, Ireland, the Isle of Man or the Channel Islands and has not been seen in Wales since the 1930s. On a worldwide basis, the Purple Emperor is found throughout most of Europe and eastwards as far as China, Korea and Japan.

Given its arboreal habits, it is difficult to determine the exact nature of Purple Emperor populations using established methods. This difficulty is compounded by the fact that this is a mobile butterfly, with adults occasionally seen flying over open fields, miles away from any known colony. It is thought that the butterfly inhabits quite large patches of habitat that can either be found in a single large wood or that are spread over a number of smaller woods. Colonies also vary in size, from those that are very small with just a dozen or so adults, to the largest with over 100 adults recorded at the peak of the flight period. In 2018 Matthew Oates and Neil Hulme, two authorities in this species, were regularly observing counts of over 200 at the Knepp Castle Estate in Sussex, with a record of 388 sightings on one very special day.

HABITAT

The Purple Emperor is found in sallow-rich landscapes, typically those found in woodland or mixed scrub, where Goat Willow and its hybrids are the most widely used larval foodplants. Confusingly,

Sallow-rich oak woodland at Fermyn Wood in Northamptonshire

Sallow-rich scrub at the Knepp Castle Estate in West Sussex [MO]

the term 'sallow' is used to describe the subset of willows that are broad-leaved, such as the Goat and Grey Willows. The butterfly is associated with large expanses of semi-natural and sallow-rich deciduous woodland, often where oak is the dominant tree, such as the woodlands of Bentley Wood on the Wiltshire/Hampshire border and Fermyn Wood in Northamptonshire, although those dominated by other tree species such as Ash and Beech are used on occasion. Sites that comprise dense and sallow-rich scrub are also used, and this is the dominant habitat for the Purple Emperor on the Knepp Castle Estate, the location of a pioneering rewilding project. The butterfly therefore requires neither woodland nor oaks for its survival, contrary to established thinking (Matthew Oates, pers. comm.).

STATUS

The Purple Emperor suffered a significant contraction of its range during the 20th century, when it was previously known as far north as the Humber, in parts of central Wales, and in East Anglia. Fortunately, there has been a reversal of fortunes in recent years, with the butterfly expanding into both new and former areas to some degree, although there is some recording bias as observers become more adept at finding this species. The long-term trend shows a moderate decline in distribution and a significant increase in abundance, with the more recent trend over the last decade showing a significant increase in distribution and a moderate decline in abundance. Using the IUCN criteria, the Purple Emperor is categorised as Near Threatened (NT) in the UK.

The greatest threat to this butterfly is the loss of the sallow-rich habitat on which it depends, and it is unfortunate that sallows, which have no commercial value, are routinely targeted by foresters and farmers that see them as no more than a useless weed. Fragmentation of sites is also a threat, despite the butterfly's mobility.

LIFE CYCLE

The butterfly has a single generation each year, with adults flying for a six-week period, the timing of which is dependent on location. At 'early' sites adults will start emerging in the middle of June and fly until the latter part of July, with adults flying until at least the end of July at 'late' sites, with a few stragglers usually turning up in early August.

Imago

The best time to get 'up close and personal' with a Purple Emperor is mid-morning and late afternoon/early evening in the first third of the flight season, before all of the females have been mated, when the males will come down to the ground to feed on animal droppings and collect minerals and moisture from damp earth (females will also occasionally descend to collect moisture in hot weather). Males will also feed from mud and other debris that has accumulated on car tyres and they are also partial to sweat and readily land on observers. They have also been observed feeding from hydraulic fluid leaking from forestry vehicles (Mark Colvin, pers. obs.). Most literature on this butterfly also states that the butterfly feeds on aphid honeydew as a primary source of nourishment, although this seems to be an assumption rather than an observation-based fact, and this is an area worthy of more study.

The male has a characteristic flight pattern when he comes down from the treetops, spiralling around a particular spot before eventually landing and crawling to his chosen substance. The males will sometimes spend over an hour feeding in the same place, when each displays its characteristic yellow proboscis. Males seem quite oblivious to their surroundings when feeding and, once they have started feeding in earnest, can be approached quite closely. I once found a male on a main ride at Bentley Wood that was so engrossed while feeding on a fox scat that it remained in place and unharmed as a four-wheel drive vehicle passed slowly over it.

In late morning, the males will fly off to establish a territory or lek, often congregating at a feature that has traditionally been called a 'master tree' and which provides a vantage point for intercepting passing females, although other insects and even birds are also pursued. These trees are typically at a high point in the wood where males and females can meet, often those growing on the summit of a hill, and the same trees are sometimes used year after year. However, in a 2008 *British Wildlife* article entitled 'The Myth of the Master Tree: Mate-location strategies of the Purple Emperor butterfly', Matthew Oates determined that the term 'master tree' is a misnomer since the lek may not be a single tree, but a group of three to 10 trees that may be occupied by more than two males at a time. Males in these 'sacred groves' (as Oates calls them) can change their perch within their territory based on the position of the sun and the amount of shelter offered. Seeing the males battle it out for the best vantage points, with flashes of purple as the light hits their wings, is quite breath-taking. Oates also describes two patrolling strategies undertaken by males when searching for females. The first is 'sallow searching', which takes place in mid- to late-morning, when stands of sallows are investigated for newly emerged females that have yet to make their way to the canopy. The second is 'oak-edging' when males patrol the woodland edge in search of females.

A mating pair that drifted down from the canopy, ♀ on left [NH]

When a virgin female is located by a male, she will lead him to a spot high in the canopy, possibly several hundred metres away, where they settle and mate. If the female has already mated, however, then she has the curious habit of drifting towards the ground (known as the 'rejection drop') while the male circles frantically around her, before he ultimately gets the message and returns to his perch. In *The Butterflies of Sussex*, Neil Hulme writes of his experiences while watching accosted females landing and crawling deep into bramble or nettle patches to escape the attention of persistent males.

Females lay their eggs early in the afternoon when the female 'strikes' the sallow, as it is known, when she enters the tree and disappears within it as she flits around before laying a single egg on the upperside of a sallow leaf that is in half-shade, either on the shaded side of the tree or deep within the crown. A good amount of shade is a prerequisite for successful development of the larva—I was very fortunate to spend the 2013/2014 season in the company of Matthew Oates while he monitored immature stages of the Purple Emperor at Savernake Forest in Wiltshire and, on several occasions, we came across an exposed larva that had unfortunately desiccated and died. Oates documented the initial findings from his studies in a 2012 *British Wildlife* article entitled 'Adventures with caterpillars: The larval stage of the Purple Emperor butterfly'.

Females seem to undergo bouts of egg-laying lasting 10 minutes or so, during which time half a dozen eggs will be laid at various heights on the tree. The female also takes some time to lay each egg as mentioned by F.W. Frohawk in his *Natural History of British Butterflies* in 1924: "*In depositing the butterfly settles on a leaf, then lowers the abdomen so that the extremity rests on the upper surface, and, after remaining motionless for eight or ten seconds, lays an egg and flies off*". Goat Willow (including hybrid plants) is the most widely used foodplant although the butterfly also uses Grey Willow and, on the rare occasion, Crack-willow. Eggs are laid on a wide range of tree sizes, from medium-sized shrubs to tall canopy trees, and a range of ages, although trees younger than five years old are not used.

Ovum

The 1-mm high dome-shaped egg is blue-green when first laid but, after a few days, a ring of purple forms at its base that gradually moves towards the top of the egg as the larva develops, ultimately contributing to the head capsule of the larva. The egg also has 14 ribs that run from its top to its base.

The newly laid egg is blue-green [GP]

After a few days a purple ring forms at the base of the egg

The larva can be clearly seen just prior to the egg hatching

Larva

After between two and three weeks, depending on temperature, the **first instar** larva eats the eggshell just below the crown which forms a lid as the larva pushes its way out. The larva is just over 3 mm long and eats the majority of the eggshell with the exception of the base, leaving a tell-tale 'halo' where the egg was positioned, before setting up home at the very tip of a leaf, almost always the one on which the egg was laid. Here it lays down a silk pad on which to rest, where it may occasionally be found completely immersed in water following a period of rain. The larva always faces the leaf stalk when resting but will raise the front of its body off the leaf in order to reduce the size of any silhouette (when viewed from underneath) that might attract a predator.

The larva feeds either side of its resting place, leaving characteristic feeding damage with only the midrib intact in places, but always returns to its resting pad after each meal. Even at this young age the larva sports a series of oblique pale yellow stripes along its sides that mimic a sallow leaf. The key distinguishing feature of this instar is that its head is quite 'globular' and does not exhibit the pair of horns found in later instars. When fully grown, the first instar is just under 5 mm long.

1st instar: the larva does not yet possess the characteristic horns of later instars

1st instar: a fully grown larva with the front of its body raised off the leaf

1st instar: even at this young age there is characteristic feeding damage

After around 10 days, the larva moults into the **second instar** and eats most of its cast skin. The larva is now in possession of two horns that extend forward from the head capsule in the shape of a letter 'V' and each horn is slightly divided at its tip—a feature found in this and all later instars. The larva rests with its head-down, facing the leaf surface, when the two horns project forward parallel with the surface of the leaf. When in motion, however, the larva raises its head, when the appearance of the horns and tapered body are reminiscent of a slug. The fully grown second instar is just under 8 mm long, excluding its 2-mm long horns.

2nd instar: the larva continues to raise the front of its body when at rest

2nd instar: the larva is now in possession of a pair of horns on its head

2nd instar: the newly moulted larva is tiny when shown in context at the tip of a leaf

Life Cycles of British & Irish Butterflies

3rd instar: a pre-hibernation larva

After around three weeks, the larva moults into the **third instar**, the instar in which it will overwinter, and it is now around 10 mm long, excluding the 2.5-mm long horns. As the larva moves it lays down silk on the leaf surface in order to establish a secure foothold and it continues to rest with its front half raised, especially when alarmed.

The third instar gradually darkens in colour during October and, by early November, matches its surroundings. With the onset of autumn there is a high probability of leaves falling and larvae habitually use silk to attach the leaf stalk to the stem before they start feeding on it.

The larva adopts one of three positions when overwintering—next to an unopened sallow bud, in the fork of a branch or in a crevice in the bark. Whichever is chosen, the larva is well camouflaged since it changes colour to match the surface on which it hibernates and always lies flush against it.

3rd instar: a larva overwintering next to a sallow bud

3rd instar: a larva overwintering in the fork of a branch

3rd instar: a larva overwintering in a crevice in the bark [NH]

3rd instar: the larva develops a greener hue post-hibernation

4th instar: the larva now has a brighter appearance

With the onset of warmer weather, usually in early April, the larva starts to develop a noticeably greener hue that perfectly matches the developing sallow buds and its colour is restored to its pre-hibernation shade of green as it matures, although the yellow stripes on its sides are now white. The larva usually recommences feeding in the second half of April and initially nibbles on the unfurling sallow leaves and, as soon as the leaves become fully developed, will create a silk pad on a chosen leaf, known as its 'seat leaf', where it rests. The seat leaf is always left largely intact, since the larva moves to nearby leaves to feed. The lack of any significant feeding damage on its resting leaf ensures that the larva is better camouflaged and is a strategy also used in subsequent instars. The larva also continues to rest with its front raised to minimise any silhouette. When fully grown, the third instar is just under 13 mm long, excluding the horns.

While superficially similar to the third instar in terms of both appearance and behaviour, the **fourth instar** is more brightly coloured and has relatively slender horns that now exhibit a bluish tinge. When fully grown, the fourth instar is around 20 mm long, excluding the 3.5-mm long horns.

After between two and three weeks, the larva moults into the final and **fifth instar**, and does not eat its cast skin, but lets this drop to the ground. Its 6-mm long horns continue to give it a slug-like appearance, but since these do not increase in size as the larva grows (they are part of the head capsule), they are proportionally much longer relative to the length of the body in a larva that has just moulted. In this instar the chocolate-tipped horns now have an extensive blue tinge.

The larva continues to create a silk pad on its seat leaf and moves to nearby leaves to feed. When fully grown, the larva is

5th instar: the chocolate-tipped horns now have an extensive blue tinge

38 mm long when at rest and 44 mm long when crawling. It is a sizeable beast and, unsurprisingly, its weight causes whichever sallow leaf it is resting on to lie vertically. As the larva matures it tends to rest with its head flat against the leaf surface and, although it no longer raises the front of its body away from the leaf surface, is still perfectly camouflaged against the sallow leaf. One surprising characteristic of a mature larva is that it is often possible to determine its sex by looking at its 8th segment. If the larva is male, then this segment shows two pale patches that are its developing testes.

When ready to pupate, the larva turns a much paler green and may travel some distance, often to the top of a tall sallow, to find a suitable pupation site. The larva takes up position on the underside of a chosen leaf where it creates a silk pad from which the pupa will be suspended. The larva rests on the underside of the leaf for a couple of days with its head facing the leaf stalk but, as the time for pupation nears, the larva reverses this position, with its head now facing the leaf tip and its tail end positioned over the silk pad. After another couple of days, the larva pupates, having spent around 300 days in total in the larval stage, from August to early June.

5th instar: the larva perfectly mimics a sallow leaf

5th instar: a larva creating the silk pad from which the pupa will be suspended

5th instar: a pre-pupation larva

Pupa

The 30- to 35-mm long pupa is arguably the most difficult stage to find in the wild since, like the larva, it is perfectly disguised, matching the shades of green of the sallow leaf from which it is suspended upside-down. The pupa has a curious shape, with a flattened appearance when seen side on, but appearing relatively narrow when seen from the back. It will also shake violently from side to side when disturbed, which is clearly a defence against predators. After two to four weeks, depending on temperature, the pupa darkens slightly before this most majestic of butterflies emerges.

The pupa mimics a sallow leaf

The pupa appears relatively narrow when seen from the back

Life Cycles of British & Irish Butterflies

Red Admiral
Vanessa atalanta

❖ RED ADMIRAL IS SUPPORTED BY HOLLY CHAPPLE, WENDY GREENAWAY & VINCE MASSIMO ❖

	January	February	March	April	May	June	July	August	September	October	November	December
IMAGO												
OVUM												
LARVA												
PUPA												

The **Red Admiral** is one of our most iconic butterflies and can turn up just about anywhere, although its presence each year is largely dependent on the strength of an annual migration from the continent. The butterfly is unmistakable, with velvety black wings intersected by striking red bands, together with white spots on the apex of each forewing. As Richard South puts it in *The Butterflies of the British Isles* of 1906: "*The vivid contrast of black and scarlet in this butterfly will certainly arrest the attention of even the least observant*". Male and female have a similar appearance but can sometimes be distinguished based on the size and shape of the abdomen, which is shorter and wider in the female due to the eggs she carries. Both sexes have multicoloured undersides that are surprisingly cryptic, allowing the butterfly to blend in perfectly with its background when it is settled, closed-winged, on a suitable surface such as a tree trunk or stony ground.

The ♀ is slightly larger with a wingspan of between 70 and 78 mm

The ♂ has a wingspan of between 64 and 72 mm

In his *Musei Petiveriani* of 1699, James Petiver named this butterfly the 'Admiral', which is thought to be a reference to the Royal Navy ensign that was hoisted when an admiral was onboard ship, and which has a plain background colour with the red, white and blue tricolour of the Union Jack positioned in the top left quadrant. Similarly, the forewings of the Red Admiral are a plain black with a red, white and blue tricolour found toward the apex. Unfortunately, Benjamin Wilkes misinterpreted the intent of the word 'admiral', naming this butterfly the 'Admirable' in his *The English Moths and Butterflies* of 1749, and this was then used widely in entomological circles with Edward Donovan, in his *The Natural History of British Insects* of 1799, going on to name the butterfly 'Red Admirable'. Most subsequent authors reverted back to Petiver's original spelling, including Adrian Haworth in his *Lepidoptera Britannica* of 1803, giving the butterfly the name 'Red Admiral' that we use today.

♂ and ♀ have similar undersides

Life Cycles of British & Irish Butterflies

DISTRIBUTION

Most Red Admirals are unable to survive our winters and the presence of this butterfly is therefore largely dependent on an annual influx of migrants from the continent. The distribution and abundance therefore fluctuate from year to year—in some years this butterfly can be widespread and common, and in others rather local and scarce, depending on the strength of the migration. In most years, however, the butterfly can turn up anywhere in Britain and Ireland, including Orkney and Shetland. On a worldwide basis, the butterfly is found across Europe to western Asia, in parts of North Africa and in parts of North America. It has also been recorded near the Arctic Circle, with some records from the Faroe Islands and Iceland.

HABITAT

This widespread butterfly can be found in every habitat, from the seashore to the tops of the highest mountains. It is also a familiar sight in gardens and, in autumn, can be found in orchards where it feeds on fruits that have fallen to the ground. One youngster was entranced at the sight of tens of Red Admirals feeding on rotting fruit beneath a particularly productive plum tree that grew in his uncle's allotment near his Cheltenham home. How much influence that had on the author becoming a lepidopterist is hard to say, but the Red Admiral must take some credit for the book that you are currently reading.

Adults can be found in orchards in the autumn [CK] **feeding on an overripe fruit** [inset CK]

STATUS

This widespread and often common butterfly is not a species of conservation concern. The long-term trend shows a moderate increase in distribution and a significant increase in abundance, which may indicate that our increasingly mild climate is better suited to this butterfly. The more recent trend over the last decade shows a stable distribution and a moderate decline in abundance. Using the IUCN criteria, the Red Admiral is categorised as Least Concern (LC) in both the UK and Ireland.

LIFE CYCLE

A simple summary of the timing of the each of the life cycle stages throughout the year (the species 'phenology') is hard to pin down due to the unpredictability of the butterfly's migration patterns and its ability to overwinter, and these two topics have received an enormous amount of attention over the years.

Britain and Ireland experience several migrations, not just one, and the timing of these varies to some degree from year to year. The first migration, that is believed to originate in North Africa and southern Europe, contains the earliest migrants to reach us. A second is from Spain and Portugal that reaches us in May and June, and a third is from central Europe that reaches us in August. The butterfly also undertakes a reverse migration by flying south in the autumn, which is triggered by falling temperatures and therefore starts earlier in cooler years. T.G. Howarth writes of one example in his *South's British Butterflies* of 1973: "… *we rarely hear of butterflies moving about at night, but this species as well as the Painted Lady, are known to do this, particularly so since the advent of the use of mercury-vapour light for capturing moths in the early 1950s. Evidence of a southward migration was observed in the Isle of Wight on October 1955 when many specimens came to this type of light trap and when released the following morning were all observed to fly off in a southerly direction*". Prior to the emigration, adults can be found gorging on available nectar sources.

The butterfly is also now resident to some degree and not purely a migrant, but only in certain parts of southern England. Whilst there are records of some adults overwintering in most years, even from a century ago, successful overwintering is now being observed more frequently thanks to a milder climate, with adults hiding away in outhouses, rabbit holes, hollow tree trunks and other suitable spaces. However, the butterfly does not hibernate as such and can be found flying on warm winter days, unlike the Brimstone, Comma, Peacock and Small Tortoiseshell that become fully torpid during the winter months.

Immature stages are also regularly overwintering, although this is by no means a widespread phenomenon. Led by Vince Massimo, contributors to the UK Butterflies website are encouraged to monitor Red Admiral immature stages each winter, and there is an increasing body of evidence to show that eggs laid in October or later in southern England can produce adults the following year, although development is retarded due to the cold weather. Summarising observations of overwintering immature stages from 2014 to 2018, Massimo provides a general rule that eggs hatching in November and early December produce adults (if development is successful) by the third week of May the following year, which correlates with the findings of other enthusiasts who have monitored the overwintering of this species. He also notes that females have been observed laying eggs in every winter month in recent years.

The butterfly will feed on animal dung on occasion [MC]

Imago

The butterfly is one of the first to fly in early morning, and one of the last to go to roost, when it will settle head-down on the trunk or lower bough of a tree, where it is concealed by its cryptic undersides. Adults also feed on a wide variety of flowers and will use whatever nectar sources are available. The flowers of Blackthorn and hawthorns are used early in the year, whereas brambles, Buddleja, Common Fleabane, Michaelmas-daisies, thistles, Wild Privet and Wild Teasel are used in the summer months. At the end of the year the butterfly is particularly fond of Ivy blossom and Hemp-agrimony. In addition to rotting fruit, adults are especially fond of sap runs, as noted by Richard South: "*The seductive fluid obtained from such trees is evidently more potent than the nectar from flowers, as under its influence the insect is so listless that it may be taken up between the finger and thumb*". The butterfly will also feed on aphid honeydew found on leaves, and also obtain minerals from dung and damp ground.

The male will establish a territory on a hilltop or some other prominent feature in the landscape, such as a sheltered and sunny clearing, where he will perch on a twig or branch and wait for a passing female, flying up to intercept anything that looks like a Red Admiral, including Commas and Peacocks. Rival males are chased off before the resident male, or occasionally the rival, returns to take up position in the territory. Female encounters and mated pairs are rarely observed, at least in Britain and Ireland. This may be because the vast majority of adults that we see are either already-mated migrants or are newly emerged and home-grown adults whose main objective is to head south. Alternatively, it could simply be that mated pairs are elusive and hard to spot—the few lucky observers that have witnessed both courtship and pairing suggest that the female leads the male to a spot high up in the shaded part of a tree, where mating takes place.

An egg-laying female is very easy to spot as her typically powerful flight is replaced by one that is slow and meandering as she flies over the foodplant, depositing an egg on the upperside of a young leaf before flying off, with bouts of egg-laying interspersed with periods of feeding and resting. The primary larval foodplant is Common Nettle although Hop, Pellitory-of-the-wall and Small Nettle are used occasionally. The 1997 Butterfly Conservation booklet *The Red Admiral Butterfly*, written by Mike Tucker, gives an indication of the number of eggs laid by a female: "*There is a recent record of a captured female which laid over 700 ova in captivity, and this was only part of her full potential*".

A lush bed of nettles next to a south-facing wall in Crawley, Sussex [VM]

Eggs laid on young nettle growth [VM]

Ovum

The 0.8-mm high, shiny and barrel-shaped egg has nine or 10 raised ribs that run its length, and these appear transparent and delicate when seen close up. The egg is light green when first laid but turns a darker grey as the larva develops, with the black head capsule of the larva showing clearly through the eggshell just prior to its emergence.

The egg has nine or 10 delicate ribs along its length

The black head of the larva can be seen just before it emerges from the egg

Larva

After between five and 10 days, depending on temperature, the 1.5-mm long **first instar** larva makes its exit from the egg by eating a hole in the crown, but it does not eat the remainder of the eggshell. The influence of temperature is not to be underestimated—Vince Massimo has found that, while eggs may hatch in any month of the year, some successfully overwinter on occasion, with an incredible observation of a few eggs that were laid in November 2017 hatching in February 2018. Given this variability, the durations provided here are indicative of immature stages found in the summer months, rather than those developing in early spring, late autumn or over the winter.

Once emerged, the first instar immediately crawls to the base of its nettle leaf where it creates a protective shelter, under which it feeds. The shelter is usually a simple affair, with silk spun across

1st instar: a larva emerging from its egg

1st instar: a larva creating a shelter at the base of a leaf [VM]

1st instar: a newly emerged larva

1st instar: a larva preparing to moult

the basal lobes of a single leaf, and the resulting white patch at the leaf base is relatively easy to spot against the green of the leaf. That being said, the construction of the shelter is quite variable, and may be formed in the crown, involve two leaves, be created on the underside as well as the upperside and, on occasion, even result in a leaf edge completely folding over on itself.

The first instar has a shiny black head, a greenish-yellow ground colour and is covered in 10 rows of dark warts that each bear a long and curved black hair. The larva darkens as it matures and, just prior to moulting, when 3 mm long, may exhibit pairs of white blotches on its back on the 5th, 7th and 9th segments that provide a hint at the colour of the spines of the second instar larva.

After five days or so, depending on temperature, the larva moults into the **second instar** and is now covered in seven rows of spines, with each spine clothed in several bristly hairs. The ground colour is now olive-brown that accentuates pairs of pale yellow spines found on the 5th, 7th and 9th segments, although the colour of these spines is quite variable and the spines on some larvae are completely black. The larva may also have a faint pale yellow marking at the base of each side, from the 4th to 11th segments, each of which is interrupted by a dark spine, and this is a feature that becomes more prominent with each successive instar. The larva continues to live under a protective shelter, although this is more substantial than that of the first instar and usually involves several leaves in its construction. When fully grown, the second instar is 5 mm long.

2nd instar: a larva with pale yellow spines on the 5th, 7th and 9th segments

2nd instar: the protective shelter is more substantial in this instar

If temperatures are sufficiently high then, after only a few days, the larva moults into the **third instar**. The larva now has a darker ground colour and is covered in numerous white spots. The arrangement and colour of the spines is the same as in the second instar, although the structure of the spines is now slightly more complex and branching. Particularly pale individuals may also have additional pale yellow blotches at the base of some spines, not just those found on the 5th, 7th and 9th segments.

3rd instar: a recently moulted larva

3rd instar: a larval shelter with feeding damage at its tip [VM]

The larva now creates a shelter by neatly silking together the two edges of a leaf, starting at the base of the leaf, causing the leaf edges to fold upwards as the silk dries and contracts. The resulting construction is so neat, and so visible when in a patch of nettles, that it is not only easy to find in the field but is also clearly identifiable as a Red Admiral construction (although Small Tortoiseshell larvae will create similar structures when moulting), rather than that of nettle-feeding moths which typically look like a rolled-up cigar. The larva feeds on the tip of this shelter and subsequently uses silk to seal up the end after it has finished its meal. The larva abandons its shelter when it is no longer sufficient for its needs, moving to a larger leaf where it creates a new shelter. The Red Admiral larva is one of the easiest to find in the wild since, once it has reached the third instar, its presence is given away by a tell-tale series of discarded shelters that are particularly noticeable within their patch of nettles, with the larva inhabiting the largest shelter. When fully grown, the third instar is 11 mm long.

After only another five days or so, the larva moults into the **fourth instar**. The larva is almost identical in appearance to the third instar and the two instars are best distinguished by size—when fully grown, the fourth instar is 16 mm long. The larva continues to exhibit a high degree of variability in terms of its ground colour—some larvae are almost black, whereas others appear much paler due to the extent of the white markings all over their body.

4th instar: a pale larva with numerous white spots on its body

4th instar: a relatively dark individual [VM]

In just under a week, depending on temperature, the larva moults for the last time into the **fifth instar**. While the larva continues to possess seven longitudinal rows of branched spines, these are often the same colour along the length of the larva, including those found on the 5th, 7th and 9th segments. That being said, the colour of the fifth instar is very variable, ranging from black through to a very pale yellowish-green. Some have suggested that this is a defence against avian predators that might 'lock on' to one particular search image, especially since larvae are found out in the open on occasion, usually when moving to create a new larval tent. The tent is a much more substantial affair

5th instar: a typical larval tent [VM]

5th instar: an almost black larva

5th instar: a pale yellow larva

5th instar: this colourful larva has spines with an orange base in this instar and watching one being constructed is fascinating—the larva ascends to the top of a nettle sprig, fells the main stem and then draws several leaves together with silk, thereby creating a tent within which it lives and feeds. When fully grown, the larva is 35 mm long.

When preparing to pupate, the larva creates a new tent within which it creates a silk pad from which it hangs upside-down in a J-shape. The final tent of a mature Red Admiral larva is easy to spot in nettle patches and this is therefore one of the easiest pupae to find in the wild.

In the winter of 2016–2017, Vince Massimo closely monitored two eggs that were found on 31 October and one larva that survived eventually reached its fifth instar on 14 March. Much to his surprise, the larva moulted again on 27 March into a sixth instar, which was similar in appearance to the fifth instar, although larger, with the larva pupating on 3 April and a female emerging on 23 April. This 'plasticity' of larval instars has been observed in other species too, notably in the Glanville Fritillary, where larvae that have overwintered but are under-developed (based on their body mass) occasionally produce an extra instar.

5th instar: a larva preparing to pupate [VM]

Pupa

After 10 days in the final instar, the entire larval stage lasting only a month, the larva pupates, and the resulting 23-mm long pupa is attached to the silk pad by the hooks on the cremaster. Unlike several closely related species, the head of the pupa is quite blunt without any horns. The pupa is a plain greyish-brown in colour with some noticeable golden flecks on its back that may have been the inspiration for the term 'aurelian', from the Latin '*aureus*' meaning 'golden' or 'gilded', that was once used to describe lepidopterists. The strikingly coloured adult butterfly emerges after two or three weeks.

The pupa has golden flecks on its back

The wings can be seen through the pupal case just prior to emergence [VM]

Life Cycles of British & Irish Butterflies

Painted Lady
Vanessa cardui

❖ PAINTED LADY IS SUPPORTED BY GAVIN DEANE & BILL STONE, SUFFOLK BUTTERFLY RECORDER & IN LOVING MEMORY OF KATHY COTTLE 1957–2015

	January	February	March	April	May	June	July	August	September	October	November	December
IMAGO												
OVUM												
LARVA												
PUPA												

This butterfly is one of the greatest migrants of the butterfly world and its ability to fly long distances has resulted in the largest distribution of any butterfly. Migrants arrive in Britain and Ireland every year and, every 10 years or so, the migration can be spectacular. The most recent '*cardui*' year was in 2009 when the butterfly literally swarmed over the southern half of Britain, even grabbing the attention of the national press. Like many observers, I recall butterflies streaming purposefully in a northerly direction over my garden and, at their peak on 24 May, I was counting 20 per minute or more for several hours.

The corresponding egg lay was huge. In late June of that year I remember taking a walk in the countryside and there was hardly a thistle (the larval foodplant) that did not contain at least one Painted Lady larva. In '2009: The Year of the Painted Lady *Vanessa cardui*', published in *Atropos* in 2010, Richard Fox describes several instances of large numbers of larvae: "*Thousands of larvae were found on thistles and nettles at Barbury Castle and Markham Banks, both close to Swindon. These aggregations were eclipsed by an estimated 500,000 Painted Lady larvae discovered by Malcolm Lee and Barry Ofield in two fallow fields, totalling a mere 7.5ha, close to Port Isaac, Cornwall. Both here and at Barbury Castle observers counted 20 or more caterpillars per thistle plant*". Despite the real potential of starvation with such large numbers of larvae, the resulting emergence was impressive, with hundreds of thousands of adults estimated at several sites at the end of July and beginning of August, when new immigrants were still seen coming in off the sea.

Painted Ladies nectaring in Sussex during the 2009 invasion [VM]

While the butterfly undertakes similar migrations throughout its worldwide range, the migration to our shores has been surrounded in mystery. Where do the butterflies disappear to in the autumn? Where do the butterflies that we see originate? How far is an individual butterfly able to travel? What triggers the migration? Thankfully, research into this butterfly has provided some answers.

In 'Migration of the painted lady butterfly, *Vanessa cardui*, to north-eastern Spain is aided by African wind currents', published in *Journal of Animal Ecology* in 2007, Constanti Stefanescu and colleagues show a strong correlation between wind patterns and the ability of the butterfly to migrate large distances in a corresponding direction, concluding that the butterfly hitches a ride on the wind currents in the upper levels of the atmosphere when undertaking long-distance flights. In their 2008 paper 'Mysteries of Lepidoptera Migration Revealed by Entomological Radar', published in *Atropos*, Jason Chapman and Rebecca Nesbit confirmed this theory by using vertical-looking radar (VLR) to show that different insects, including the Painted Lady, regularly fly at altitude and out of sight at a height of up to 1,200 m.

Radar signatures were correlated with insects caught in a net suspended below a helium-filled balloon that was attached to a motorised winch, making subsequent radar detection surprisingly accurate, with an ability to distinguish a Painted Lady from the equally migratory Silver Y moth *Autographa gamma*, and for their direction of travel to be recorded. Based on their data, Chapman and Nesbit were able to determine that 11 million Painted Ladies reached us in the spring of 2009, with an incredible 21 million butterflies emigrating in the autumn, with most butterflies flying at a height of between 200 and 400 m. Up until this study, evidence of a reverse migration was somewhat anecdotal, based on a few sightings of butterflies heading out to sea from the south coast in the autumn. It is also thought that the butterfly flies at night as well as during the day, explaining those individuals that occasionally turn up in moth traps.

The ability for an individual butterfly to travel long distances was confirmed in another study by Constanti Stefanescu and colleagues, as described in their paper 'Long-distance autumn migration across the Sahara by Painted Lady butterflies: exploiting resource pulses in the tropical savannah', published in *Biology Letters* in 2016. This study used measurements of hydrogen isotopes from parts of the wings to prove unequivocally that the butterflies found in sub-Saharan Africa in autumn originate from Europe, confirming that the butterflies traverse one of the most inhospitable regions of our planet—the Sahara Desert—covering a distance of over 1,000 miles. A similar study by Gerard Talavera and colleagues and described in their paper 'Round-trip across the Sahara: Afrotropical Painted Lady butterflies recolonise the Mediterranean in early spring', published in *Biology Letters* in 2018, showed that the reverse is true, and that butterflies found in the Mediterranean region in spring originate from sub-Saharan Africa. This evidence adds some weight to the belief that the first arrivals on our shores in early spring originate directly from north-west Africa, which is also a distance of just over 1,000 miles. Chapman and Nesbit also write: "*Wind speeds 500m above the ground are often four or five times faster than the butterflies' flight speed, so that if they were to fly at these heights they could reach speeds of almost 100km/hr (60 mph)*".

In summary, it is now thought that the butterflies that reach our shores originate from the breeding grounds in northern Africa and Arabia, reaching us both directly (the first arrivals from north-west Africa) and indirectly from continental Europe (the offspring of subsequent broods), with a proportion of individuals flying out of sight at altitude on favourable wind currents. The butterfly then undertakes a reverse migration out of Europe in the autumn when, confirmed by the paucity of sightings during this event, flying at altitude is more fully exploited. In 'Multi-generational long-distance migration of insects: studying the painted lady butterfly in the Western Palaearctic', published in *Ecography* in 2013, Stefanescu and colleagues provide a further thought-provoking fact: "*The migratory cycle in this species involves six generations, encompassing a latitudinal shift of thousands of kilometres (up to 60 degrees of latitude)*"—this is a round-trip of over 13,000 km with the butterfly reaching far above the Arctic Circle. The movements of this butterfly are therefore on a par with, and even exceed, the well-documented long-distance migrations of the Monarch. One mystery that remains is the trigger for the migration and suggestions include an increasing day length, increasing temperatures, lack of suitable nectar sources and larval foodplants, and an increasing density of individuals (that could result in their offspring overwhelming the foodplant with many larvae perishing as a result). Similar theories exist for the autumn emigration, such as a decreasing day length and falling temperatures.

Like many of the 'vanessids', male and female are notoriously difficult to tell apart, although the male (whose wingspan is between 58 and 70 mm) is slightly smaller than the female (whose wingspan is between 62 and 74 mm). The two sexes also have similar undersides and are remarkably well camouflaged when resting with wings closed on the ground.

The butterfly was given its name by James Petiver in his *Musei Petiveriani* of 1699, with William Lewin, known for his brevity, naming this butterfly the 'The Thistle' in his *The Papilios of Great Britain* of 1795. Its specific name, *cardui*, also recognises the thistle genus *Carduus*.

♂ and ♀ are similar in appearance

The cryptic undersides provide excellent camouflage [JA]

DISTRIBUTION
This butterfly can be found anywhere in Britain and Ireland, including Orkney and Shetland where it has also bred, although its distribution in any given year is wholly dependent on the strength of the annual migration, wind patterns and conditions that allow the butterfly to breed. The butterfly can be remarkably common in good years, such as 2009, and extremely scarce in others. Given its migratory tendencies, the butterfly is found all over the world, with the exception of South America. Also, some authorities consider the Australian population to comprise a different species, *Vanessa kershawi*, where it is known, quite simply, as the Australian Painted Lady.

HABITAT
Given the nomadic nature of this species, it can be found in just about any open area, from the seashore to town gardens to the tops of the highest mountains. This is also one of the few species that can breed in intensive farmland since even these sites typically contain a patch of thistles, the primary larval foodplant.

STATUS
The long-term trend shows a moderate increase in distribution and a significant increase in abundance, with the more recent trend over the last decade showing a moderate decline in distribution and a significant decline in abundance, although these trends are simply a reflection of the strength of immigrations over these periods. Using the IUCN criteria, the Painted Lady is categorised as Least Concern (LC) in both the UK and Ireland.

Thistles act as both a nectar source and larval foodplant

LIFE CYCLE
Adults are seen from late March as they start to arrive on our shores, with numbers building up in May and June as further migrants arrive from the continent. This influx gives rise to the next generation that peaks in early August and that may go on to produce another brood if conditions are favourable, with the butterfly completing its life cycle in six weeks or so.

This continuously brooded butterfly has no natural diapause, and this is almost certainly the root cause of all stages perishing in Britain and Ireland over the winter. There are always exceptions to the rule and, as documented in *Atropos* in 1998, John Wacher provides concrete evidence of one marked individual successfully overwintering at Hayle in Cornwall in the winter of 1997–8. In general, however, winter sightings of this species are thought to be of new immigrants.

Imago
This very active butterfly is often found refuelling at its favourite nectar source of thistles, but will use a variety of other plants, including brambles, Buddleja, Bugle, Common Bird's-foot-trefoil, Common Fleabane, Devil's-bit Scabious, hawkweeds, heathers, Hemp-agrimony, Ivy, knapweeds, ragworts, Red Clover, Wild Marjoram and Wild Privet.

When looking for a mate, the male will perch in a warm spot such as bare earth or a rock where he waits for a female, flying up to investigate any passing object. Both sexes will also undertake 'hill-topping' where they fly to a high point in the landscape in search of a mate. Whatever the strategy, a receptive female that is encountered is soon mated.

Egg-laying females are easy to spot as the fly around tall thistles that are selected not for their flowerheads, but for their lush leaves. They will land on a leaf, carefully avoiding the spines, and lay a single egg, usually on the leaf upperside, laying up to 500 eggs in total. Thistles are the most widely used foodplant in Britain and Ireland, although burdocks, Common Nettle, mallows, Viper's-bugloss and various cultivated plants have also been recorded, with the butterfly using a much wider range of foodplants throughout its worldwide range.

A ♀ laying a single egg on thistle [PA]

Eggs laid on the upperside of a thistle leaf

The egg has 16 ribs that run along its sides [GN]

Ovum

The 0.65-mm high and oval egg is light green when first laid but eventually turns grey as the larva develops, when its head is visible through the shell. The egg has 16 ribs that run along its sides from top to bottom, becoming progressively less prominent as they reach the egg base.

Larva

After a week, the 1.6-mm long **first instar** larva eats away at the crown of the egg so that it forms a lid, before pushing its way out. The larva does not eat the remainder of the eggshell and immediately moves to the underside of the leaf where it lays down a simple silk web, under which it eats away at the cuticle, leaving tell-tale transparent windows when the leaf is viewed from above. Frass also collects within the silk web and is rather conspicuous. The dark olive larva has a black head, a black band on the first segment and several rows of black warts that run along the length of the body, with each wart bearing a long wiry hair. When fully grown, the first instar is just over 3 mm long, when it gives a hint of the colouring of the second instar, with pale patches forming on the 5th, 7th, 9th and 11th segments.

After a week, the larva moults into the **second instar**. The dark band on the first segment is still present and the larva is now in possession of several rows of tubercles, each of which bears a long wiry hair, with several more hairs emerging from the tubercle base. Most tubercles are black, although two of those on the back of the larva on each of the 5th, 7th and 9th segments are cream, and there is a hump on the back of the 11th segment of the same colour. The larva continues to live under a web on the underside of the leaf, where it feeds on the leaf cuticle and, when fully grown, is 6.5 mm long.

In just under a week, the larva moults into the **third instar** and continues to live under a web. It now has more significant and shiny tubercles with the same arrangement as in the previous instar, although the pale tubercles on the 5th, 7th and 9th segments are now a very conspicuous yellow. Even at this young age the larva starts to show some variability and some larvae also have a yellow tubercle along the middle of the back on these segments, and at the base of the 2nd and 3rd segments. From

1st instar: the larva is dark olive with long wiry hairs all over its body

2nd instar: the larva is now covered in several rows of mainly black tubercles

3rd instar: a larva feeding under its silk web

3rd instar: the tubercles are now more prominent

the 4th to 11th segments, the larva also starts to develop pale crescent-shaped markings between each pair of segments, just below the spiracles. A dark line also runs along the middle of the back that is bordered on each side by a paler line although this, again, is a variable feature that is barely noticeable in some individuals. When fully grown, the third instar is 11 mm long.

4th instar: a newly moulted larva

In less than a week, the larva moults into the **fourth instar** and is similar in appearance to the previous instar, although the markings are much more clearly defined, especially the crescent-shaped markings along the base of each side. The larva is best distinguished by size—when fully grown, the fourth instar is around 20 mm long and almost twice the length of a third instar. The larva continues to spin silk among the foodplant but now feeds more extensively on the leaves and will completely strip them, leaving only the toughest stems and spines.

In under a week, the larva moults for the last time into the **fifth instar** and now comes in several forms, from those that are very dark and resemble the previous instar, to those that are much paler and have a more colourful appearance, although all forms retain the pale yellow crescent-shaped markings along the base of each side from the 4th to 11th segments. The larva is now covered in seven rows of spines that run along its length from the 4th to the 12th segments, with fewer spines on all other segments. Unlike the tubercles of earlier instars, the spines are all the same colour, and are usually a pale yellow. In under a week, and after just over four weeks in this stage, the larva is fully grown and around 30 mm long, when it spins a loose tent of leaves either on the foodplant or nearby vegetation, creating a silk pad from which it hangs head-down in a J-shape.

5th instar: dark form of the larva
5th instar: pale form of the larva
5th instar: a pre-pupation larva on the foodplant

Pupa

The 24-mm long pupa is formed head-down, attached to the silk pad by the cremaster, and comes in several colour forms—including grey, dark green and brown—and sometimes has a brassy appearance. Unlike several related species, the pupa does not possess any horns on the head but does have significant protrusions at the base of each wing case, and on its back on both the thorax and in two rows that run along the length of the abdomen. After 10 days or so, the adult emerges and is flushed with salmon-pink, making for a most-beautiful insect, although the colour fades rapidly with the passage of time as this transcontinental migrant goes on its way.

The grey form of pupa
The brown form of pupa

Life Cycles of British & Irish Butterflies

Peacock
Aglais io

❖ PEACOCK IS SUPPORTED BY ALEXANDRA BARBILEV, GLYN CLARKE & IMOGEN PAYNE ❖

	January	February	March	April	May	June	July	August	September	October	November	December
IMAGO												
OVUM												
LARVA												
PUPA												

The Peacock is a much-loved and familiar sight in gardens across Britain and Ireland and is unmistakable, with spectacular 'eyes' on all four wings, the most realistic of which are on the hindwing uppersides that give this butterfly its name. A flash of the eye spots is usually enough to deter any potential predator and they are put to good effect when the butterfly is vulnerable during hibernation, or when it is roosting on a surface such as a tree trunk. This is also when the butterfly's black undersides, that could not be more different to the uppersides, provide surprisingly convincing camouflage.

Like closely related species, the male and female are similar in appearance although the male, with a wingspan of between 63 and 68 mm, is slightly smaller than the female whose wingspan is between 67 and 75 mm. The butterfly was originally named 'The Peacock's Eye' by James Petiver in his *Musei Petiveriani* of 1699, although this was shortened to the name we use today by Benjamin Wilkes in his *Twelve New Designs of English Butterflies* of 1742.

♂ and ♀ are similar in appearance

The undersides of both sexes are almost black

DISTRIBUTION
This is a highly mobile butterfly that occurs throughout Britain and Ireland and has been recorded from Orkney and Shetland on occasion, although it is not found in some parts of northern Scotland. On a worldwide basis, the Peacock is found throughout most of Europe and across temperate Asia as far as Japan. In *The Butterflies of Britain and Ireland*, Jeremy Thomas writes that the furthest distance travelled by an individual is recorded as 95 km and that, while the butterfly is known to fly north in the spring and south in the autumn, it is best classified as a nomad rather than a true migrant.

HABITAT
Given its nomadic lifestyle, this butterfly can turn up just about anywhere, including woodland edges, open grassland, field margins, lanes, waste ground and gardens. If not simply passing through, then sites are chosen either for their abundance of nectar

A sunny, sheltered and nettle-rich area at Woolhampton in Berkshire

sources or the availability of lush nettles growing in a sunny and sheltered position that are used by egg-laying females.

STATUS

This common and widespread species is faring well and shows signs of colonising the few remaining areas in northern Scotland where it has not historically been found. This butterfly is not, therefore, a species of conservation concern. The long-term trend shows a moderate increase in both distribution and abundance, with the more recent trend over the last decade showing a stable distribution and a moderate increase in abundance. Using the IUCN criteria, the Peacock is categorised as Least Concern (LC) in both the UK and Ireland.

LIFE CYCLE

The butterfly overwinters as an adult and unseasonably warm weather can prematurely awaken it from its slumber, and butterflies have been recorded flying in every month of the year, even when there is snow on the ground. The majority of adults become active at the end of March and throughout April and these then pair and ultimately give rise to the next generation that emerges from the end of July to early September. The newly emerged adults feed for a short time on a variety of summer nectar sources that allows them to build up essential body fats before they overwinter, this being one of our longest-lived butterflies. While this butterfly is generally single-brooded, larvae have been found feeding in September in warm locations on occasion, and these go on to produce a second brood that emerges in early October.

Imago

The butterfly's striking appearance has developed for good reason—to deter mice, birds and other would-be predators. Aside from its underside camouflage and a flash of its eye spots, the butterfly has another trick up its sleeve, which is to make a hissing sound by rubbing its wings together, which is so loud that it is audible to the human ear. This is particularly powerful when adults are hibernating in groups in a hollow tree, wood pile or outbuilding, when their collective defences must be quite overwhelming. Some authors write of finding up to 40 butterflies hibernating together and they are often in the company of the Small Tortoiseshell and the Herald moth *Scoliopteryx libatrix* that favour similar overwintering sites. The butterflies will also use these same sites when roosting in the autumn, remaining in the vicinity before finally entering hibernation.

Much of what we know about the Peacock's behaviours, and those of the Small Tortoiseshell, comes from the work of Robin Baker and his paper 'Territorial Behaviour of the Nymphalid Butterflies, *Aglais urticae* (L.) and *Inachis io* (L.)', published in *The Journal of Animal Ecology* in 1972, makes for a fascinating read.

Pre-hibernation adults are sexually immature and spend their time feeding on flowers in order to build up their fat reserves that help them get through the winter. Betony, brambles, Buddleja, Common Fleabane, Devil's-bit Scabious, hawkweeds, Hemp-agrimony, ragworts, Water Mint, Wild Marjoram, Wild Privet, Wild Teasel and Yarrow are all used, and the butterfly will also feed from sap runs when available. Post-hibernation adults, on the other hand, spend most of the morning seeking out warm and sheltered spots to bask, and feed on whatever nectar sources are available in early spring, including the flowers of sallows, followed later by Blackthorn, Bluebell, Bugle, Cuckooflower, dandelions and Ground-ivy. A few days after emerging from hibernation the adults turn their attention to reproducing.

Each morning, after feeding, the male will wander up to 0.5 km before establishing a territory from 11.30 am. Most males are in position by 1 pm, and roost near their territories that evening. Territories are often on the sunny side of either a wood, a group of trees or a tall hedge, which Baker felt may be *en route* to a suitable egg-laying site used by females. The male waits on the ground and will fly up at any dark object entering the territory. Other males entering the territory are quickly seen off if they are merely passing through. Others may be more persistent, with the two males battling it out as they circle around one another, with clashes of wings as each tries to fly above and behind the other. They rise high into the air before dropping to the ground, with the pair either battling it out once again (resulting in a series of rises and drops that takes them up to 200 m from the original spot) or with one of the males moving off to find a new territory.

The male is particularly persistent when he encounters a female. An unresponsive female will try to lose her potential mate by flying away and, if pursued, will drop into vegetation. If she happens to fly through another male's territory, then the resulting skirmish between the two males often takes precedence over the female, who then flies off undisturbed. She may also try to escape detection by landing on the far side of a tree trunk and snapping her wings shut.

A ♂ drumming his antennae on the hindwings of the ♀ [NF]

A mating pair with the ♀ below [PA]

A more receptive female is pursued until she lands, when the male alights behind her and gently taps her hindwings with his antennae—much like the behaviour found in the Small Tortoiseshell. If the female is receptive then she will fly away and, if the male succeeds in staying with her over a prolonged period, often for two or three hours, she will lead him to a shady spot in late afternoon or early evening where they mate.

The female keeps her wings closed while she lays her eggs in batches of up to 400 eggs, a process that can take over two hours. Since she is literally 'putting all of her eggs in one basket', she takes great care when finding a suitable spot, selecting the middle of a large nettle patch that is sheltered and in full sun, often next to a woodland edge or hedgerow. Only a small

A ♀ laying her batch of eggs [VM]

Life Cycles of British & Irish Butterflies

proportion of available plants are suitable and there are reports most years of two females not only laying on the same plant, but on the same leaf, and this behaviour also extends to laying alongside a Small Tortoiseshell. Eggs are typically laid on Common Nettle, although eggs and larvae are occasionally found on Small Nettle and Hop.

Eggs are laid in untidy batches

The barrel-shaped eggs have eight to 10 ribs running down their sides

Ovum

The 0.8-mm high, barrel-shaped eggs are laid untidily on top of each other and are piled up to six deep in the centre of the egg mass. Each egg has between eight and 10 ribs that run down its sides from top to bottom and is yellowish-green when first laid but turns a darker green after a few days. Just before hatching, the dark head of the larva can be seen quite clearly through the eggshell, and this gives the entire egg batch a dark grey appearance.

Larva

After two weeks or so, depending on temperature, each 1.6-mm long **first instar** larva emerges from its egg at the same time as its siblings by eating away the crown and, without eating the remainder of the eggshell, immediately goes about contributing to a communal web, on the leaf on which the eggs were laid. The larvae feed by both day and night and, once the initial leaf has been consumed, they move to the tip of the plant where they create a new web. As the larvae grow, they will move to new leaves, with later instars even moving to new plants, building new webs along the way, and a series of discarded webs give away the presence of nearby larvae. The key distinguishing feature of this instar is that it has no tubercles or spines—it simply has four rows of long and fine black hairs that each emanate from a dark bulbous base. The larva has a black head and a pale greenish-grey ground colour when newly emerged that turns a yellowish-brown as it matures, with a reddish-brown hue developing on the thoracic segments and just below the back on each side. When fully grown, the first instar is 4 mm long.

After a week, the larva moults into the **second instar** and, following the moult, the web is decorated with shed larval skins and frass. The larva is similar in appearance to the previous instar, although conical bristle-covered black tubercles are now found at the base of the longest hairs, and the reddish-brown markings are more extensive, the larva appearing browner as a result. Unsurprisingly, for something so small, these differences can be hard to discern, although the fully grown second instar is 8 mm long and twice the length of a first instar. Larvae continue to live gregariously on a thick silk web and, when disturbed, will raise the front half of their bodies.

1st instar: newly emerged larvae are greenish-grey [VM]

1st instar: larvae turn yellowish-brown as they mature

2nd instar: fully grown larvae [VM]

After as little as four days, the larva moults into the **third instar** and now has a much darker purplish-brown ground colour that is sprinkled with white spots, with those at the base of each side clustered together to provide the illusion of a lateral line. The true legs on the thorax are black while the midabdominal prolegs (the 'claspers' on the 6th to 9th segments) are pale olive-brown. The larva is now covered in spines and those just below the back on each side are especially long. The larvae remain gregarious and their collective basking allows them to raise their body temperature, which aids digestion. When fully grown, the third instar is 14 mm long.

3rd instar: the larvae continue to live gregariously

3rd instar: the ground colour is dark purplish-brown

After four days or so, the larva moults into the **fourth instar** and now has a consistent purplish-black ground colour with numerous white spots dotted evenly all over the body. The body is now covered in long spines and those just below the back remain the longest, and the midabdominal prolegs remain a pale olive-brown and stand out against the purplish-black ground colour. When fully grown, the fourth instar is 25 mm long.

In just under a week, the larva moults into the final and **fifth instar** and, while similar to the previous instar, now has a jet-black ground colour and reddish-brown midabdominal prolegs. The body is also covered in short hairs which gives the larva a velvety appearance. The larva is also, of course, much longer, reaching around 42 mm when fully grown. Larvae continue to feed by both day and night and have several behaviours to avoid predation—when disturbed, a group of larvae will jerk their bodies from side to side in unison (which must be a formidable sight to any predator), will regurgitate a green liquid and, if necessary, will curl up in a ball and drop to the ground.

Larvae will disperse prior to pupation and will wander for several metres before finding a place to pupate. Once a suitable site has been found, the larva spins a silk pad from which it hangs head-down in a J-shape.

4th instar: the larva is now covered in long spines

5th instar: the larva is now a velvety black

5th instar: a larva preparing to pupate [WL]

Pupa

After a week or so, the larva sheds its skin to reveal a 27-mm long pupa which is attached to the silk pad by the cremaster. The pupa is covered in prominent points—two on the head, one on the back of the thorax, and in two rows that run down the back of the abdomen. There are two main colour forms, yellow and dark grey, the resulting colour depending on the site chosen for pupation. After around two weeks, the pupa colours up, when the wings of the adult are clearly visible through the pupal case, before this most colourful of our butterflies emerges.

The yellow form of pupa

The dark grey form of pupa

Small Tortoiseshell
Aglais urticae

❖ SMALL TORTOISESHELL IS SUPPORTED BY DEBBIE COOPER, VRULA DONALDSON & JOHN GREENAWAY ❖

	January	February	March	April	May	June	July	August	September	October	November	December
IMAGO												
OVUM												
LARVA												
PUPA												

The **Small Tortoiseshell** is one of our most familiar and well-known butterflies, appearing in gardens throughout Britain and Ireland. It has historically been one of our commonest species, although recent decades have seen worrying declines in some years, especially in the south, and, while the butterfly may make a partial recovery in subsequent years, this much-loved creature can no longer be considered the common butterfly that it once was.

The two sexes are almost identical in appearance, with the distinctive yellow and orange uppersides contrasting with the dark undersides that provide the butterfly a great deal of camouflage when hibernating. The male has a wingspan of 45 to 55 mm, although the female is slightly larger with a wingspan of 52 to 62 mm.

The butterfly was originally named the 'The lesser Tortoise-shell Butterfly' by James Petiver in his *Musei Petiveriani* of 1699, differentiating it from the 'The greater Tortoise-shell Butterfly' (what we now know as the Large Tortoiseshell), with Benjamin Wilkes applying the name we use today in his *Twelve New Designs of English Butterflies* of 1742. William Lewin used the name 'Nettle Tortoiseshell' in *The Papilios of Great Britain* of 1795 that recognises the key larval foodplants of Common Nettle and Small Nettle, as does the specific name of *urticae*, since *Urtica* is the genus of nettle. In Scotland the butterfly has historically been given the undesirable names of 'Devil's Butterfly' and 'Witch's Butterfly'.

♂ and ♀ are similar in appearance

The dark undersides contrast with the bright uppersides [JA]

DISTRIBUTION

This is one of our most mobile and widespread butterflies and is found throughout Britain and Ireland, including Orkney and Shetland, although there are no records of it breeding in the latter. Unlike most other species, it can also be found in mountainous areas where larval webs have occasionally been found. On a worldwide basis, the butterfly has a Palearctic distribution and can be found from the Atlantic coast in Europe and across temperate Asia to the Pacific coast, although it is not found in North Africa.

The butterfly does not form discrete colonies but is a great wanderer, often travelling several kilometres each day. As documented in 'The evolution of the migratory habit in butterflies', published in *The Journal of Animal Ecology* in 1969, Robin Baker found that the Small Tortoiseshell exhibited migrant tendencies, with adults flying NNW in spring and summer, and SSW from mid-August. The butterfly is also known to migrate across the Channel (in both directions) and studies in Germany showed that individuals were covering a distance of 150 km. Even though any immigration to our shores can be significant, this never reaches the scale of those undertaken by the Red Admiral or Painted Lady, and the vast majority of butterflies that we see are home-grown.

HABITAT

This butterfly can turn up almost anywhere, from city centres to mountain tops, and is often encountered while feeding in gardens. It is also found where lush nettles grow in abundance, such as field margins and the edges of farmland used for cattle. Many observers also come across the adult butterfly as it is hibernating, when it can be found in an outbuilding, such as a garage, shed or barn; an attic; a log pile; a hollow tree; or some other dark and sheltered site. Given the butterfly's propensity to hibernate in dark and sheltered places, they may also be found in houses where artificial heating will awaken these visitors from their slumber. In his *The Lepidoptera of the British Islands* of 1893, Charles Barrett writes: "*Mr. H. Jenner Fust, Jun., records a most curious instance of hybernation, in which three specimens settled down in a church bell, which was in regular use, and there spent the winter, undisturbed by the vibrations*".

Thames Path next to Chimney Meadows in Oxfordshire [JA]

STATUS

In *The Millennium Atlas of Butterflies in Britain and Ireland*, published in 2001, Jim Asher and colleagues write "*Despite great fluctuations in abundance, it remains ubiquitous and its future seems assured*". Unfortunately, the intervening years have not been kind to the Small Tortoiseshell, whose declines appear to be more severe than in the past. The causes of these declines are not well understood, with various theories put forward.

A *Sturmia bella* tachinid fly

One theory is that climate change has resulted in an increased presence of the tachinid fly, *Sturmia bella*, that is found commonly on the continent. The fly lays its eggs on leaves of the foodplant, close to where butterfly larvae are feeding. A tiny egg is then eaten whole by the larva and the grub that emerges from the egg feeds on the insides of their host, avoiding the vital organs. The fly's grub eventually kills its host and emerges from the pupa when it leaves a tell-tale mucous thread from which it made its exit, before pupating itself. However, studies show that the fly is unlikely to be solely responsible for the decline of the Small Tortoiseshell.

Changing weather patterns and climate change are also thought to impact the availability of lush and nitrogen-rich nettles, as well as the synchronisation of flight period with this preferred growth form. In *The Butterflies of Sussex*, Neil Hulme also considers that an earlier emergence may be reducing the proportion of adults that go into hibernation versus those that go on to produce another brood, exacerbating any declines due to a lack of suitable nettles.

The long-term trend shows a moderate decline in distribution and a significant decline in abundance, with the more recent trend over the last decade showing a moderate increase in distribution and a significant increase in abundance, although these statements do not account for the most recent years. Using the IUCN criteria, the Small Tortoiseshell is categorised as Least Concern (LC) in both the UK and Ireland.

LIFE CYCLE

Adults can be seen at any time of the year, even on warm winter days if the temperature is high enough to wake them from hibernation. Adults normally emerge from hibernation at the end of March and start of April and soon pair up, resulting in a first brood that emerges from late May, peaks at the end of June and flies until mid-July. A proportion of the offspring feed up before going into hibernation, while the remainder produce an often-larger second brood that emerges from the second half of July, with most adults entering hibernation by the end of September. In the northern part of its range, however, the butterfly emerges later and is single-brooded.

The switch from a single to a double brood is thought to be triggered by an increasing day length and sufficient temperature, although butterflies in the north of its range remain single-brooded when placed under these conditions, showing that there is some genetic variation across the butterfly's range. Some observers have also come across immature stages late in the year, and I have found third instar larvae in mid-September in southern England, suggesting that these are individuals of a third brood.

Imago

Adults are often seen when they are feeding on whatever nectar sources are available at that time of year, including Betony, brambles, Buddleja, dandelions, Devil's-bit Scabious, Field Scabious, Greater Stitchwort, hawkweeds, heathers, Hemp-agrimony, Ivy, knapweeds, Michaelmas-daisies, Primrose, ragworts, sallows, sedums, thistles, Water Mint, Wild Marjoram, Wild Privet and Wild Thyme.

A courting pair with the ♂ below [NH]

Robin Baker, an expert in this species, found that those butterflies that go on to hibernate spend their time basking, looking for a hibernation site, and feeding on and off throughout the day. Those that will go on to reproduce, on the other hand, concentrate their feeding in the morning, with few continuing past 1.30 pm. Around midday, each male will move off to set up a territory, while females move off to find a suitable area for laying their eggs when the time comes.

The male remains in his territory for about 90 minutes but will then move off to set up another territory, presumably because he has yet to find a mate, and he remains territorial until he goes to roost. Both sexes roost deep in a nettle patch, where their dark undersides provide camouflage against their background. Territories are usually in front of a hedge and close to a nettle patch, where females are likely to fly as they search out suitable spots for when they are ready to lay. Here the male rests on the foodplant or ground with his wings open, waiting for a passing female, and launches himself at any passing object. Unlike the closely related Peacock, the Small Tortoiseshell will tolerate other males in the vicinity.

A mating pair [NH]

An encounter with another male results in the pair flying in tight circles around one another, ascending as they do so, with the original male eventually seeing off his rival and returning to his territory. When a female enters the territory, however, a most curious courtship begins. The male approaches the female from behind and 'drums' his antennae on her hindwings, making a faint sound that is audible to the human ear. The female may fly a little distance, with the male following, when the process repeats. This can go on for several hours with the couple spending a good amount of time basking together and with the male chasing off any intruding male. The female will also drop into a nettle patch and crawl for a short distance, presumably to test the male for suitability and, if she loses him, will subsequently accept the advances of a different male. Eventually, usually in early evening, the female will lead the male into vegetation, often a nettle bed within the territory, and crawl between stems with the male following and, if the male succeeds in staying with her, the pair mate and remain coupled until the following morning.

After the pair separate, Baker found that the female selects a preferred nettle patch in the afternoon and roosts low down within it overnight. She emerges the following morning to bask and make short flights, before laying a batch of eggs, usually between 10 am and 2 pm. The female may lay further batches on subsequent days, with each batch comprising between 60 and 100 eggs that can take over an hour to deposit. Plants that are in full sun and at the edge of the nettle bed are selected, with eggs laid on the underside of one of the larger leaves at the top of young nettle growth. Like many other species that lay their eggs in batches, the female will occasionally lay in the presence of other females, even on the same leaf and occasionally on top of an existing egg batch. The batches themselves are untidy, with eggs laid on top of one another, as in the Peacock.

An egg-laying ♀ [VM]

Several egg batches laid on the same leaf

Eggs are piled on top of one another

Ovum
Each egg is 0.8 mm high and barrel shaped, with nine ribs that run from top to bottom. Eggs are green when first laid but turn yellowish-brown over time, and finally darken to a dull grey just prior to hatching, when the dark head of the larva is visible.

Larva
After between one and three weeks, depending on temperature, each 1.3-mm long **first instar** larva emerges from its egg by eating away the crown. Without eating the remainder of the eggshell, the larva immediately goes about contributing to the creation of a communal web, usually at the top of the nettle, from which larvae emerge to bask and feed. The larva has a shiny black head and a pale brownish-green ground colour with 10 rows of tubercles that run the length of the body, with each tubercle bearing a single black hair. Larvae feed by both day and night and, as they grow, move to new plants, building new webs along the way. This ultimately leaves a trail of webs, decorated with shed larval skins and droppings, that shows the passage of time, and allows the patient observer to trace the larvae all the way back to the plant where the eggs were laid. When fully grown, the first instar is around 3 mm long.

After around nine days, the larvae moult into the **second instar** and spin a new web, as they do after each moult. Even at this young age, the larva starts to develop a colouring that is reminiscent of the final instar. The pale yellow ground colour sports a broken line of brown markings that run along the back, a row of large blotches that run along each side that results in a chequered appearance, and a speckled line that runs in line with the spiracles on each side of the body. The larva is also covered in seven rows of dull green tubercles that each end in a long bristle, with several hairs emanating from the tubercle base. The fully grown second instar is just under 5 mm long.

1st instar: fully grown larvae

2nd instar: the black and yellow markings are clearly visible

3rd instar: the larva now has more significant tubercles [VM]

After five days, the larva moults into the **third instar** and is similar in appearance to the previous instar but has more significant tubercles. The overall colour of the larva is now quite variable—some are similar to the previous instar, whereas others are much darker and have a speckled appearance, with yellow markings only along the base of each side. The fully grown third instar is 8.5 mm long.

After another five days, the larva moults into the **fourth instar** and now has more complex and branching tubercles. The larva continues to exhibit both pale and dark forms (and intermediates), with all but the darkest form possessing two thin yellow stripes along the back, and two along each side (with one above and the other below the spiracles). When fully grown, the fourth instar is 19 mm long. The larvae continue to live gregariously but will split into smaller groups and singletons. Larvae have several techniques to avoid predation. When disturbed, a group of larvae will jerk their bodies from side to side in unison, which must be a formidable sight to any predator. Larvae will also regurgitate green fluid and will, if necessary, curl up in a ball and drop to the ground.

4th instar: larvae typically have yellow stripes along the back and sides [VM]

5th instar: the larva is covered in seven rows of significant tubercles [VM]

After four days, the larva moults into the final and **fifth instar**, when the tubercles are the most substantial of all instars. As in earlier instars, there is much variability in the colour of the larva, from those that are very pale and with prominent yellow stripes, to those that are almost completely black. The mid-abdominal prolegs (the 'claspers') are also now noticeably green, unlike the olive of the previous instar. Behaviourally, the larvae now separate and lead largely separate lives. Some larvae disperse over the patch of nettles and remain in the open, while others create a shelter in which to live by silking together the edges of a leaf, although the resulting construction is quite varied—leaves may be folded upward or downward and may be neatly folded or rather messy and asymmetrical. After five or six days, the larva is fully grown and around 30 mm long. It will wander in search of a pupation site and may remain within its nettle patch or travel several metres before suspending itself head-down from a silk pad attached to a hedge, fence, wall or some other suitable surface, before pupating after as little as 14 hours.

5th instar: a pre-pupation larva

The pupa may exhibit a metallic golden sheen [VM]

The pupa may be a dull brown

Pupa

The 21-mm long and slender pupa is formed head-down, attached by the cremaster. The pupa has two horns on its head, several protuberances on the back of the thorax and two rows of amber-tipped spines along the back of the abdomen with a raised bump between each pair. The colour of the pupa is quite variable, ranging from those that possess a beautiful metallic golden sheen to those that are a dull brown. After between two and four weeks, depending on temperature, this most familiar of butterflies emerges.

Life Cycles of British & Irish Butterflies

Comma
Polygonia c-album

❖ COMMA IS SUPPORTED BY PAUL ATKIN, NICOLA MAIN & CLIVE PRATT ❖

	January	February	March	April	May	June	July	August	September	October	November	December
IMAGO												
OVUM												
LARVA												
PUPA												

The Comma gets its name from the only white marking on its underside, which resembles a comma when looking at the right hindwing. Of course, when looking at the underside of the left hindwing, it more closely resembles the letter C, which is where the specific name of this species, *c-album*, comes from. I much prefer the vernacular name 'Jagged wing'd Comma' that was one of several names provided by James Petiver in his classic 1717 work *Papilionum Britanniae*, since this makes reference to the stunning outline of the wings that, when closed, give the appearance of a withered leaf, making the butterfly inconspicuous when it is resting on a tree trunk or when hibernating. The genus *Polygonia* is taken from the Greek 'poly' (many) and 'gonia' (angles) which also refers to the jagged wing edges found in this and related species. The wing uppersides, however, have a beautiful orange-brown ground colour that is covered in dark blotches and spots, the butterfly superficially resembling a fritillary or a tatty Small Tortoiseshell.

♂ and ♀ are similar in appearance, with a wingspan of between 50 and 64 mm

Both vernacular and specific names come from the only white marking on the underside [PC]

An adult of the pale *hutchinsoni* form

An underside of the pale *hutchinsoni* form

The Comma is known for a particular form named *hutchinsoni* that is much paler in appearance on both upperside and underside than the nominate form. This form is found throughout the butterfly's range and is normally attributed to individuals that go on to produce a second brood in the same year. Its name is a tribute to Emma Hutchinson, a renowned Victorian entomologist who reared this species extensively in captivity (often sharing livestock with other entomologists) and who ultimately discovered its double-brooded nature and the corresponding variation between broods. The name was announced by J.E. Robson in 1881 in *The Young Naturalist*: "*The Summer form is so different, and so constant in its appearance, that it ought to have a distinctive name, and we suggest it be called var.* Hutchinsoni, *in compliment to the lady whose liberality has enriched so many cabinets with specimens; whose knowledge of the species is not exceeded by that of any one living*".

DISTRIBUTION

The Comma is one of the few species that has bucked the trend by expanding its range considerably in the last few decades and it is now a familiar sight throughout England, Wales and the Channel Islands. It also reached the Isle of Man in the 1990s and Scotland in 2000, where it hadn't been seen since 1870. The news from Ireland is equally positive with the butterfly first recorded at Portaferry, County Down, in the late 1990s, and regular sightings also now come from south-east Ireland in Counties Wexford, Carlow and Wicklow, with records of immature stages confirming that it is resident. The butterfly has also been recorded in Counties Kildare, Dublin and Tipperary. On a worldwide basis, this butterfly is widely distributed in Europe, extending through Asia to Japan, and is also found in a few countries in North Africa.

Such a positive outlook is a welcome respite for a butterfly that was once a cause of much concern. This butterfly was formerly widespread over most of England and Wales, and parts of southern Scotland, but by the middle of the 1800s had suffered a severe decline that left it confined to the Welsh border counties, especially Gloucestershire, Herefordshire and Monmouthshire. It is thought that the decline, especially in a former stronghold of Kent, may have been accelerated by a reduction in Hop farming, a key larval foodplant at the time, as well as the ritual burning of Hops after the harvest when they played host to Comma larvae and pupae. However, this does not explain the degree of decline that was experienced and changes in climate and weather patterns may have been the root cause. Happily, since the 1960s, this butterfly has made a spectacular comeback, spreading north at a rate of 20 km per year, with Common Nettle emerging as the primary larval foodplant. Currants, elms and willows are also used, and larvae have also been found on other foodplants, such as Hazel, on occasion.

HABITAT

The Comma is primarily a woodland butterfly, where it can be seen along woodland rides and in clearings and glades. However, especially in late summer as the butterfly builds up its fat reserves before entering hibernation, the butterfly is frequently seen in gardens, parks and churchyards where it feeds on various nectar sources including Buddleja and other flowers, and in orchards where it feeds on rotting fruit such as fallen plums. In other habitats, adults nectar primarily on thistles, brambles, knapweeds, Wild Privet, and Ivy—the latter acting as an important resource in autumn when many other plants have gone over. Adults looking to hibernate eventually search out a suitable overwintering location such as a tree trunk, branch, hollow tree or log pile, where they are almost invisible with their wings closed.

The Comma can be found in woodland rides and clearings at Pamber Forest in Hampshire

STATUS

As already mentioned, the Comma is one of the few species that is thriving, and it is believed that this change in fortunes is linked to climate change as well as the adjustment it has made in preferring Common Nettle over Hop as its primary larval foodplant. Its range has been continually expanding over several decades and this species is not, therefore, a species of conservation concern. The long-term trend shows a significant increase in both distribution and abundance, with the more recent trend over the last decade showing a moderate increase in distribution and a moderate decline in abundance. Using the IUCN criteria, the Comma is categorised as Least Concern (LC) in the UK.

LIFE CYCLE

Like other species that hibernate as an adult, the Comma can be seen at any time of the year and may even put in an appearance on warm winter days. However, the majority of adults emerge from their slumber in March and are usually one of the first species to reappear in the spring when they pair and ultimately give rise to the next generation which appears at the end of June and start of July. A proportion of the resulting adults are of the typical dark form and will ultimately go on to hibernate and survive for around 10 months, but others are of the paler *hutchinsoni* form that go on to produce another brood in the same year (that results in an emergence of fresh adults in late summer) and these adults survive only a few weeks. F.W. Frohawk, in his *Natural History of British Butterflies*, has an elegant way of making the point about the overlapping broods: "*The* Hutchinsoni *examples pair at once and produce the second brood, which starts emerging in August and continues until October; all of the ordinary type hibernate and pair in the spring with the hibernated examples of the second brood. It will be seen that after hibernation copulation take place between uncles, aunts, nephews and nieces*".

So what factors determine if a given individual will produce the dark form of the adult or the pale *hutchinsoni* form? Many theories have been put forward over the years based on factors such as when the egg was laid, temperature and the quality of the foodplant. While all of these may have some impact, this particular nut was cracked by Sören Nylin of Stockholm University who determined in the laboratory that the key factors are photoperiod (the number of hours of daylight) and how that changes over time. It is now known that if the number of daylight hours is sufficient and day length is increasing (that is, before midsummer's day) as the larva develops, then the adult will be of the *hutchinsoni* form that goes on to produce another generation, whereas if the number of daylight hours is small or the day length is decreasing, then the adult will be of the regular dark form that enters hibernation. The assumption is that a good spring allows for an earlier emergence and more rapid larval development and this results in a higher proportion of *hutchinsoni* adults that can then comfortably fit in another brood. Of course, there are always exceptions to the rule and individuals of the dark form may also go on to produce a second brood, and individuals of the pale *hutchinsoni* form may overwinter.

Imago

After emerging from hibernation, both sexes search out whatever nectar sources are available at this time of year, such as sallow catkins or Blackthorn blossom. Adults also spend a good amount of time basking on warm surfaces such as tree trunks, wood piles, dead Bracken and fence posts.

The male butterfly sets up a territory, often on a perch one or two metres up on the sunny side of a woodland margin or ride. Here he will sit waiting for a passing female and will fly out to investigate any passing insect. The male will also make short flights in search of a female—always returning to the same perch if none is found. This territorial characteristic is an advantage to photographers since, even when disturbed, the male will fly off for several metres or so before predictably returning to the same spot. Mating pairs are rarely seen, and it is thought that, in general, they pair high up in the tree canopy, although there are occasional reports of pairs that have been found much lower down, less than a metre off the ground.

A rarely seen mating pair [NH]

When egg-laying, the female makes short fluttering flights over the foodplant, landing every few feet to test it for suitability before she lays a single green egg, each female laying around 250 eggs in total. Eggs are usually laid toward the edge of a leaf upperside, but there are records of females laying on the leaf underside and on stems, especially on Hop. Favoured plants are those that are in sheltered and sunny locations, such as the margins of woods, in woodland glades and rides, or next to a south-facing wall or hedgerow, where eggs can be quite easy to find. Less sunny areas may be used on occasion and I once found several larvae on an elm tree near Stockbridge Down in Hampshire, feeding on the underside of leaves on the shaded north-facing side of the tree.

Ovum

The beautiful almost-spherical egg is just under 1 mm high and has 10 or 11 conspicuous ribs running from top to bottom. Eggs are green when first laid but eventually turn yellow and ultimately grey with the head of the larva appearing under the surface of the crown just before hatching. Those of the first brood are laid in April and May and those of the second brood in July and August.

A Comma egg laid on the upperside of a leaf of Common Nettle

Larva

After two to three weeks, depending on temperature, the 2-mm long **first instar** larva emerges from the egg and, without eating its eggshell, moves to the underside of a leaf where it spins a silk pad and starts to feed, leaving holes in the centre of the leaf. As the larva grows it moves to feed on the leaf edges. The newly emerged larva is pale green and has several longitudinal rows of hairs that each end in a dark bulbous base. The larva soon develops a banded appearance with the 2nd, 3rd, 5th, 7th and 9th segments appearing lighter. The larva ultimately develops a brown ground colour and just prior to moulting, when the larva is just over 3 mm long, the lighter markings are quite conspicuous, especially along the back of the larva, and hint at the white markings found on the second instar. Larvae may also be found in small groups that are scattered over a small area and these are almost certainly the offspring of a single female that has chosen to lay her eggs in close proximity.

After just over a week, the larva moults into the **second instar**. It continues to feed on the underside of the leaf but is now covered in seven rows of spines that run the length of the body, except for the first segment. All of the spines are black with several exceptions—the dorsal and sub-dorsal spines of the 5th, 7th and 9th segments, and the sub-dorsal spines of the 2nd, 3rd and anal segments, are ivory white. When fully grown, the second instar is approximately 6 mm long.

1st instar: a newly emerged larva

1st instar: a fully fed larva with pale blotches appearing on its surface

2nd instar: a larva showing the distinctive white spines on the 5th, 7th and 9th segments

3rd instar: a larva showing the two-tone colouring between front and rear segments

After a week or so, the larva moults into the **third instar**, and now has an appearance that remains relatively consistent until the larva becomes fully grown—the 1st segment has no spines, the 2nd to 5th segments have amber-yellow spines, and the 6th segment through to the anal segment have white spines, and the spines on the 2nd, 3rd, 5th, 7th and 9th segments are particularly long. The 6th to last segment are also white along the back, resulting in an overall two-tone colouring that is quite distinctive, giving the appearance of a bird dropping, making it easy to identify this as a Comma larva when found in the wild. The larva now feeds increasingly on the upperside of a leaf and, when fully grown, the third instar is 11 mm long. There is some variation between individuals and some are much paler than others, for example, although the resulting adults are normal.

4th instar: the larva now has spines that are of an equal height

4th instar: individuals can vary in appearance and this larva has white rather than amber-yellow spines [VM]

4th instar: some late-instar larvae can be quite pale [VM]

After around two weeks, the larva moults into the **fourth instar**. The overall appearance is similar to that of the third instar, although the spines along the back of the larva are now equal in height. All other features are so variable from individual to individual that it is difficult to make generalisations that always hold true in the field. When fully grown, the fourth instar is 16 mm long.

5th instar: the final instar has distinctive bold markings

5th instar: the larva hangs head-down in a J-shape prior to pupation [VM]

After a week or so, the larva moults for the last time into the **fifth instar**. The larva now has a quite spectacular appearance and is unmistakable, with much bolder and continuous markings than previous instars and a clear distinction between the orange-brown of the 2nd to 5th segments and the white of the 6th to anal segments. The sides are also adorned with orange-brown markings that match the colour found at the front of the larva. Prior to pupation, when it is around 34 mm long, the larva spins a dense silk pad low down either on the foodplant, on surrounding vegetation or on some other suitable platform from which it hangs head-down, forming a J-shape.

Pupa

The pupa is suspended head-down, attached by the cremaster, and is just over 20 mm long. It varies in colour from a quite beautiful mix of greens and browns, through to a more uniform shade of brown. It has, however, a very curious shape, with a distinctive hump on the back of the thorax, giving an overall impression of a withered leaf. The pupa also has a small number of subtle silver spots on its back that give the impression of water droplets. After around two weeks, the adult butterfly emerges to brighten up woodlands and gardens throughout its range.

The pupa is often a subtle mix of browns and greens

A pupa that is of a more uniform brown

The pupa has several silver spots on its back [VM]

A pupa just before the adult emerged [VM]

Life Cycles of British & Irish Butterflies

Marsh Fritillary
Euphydryas aurinia

❖ MARSH FRITILLARY IS SUPPORTED BY MIKE GIBBONS, THE GIBBS FAMILY & THE MAKIN FAMILY ❖

The **Marsh Fritillary** has the most colourful upperside of all of our fritillaries, with a highly variable chequered pattern of orange, brown and yellow markings. The bright colours fade after a few days when the butterfly appears to have a sheen on its wings as a result of lost scales, causing Moses Harris in his *The Aurelian* of 1766 to name this butterfly 'The Dishclout, or Greasey Fritillaria'. The name 'Marsh Fritillary' was provided by William Lewin in his *The Papilios of Great Britain* of 1795 in recognition of one of the butterfly's key habitats of damp grassland. James Rennie in his *A Conspectus of the Butterflies and Moths found in Britain* of 1832 gave the butterfly the name 'The Scabious' in recognition of its primary larval foodplant, Devil's-bit Scabious, although this name never took on.

The undersides of the two sexes are similar [IL]

The Marsh Fritillary is one of our most threatened species and has seen a dramatic decline over several decades across Europe and, even though this decline has been felt in Britain and Ireland, our shores are now considered one of the few strongholds of this butterfly.

The ♂ has a wingspan of 30 to 42 mm
The ♀ is larger than the ♂ with a wingspan of 40 to 50 mm

DISTRIBUTION

The butterfly is found in the south-west of England, the islands of south-west Scotland and the adjacent mainland, the north-west and south-west of Wales and in scattered colonies across Ireland. It is not found in the Isle of Man or Channel Islands. On a worldwide basis, the Marsh Fritillary is found across Europe and temperate Asia as far as Korea, and in parts of North Africa.

The butterfly experienced significant declines in Cumbria throughout the second half of the 20th century and, in 2004, it was decided to bring the last 155 known wild larvae into captivity, which were thought to be the offspring of a single female. A pure Cumbrian line and a hybrid line (using 95 larvae from the nearest sites in western Scotland) were reared in captivity. Unfortunately, the pure Cumbrian line failed after two generations, with no fertile eggs laid, although the rearing of the hybrid line allowed 42,400 larvae to be released in 2007 at four locations in north and west Cumbria, with reintroductions succeeding at three locations.

This butterfly forms discrete colonies and even the slightest barrier, such as a hedge or a river, will hamper dispersal and studies have shown that most adults fly no more than 100 m, with subsequent broods breeding in the same grassland areas year after year. However, some individuals are occasionally found several kilometres from the nearest colony, especially late in the flight period. Irrespective of this limited dispersal, each colony is normally part of a metapopulation—a network of colonies where there is some degree of interchange. On an even grander scale, however, metapopulations themselves are often geographically separated, giving rise to several local races that differ in appearance in this very variable butterfly and that have, in some taxonomic listings, resulted in named forms and even subspecies. The population found in Ireland, especially, is thought to have a greater contrast between the orange ground colour and cream markings and has been given the name *hibernica* by some authors in recognition of its unique appearance.

The abundance of this butterfly fluctuates significantly from year to year in any given region—it may be present in large numbers one year when there tends to be more dispersal, before the population crashes in a following year, with the butterfly's range contracting to its core sites, before it recovers again just as unexpectedly. Three factors have been put forward to explain these 'boom and bust' cycles, which may operate in combination.

The first factor is related to the weather, where poor conditions during the flight period delays emergence, resulting in more predation of pupae, the inability of the butterfly to find a mate and, in the case of the female, difficulty in quickly finding a suitable Devil's-bit Scabious plant on which to lay a batch of eggs. Conversely, good weather allows the immature stages to develop more quickly, resulting in less predation, and provides the preferred conditions for adults to produce the next generation.

The second factor is the implications of a 'boom' year on the availability of the larval foodplant. In 2011 I experienced one of the downsides of a 'boom' year when I visited Strawberry Banks, the last remaining site for this species in Gloucestershire. A mass emergence of adult butterflies had resulted in such large numbers of larvae that they had outstripped their supply of foodplant. Every leaf of every Devil's-bit Scabious plant had been nibbled down to its base, resulting in some larvae either switching to less-desirable Honeysuckle as a food source or wandering away from the site in search of sustenance. Any butterflies that did make it through to adulthood were then faced with a paucity of suitable foodplants on which eggs could be laid and any resulting larvae potentially starved before they could make it to their fourth instar when they go into hibernation. Suffice to say, numbers crashed the following year.

These 'boom' years were known to early lepidopterists. In his *A catalogue of the Lepidoptera of Ireland* of 1901, W.F. de Vismes Kane writes: *"This butterfly has been known to increase so prodigiously that whole fields and roads became blackened by the moving myriads of larvae. An instance of this was observed by the Rev. S.L. Brakey, near Ennis, Co. Clare, where he drove out to see a reported "shower of worms", and found the larvae so multitudinous in some fields that the black layer of insects seemed to roll in corrugations as the migrating hosts swarmed over each other in search of food"*.

A dead final instar larva with a newly emerged *Cotesia bignelli* wasp

The third factor that can have a bearing on the fortunes of this butterfly is parasitism. In Britain and Ireland, the Marsh Fritillary is parasitised by both *Cotesia melitaearum* (which also parasitises the Glanville Fritillary) and *Cotesia bignelli* (whose sole host is the Marsh Fritillary). These wasps lay their eggs inside a larva, with the resulting grubs eating away its insides, leaving the vital organs until last, ultimately killing their host before breaking through the skin of the larva and forming cocoons within a mass of silk. The wasps can also fit in three generations for every Marsh Fritillary generation, thereby magnifying their impact on the butterfly, with some suggesting that small colonies are prone to extinction as a result of the levels of parasitism.

Given the three factors mentioned above, then it is easy to appreciate how small colonies, especially, can go extinct. A sequence of years with poor weather, a decline in numbers due to an outstripped supply of larval foodplant in the previous 'boom' year and an abundance of parasitic wasps can all sound the final death knell for a colony of Marsh Fritillary. It is somewhat ironic that the loss of the butterfly from a site immediately removes two of the factors that may have caused its demise, since the foodplant will recover and the parasitic wasps will go extinct without their host.

But all is not lost and the importance of metapopulations in any recovery is spelled out in Caroline Bulman's 2001 PhD thesis on 'Conservation biology of the Marsh Fritillary butterfly *Euphydryas aurinia*'. In essence, the Marsh Fritillary requires a network of colonies—a metapopulation—where any local extinctions can be reversed by recolonisations. The implication is that the Marsh Fritillary requires suitable habitat over a large area in order to maintain the network that is so important to the butterfly's survival.

HABITAT

The butterfly uses two main habitat types in Britain and Ireland. The first is damp neutral or acid grassland (including woodland clearings) that is used throughout most of the butterfly's range. These sites are usually dominated by grass tussocks with open areas used for breeding, although these may be sheltered by scattered scrub or a woodland edge. Examples include the Culm grasslands of Devon, the Rhôs pastures of South Wales and Dartmoor, and the damp grasslands in Argyllshire. The butterfly is also found on dry calcareous grassland that is found mainly in central southern England where south- and west-facing slopes are typically used, where the higher temperature aids larval development. An example is the chalk downlands of Dorset and Wiltshire, including Salisbury Plain. Whatever the habitat type, management focuses on producing a mix of short and long vegetation, using appropriate regimes such as grazing.

Damp grassland at North Bull Island in Dublin, Ireland

Chalk grassland at Salisbury Plain in Wiltshire

STATUS

The Marsh Fritillary has suffered severe declines over several decades, especially in eastern England and eastern Scotland. Although widespread in south-west England and Wales, Martin Warren estimated in 1994 that the butterfly was declining at a rate of 11.5% per decade. This species has therefore been a priority for conservation efforts for some time. Three reasons have been suggested for the declining fortunes of this species. The first is the draining of damp neutral and acid grassland for agriculture, and the loss of dry calcareous grassland to agriculture and development. The second is inappropriate habitat management at existing sites, especially over-grazing or under-grazing that results in unsuitable larval foodplants and vegetation structures. The third is that fragmentation of remaining habitats results in smaller and more isolated colonies, which leads to local extinctions, infrequent recolonisation events and a breakdown in the proper functioning of the metapopulation.

The long-term trend shows a significant decline in distribution and a stable abundance, with the more recent trend over the last decade showing a moderate decline in distribution and a significant decline in abundance. Using the IUCN criteria, the Marsh Fritillary is categorised as Vulnerable (VU) in both the UK and Ireland.

LIFE CYCLE

The Marsh Fritillary is single-brooded, flying between the start of May until the start of July, with a peak at the end of May and start of June. The butterfly emerges later further north, at the end of May, and flies until the middle of July. Males emerge several days before females at all sites.

Imago

Both sexes are avid nectar feeders and will use a variety of flowers, favourites including Betony, Bugle, buttercups, Cuckooflower, dandelions, hawkweeds, knapweeds, Ragged-Robin, thistles and Tormentil. When roosting overnight and during inclement weather the butterfly will remain deep within a grass tussock.

Males set up small territories centred on a particular plant or flower, from which they dart up to investigate any butterfly passing nearby. They will also patrol suitable areas, in the hope of finding a newly emerged female crawling up a plant stem after her wings have dried. Once a female is found, the male lands next to her and flutters around her before mating takes place, the pair remaining together for a couple of hours. Before separating, the male seals the genital opening in the female with a substance that prevents another male from mating with her and which, to all intents and purposes, acts as a 'chastity belt'.

The female will search out large foodplants when egg-laying, typically choosing one of the larger leaves on which to lay. She is quite conspicuous as she makes her slow flight, looking for suitable plants on which to lay, weighed down by her load of eggs. While the primary larval foodplant is Devil's-bit Scabious, the butterfly is also known to use Field Scabious and Small Scabious on calcareous grassland.

A mating pair with the ♀ on the left [PH] | Two ♀s laying eggs on the same leaf of Devil's-bit Scabious

In 2013, I studied the Marsh Fritillary population on North Bull Island in Dublin and was astonished to find two females laying on the same leaf, with another laying on an adjacent plant, in a large area that is approximately the size of a football pitch. The chances of the same Devil's-bit Scabious leaf being used among the thousands of plants available would appear to be extremely small and this co-location cannot be put down to coincidence since others have witnessed the same phenomenon. The most likely explanation is that a given plant may be in a favoured position for the site, possibly based on a local microclimate. Like most butterflies, a female Marsh Fritillary will select a surprisingly small subset of available foodplants when egg-laying, choosing those growing in optimum conditions for their larvae. A deliberate strategy of co-locating egg batches was also considered plausible on the basis that the gregarious larvae of the Marsh Fritillary can thermoregulate more efficiently when numbers are higher, when they raise their temperature to around 35°C which aids digestion. More convincingly, their prickly bodies are a greater deterrent to birds, mice and other predators, especially when the larvae jerk from side to side in unison and *en masse* when disturbed, when they also make a sound as their bodies brush against any dead leaves.

A return trip to North Bull Island later that August lent weight to the theory of deliberate co-location in order to provide safety in numbers, since the two egg batches I was following had resulted in a single and relatively large larval web. During a field trip by the Dublin Naturalists' Field Club on 17 August, an estimated 200 larval webs were found. Many of these webs, together with the trail of previously consumed foodplant and associated webbing, resulted in enormous structures which, in some cases, exceeded two metres in length. These structures were clearly the work of larvae whose associated egg batches were laid in close proximity.

Ovum

Females emerge with between 300 and 400 mature eggs that are laid in a single batch on the underside of a leaf of the foodplant, the whole process taking several hours. This normally happens within a day of mating and may even happen on the day of emergence. Having laid the initial batch of eggs, additional eggs develop in the female as she feeds, and these are subsequently laid in smaller batches that vary in size from 50 to 200 eggs. Eggs are laid in neat formations of three or four layers and are pale yellow when first laid, turning dark red and then leaden grey just before hatching, when the head of each larva can be seen quite clearly through its eggshell. Each egg is around 0.8 mm high and has around 20 ribs that run from the crown to halfway down the egg, where they then branch and eventually disappear before reaching the base.

A newly laid batch of eggs on the underside of a Devil's-bit Scabious leaf

Eggs are pale yellow when first laid

Eggs turn dark red as they mature

The heads of the larvae can be seen through the eggshell prior to hatching

Larva

After four or five weeks, the 1.2-mm long **first instar** larvae hatch from their eggs *en masse* and immediately start creating a communal web in which they live, feed and moult by binding together two leaves of the foodplant. As well as the obvious protection afforded by the web, it is also thought that the web helps maintain a humid environment that prevents larvae from desiccating. The first sign of feeding is brown patches on the leaf surface that result from the lower epidermis being eaten. The larva has a pale yellowish-red ground colour and six rows of long white hairs that run the length of its body. It is fully grown after around 12 days when it is just over 3 mm long, having developed a cream ground colour with several reddish-brown markings along its body, together with a series of white spots that encircle every segment apart from the first, which sports a dark band.

1st instar: larvae hatching from their eggs

1st instar: a larval web under construction

1st instar: fully grown larvae have reddish-brown markings and white spots

After two or three weeks, the larva moults into the **second instar** and is now covered in seven rows of conical spines that run the length of its body, with each spine bearing a number of off-white bristles. The larva has a pale yellowish-brown ground colour and is covered in dark brown markings, with a pale stripe that runs along each side in line with the spiracles, although the colour of the larva is somewhat variable between webs, with much darker markings present in some larvae. After moulting, the second instar larvae immediately build a new and more substantial web that is now spun over the foodplant, and larvae can sometimes be found wandering over or basking on the

2nd instar: the larvae continue to live within a dense silk web

2nd instar: the larvae are now covered in seven rows of spines

3rd instar: the larva has longer spines with two that project away from the anal segment

surface of the web, before they return to its protection by crawling through one of the small tunnels that lead to its interior. When fully grown, the second instar is just under 6 mm long.

After another two or three weeks, the larvae moult into the **third instar**. The larva is similar in appearance to the previous instar, although its spines are much longer with yellowish rather than off-white bristles, and the pale stripe that runs along each side is more prominent. In this instar the spines on the anal segment are particularly noticeable, looking like two horns projecting from the back of the larva. As in the previous instar, the colour of the larvae from web to web is quite variable with some much darker than others. In this instar the larval web is more substantial still and is relatively easy to spot on the foodplant. Larvae continue to feed under the web for the most part but can be found feeding at the edge of the web when more mature, and larvae can also be found basking on the surface of the web on sunny days. When fully grown, the third instar is just under 8 mm long.

After just over two weeks, the larvae build a new web within which they undergo their moult into the **fourth instar** and, without feeding, immediately go about constructing a dense web low down in vegetation in which they overwinter. The larvae are now jet black, rather than the brown of earlier instars, and are covered in numerous black spines. The larva also has a dull speckled stripe along each side in line with the spiracles and some dull speckling on its back.

Larvae emerge from their web in early February and will bask in the sun, retreating back into the hibernaculum when the sun disappears. The larvae eventually split into smaller groups and will bask openly together, allowing them to keep their body temperature relatively high, even on cool days, and thermal images of these groups show, quite clearly, the difference in temperature between these groups and the surrounding area. The resulting groups of larvae vary in size with the largest containing around 150 individuals, and they are extremely easy to find—there have been numerous occasions where, standing on the same spot at a good site, I could make out over 20 larval groups at some distance. The larvae do not recommence feeding immediately after hibernation and it may be over a week before they start to nibble on new growth of the foodplant.

4th instar: a dense web of overwintering larvae

4th instar: four groups of black larvae

4th instar: a fully grown larva

In the second half of March and early April, the fully grown 11-mm long fourth instar larvae moult into the **fifth instar**, either in a curled up and dead leaf, or in a specially constructed web. The fifth instar is similar in appearance to the previous instar, although the speckling along each side and on its back is much more prominent, and its underside is noticeably reddish-brown. A distinguishing factor, however, is its behaviour—larvae now live in much smaller groups of up to 20 individuals and become increasingly solitary. They also no longer rest on silk webbing but continue to bask in the sun, often on a dead leaf or other dark surface. When fully grown, the fifth instar is just over 15 mm long.

5th instar: a small group of larvae

5th instar: the larva starts to lead a more solitary existence

After a couple of weeks, the larva moults into the final and **sixth instar**. It is similar in appearance to the previous instar and, apart from being larger with relatively smaller spines, now leads a wholly solitary existence. Larvae can often be found wandering rapidly across open ground some way from their egg site looking for their next meal and, if there is a shortage of foodplant, the larva is known to feed on alternative food sources, such as Honeysuckle growing in hedgerows. The larvae will also fall from the plant and roll up into a ring when disturbed, remaining tightly closed for a minute or so. When preparing to pupate the larva finds a suitable site low down in vegetation where it creates a silk pad from which it suspends itself head-down in a J-shape. When fully grown, the larva is between 26 and 30 mm long.

6th instar: a larva feeding on small leaves of Devil's-bit Scabious

6th instar: a larva preparing to pupate

The pupa is quite beautiful

A coloured-up pupa one hour before the adult emerged

Pupa

The 14-mm long pupa is attached to the silk pad by the cremaster and is possibly one of the most beautiful pupae of all our species. The pupa is white, with a striking mix of black, brown and orange markings. Depending on temperature, the adult butterfly emerges in around three weeks.

Glanville Fritillary
Melitaea cinxia

❖ GLANVILLE FRITILLARY IS SUPPORTED BY MIKE GIBBONS, S JOHNSON & BEN RICHARDSON ❖

	January	February	March	April	May	June	July	August	September	October	November	December
IMAGO												
OVUM												
LARVA												
PUPA												

Although formerly more widespread, the Glanville Fritillary is now one of our most local species, where it is confined to the Isle of Wight and can be found on the south coast of the island and along its central chalk ridge. In his *Gazophylacium naturae et artis* of 1702, James Petiver hints at a more extensive distribution when he gave this butterfly the name of 'Lincolnshire Fritillary' and says: *"First observed there, and given me by Madam Glanvile"*. The lady in question is Eleanor Glanville, a 17th-century Lepidopterist who originally discovered this species. After her death in 1709, James Dutfield renamed this butterfly to 'The Glanvil Fretillary' in her honour in his unfinished work *New and Complete Natural History of English Moths and Butterflies* of 1748.

The ♂ has a wingspan of 38 to 46 mm

The ♀ has a wingspan of 44 to 52 mm [JA]

Following Eleanor's death, one of her sons contested her will on the grounds of lunacy, as described by Moses Harris in this often-cited passage from *The Aurelian* of 1766 (where he uses the name 'Glanvil Fritillaria'): *"This Fly took its Name from the ingenious Lady Glanvil, whose Memory had like to have suffered for her Curiosity. Some Relations that was disappointed by her Will, attempted to let it aside by Acts of Lunacy, for they suggested that none but those who were deprived of their Senses, would go in Pursuit of Butterflies"*.

♂ and ♀ undersides are similar [IL]

DISTRIBUTION

This butterfly was formerly found in many colonies in south-east England and with records as far north as Lincolnshire but, by the mid-19th century, it was confined to the Isle of Wight and the Kent coast, with the Kent colonies going extinct in the 1890s. It is now only found in any reasonable numbers at around a dozen sites on the south coast of the Isle of Wight, with its strongholds in the south-west. A number of colonies have appeared on the mainland in recent times, although all are known to have been unauthorised introductions. This butterfly is also found on Guernsey and Alderney in the Channel Islands and, on a worldwide basis, the butterfly has a Palaearctic distribution, occurring across most of Europe and into Asia, and in north-west Africa. While the butterfly requires the relatively warm south-facing coast and chalk ridge of the Isle of Wight in Britain, it is able to colonise inland areas on the continent, where it can be found flying in meadows and woodland glades up to 2,000 m. Surprisingly, given its distribution, the butterfly is not at the northern limit of its range in Britain—in some parts of Scandinavia it flies at the same latitude as Shetland, thanks to warm summer temperatures.

This butterfly forms discrete colonies that fluctuate wildly in numbers, although there is some interchange between them, with a small percentage of adults dispersing up to 500 m between chines and along undercliffs, with individuals occasionally turning up several miles from any known colony.

HABITAT

The Glanville Fritillary is found in warm and sheltered south-facing sites where regular disturbance of the ground allows Ribwort Plantain, the primary larval foodplant, to grow and flourish. Away from the colonies on the island's central chalk ridge, coastal colonies occupy three different zonal habitats. The first zone is the undercliff, such as that found at Compton Bay, where frequent slippage provides the butterfly with ideal habitat and allows the butterfly to colonise new areas as the existing habitat becomes overgrown. The second zone is a 'chine' (a steep-sided river valley that flows to the sea), such as Shepherd's Chine, where slippage at its coastal end encourages growth of the larval foodplant. The third zone is a cliff top, which is usually an extension to an undercliff or chine, where the butterfly is able to use relatively overgrown areas that are no doubt becoming suitably warm due to climate change.

The undercliff at Compton Bay

The coastal river valley at Shepherd's Chine

STATUS

Like many of our butterfly species, the Glanville Fritillary has suffered a long-term decline, although it shows a modest increase in distribution in the short term. The long-term trend shows a significant decline in distribution and a moderate decline in abundance, with the more recent trend over the last decade showing a moderate increase in distribution and a significant decline in abundance. Using the IUCN criteria, the Glanville Fritillary is categorised as Endangered (EN) in the UK.

A lack of suitable foodplant is often cited as the primary reason for the periods of 'boom and bust' that this butterfly seems to experience. One cause of this is that sites become too stable, resulting in a reduction of the disturbance needed to encourage growth of the larval foodplants. Another cause is that, in good years, large numbers of larvae decimate the available foodplant and, if the larvae do not starve, then the resulting females do not have sufficient foodplant on which to lay their eggs, and the following year is correspondingly poor.

LIFE CYCLE

Adults typically emerge throughout May and reach a peak at the end of the month and at the start of June. There is a single generation each year on our shores with no second brood, although an additional brood is the norm in southern Europe, which flies in August.

Imago

At its best sites, the Glanville Fritillary is one of the most conspicuous butterflies as it flies from flower to flower, especially those of Thrift, Common Bird's-foot-trefoil, dandelions and hawkweeds. The butterfly can also be surprisingly difficult to follow, especially if it is battling the onshore breeze that is often present at its coastal locations. The butterfly is best sought during bright sunshine when the adults are most active since, during cloudy and inclement weather, they will remain settled with their wings closed and can be difficult to find. The adults roost, often communally, on flowerheads of plantains, Thrift and grasses.

Males are the more active of the two sexes and can be found patrolling the breeding grounds, where they intercept any other Glanville Fritillary and investigate any brown object in the vegetation, in the hope of finding a virgin female. When successful in his quest, the male immediately mates with the female without any discernible courtship, the pair remaining coupled for several hours. The female is definitely the boss in this relationship—even while mating, the female may continue to fly between flowers to feed, with the male hanging on by his claspers.

The female takes great care when choosing an egg-laying site which is not surprising since, while she may lay several batches of eggs (often because she is disturbed when laying), she literally 'puts all of her eggs in one basket' on occasion. Egg batches typically contain between 50 and 200 eggs and are laid untidily on the underside of a leaf of the foodplant and towards the leaf tip. Favoured sites

A mating pair with ♀ at top

A ♀ laying a batch of eggs

are sheltered and warm with plenty of bare ground, and where relatively young and small-leaved foodplants grow vigorously, often at the base of an undercliff or in a sheltered hollow in a chine. There are typically very few suitable plants at any given site and it is not unusual to find more than one egg batch on a single plantain leaf, indicating that more than one female has visited this spot. While Ribwort Plantain is the primary larval foodplant, Buck's-horn Plantain is used occasionally as a secondary foodplant by more mature larvae. In the spring of 2018 at Compton Bay I found a web of larvae that were using both plants simultaneously, even though there was no shortage of Ribwort Plantain. Larvae have also been found feeding on Bristly Oxtongue, Sea Beet and Wild Teasel on occasion (Andy Butler, pers. comm.).

Ovum

Eggs are almost spherical with a flattened base and 20 or so ribs that run regularly from the top to halfway down the egg, before branching towards the base. They are almost white when first laid, turning a light primrose-yellow as they mature, and then turning slightly paler just before hatching, when the black head capsule of the larva can be seen through the shell.

Given the number of eggs that are laid by a single female (F.W. Frohawk records up to 300), then it is no surprise that the eggs are very small for the size of butterfly, being only 0.5 mm high. A study by Enrique García-Barros that was documented in 'Egg size in butterflies (Lepidoptera: Papilionoidea and Hesperiidae): a summary of data', published in *Journal of Research on the Lepidoptera* in 2000, shows that the ratio of the volume of the egg relative to the wingspan of the Glanville Fritillary is only 0.46%, which is one of the smallest ratios of all of the butterflies of Britain and Ireland (the largest ratio is that of the Scotch Argus, which is an incredible 3.1%, and the smallest ratio is that of the Meadow Brown which is a mere 0.31%).

A mass of eggs laid on the underside of a leaf of Ribwort Plantain

The eggs turn a light primrose-yellow a day or so after being laid

The black head capsules of the larvae can be seen just before the eggs hatch

Larva

After between two and three weeks, the **first instar** larvae emerge from the egg batch *en masse* and immediately start to build a web, usually on the underside of a leaf and close to the egg site, in which they live and feed. Given the size of the egg it is no surprise that the larvae are correspondingly small, being only 1.25 mm long. The larva has a pale yellow ground colour, a black head and a number of hairs scattered over its body. The fully grown first instar is 2 mm long and has a very different appearance, with a light brown ground colour and a covering of white spots that give a hint of the tubercles found on the second instar.

1st instar: the newly emerged larvae immediately get down to the task of building a web

1st instar: the black-headed larvae have a pale yellow ground colour

1st instar: the fully grown larva has a light brown ground colour and is covered in white spots

After around two weeks, the larva moults into the **second instar** and larvae continue to feed and rest within a dense web spun on the foodplant. The larva now has a pale brown ground colour and is covered in rows of white tubercles, each of which bears several white hairs that look quite wiry when seen close up. When fully grown, the larva is 3.5 mm long, has a much darker ground colour, and the placement of the white tubercles now gives the larva a chequered appearance.

2nd instar: a newly moulted larva

2nd instar: the fully grown larva has a chequered appearance

3rd instar: the larva now has a black ground colour and is covered in pale olive tubercles

After between two and three weeks, the larva moults into the **third instar** and continues to live within a dense web. The larva now has a black ground colour that is speckled with white flecks, and it continues to possess several rows of pale olive tubercles, each of which is adorned with whitish bristles. There is some variation in the colour of the tubercles and, on several occasions, I have found third and fourth instar larvae that are completely black. The fully grown third instar is 5 mm long.

After around a week, the larva moults into the **fourth instar**. The tubercles are now olive-brown, giving the newly moulted larva a brown appearance. As the larvae feed and grow, the tubercles are placed further apart, and the larva appears much darker as the dark ground colour of its body is exposed. When fully grown, the fourth instar is around 6.5 mm long.

4th instar: larvae continue to live within a dense web

4th instar: the olive-brown tubercles give newly moulted larvae a brown appearance

4th instar: more mature larvae appear much darker

After between one and two weeks, the larva moults into the **fifth instar**. The larva now looks like a miniature version of the final instar larva, with a rust-red head, brown claspers, a black ground colour that is covered in white specks, and numerous spines. The white specks are found between segments and, in his *The Butterflies of Britain and Ireland*, Jeremy Thomas provides a theory for these that emerged from his studies of this species with David Simcox: "*Groups of caterpillars regulate their temperatures at 33–34°C with remarkable precision, and take care not to overheat. If this maximum is exceeded, they retreat into shade or extend their bodies to expose silvery connecting sections between each segment, which we believe reflect rather than absorb the sun's rays*".

After feeding for a short time, the larvae construct a dense ball of silk that acts as their hibernaculum over the winter. It is formed low down within grasses and other vegetation, which provide a level of camouflage for this otherwise highly conspicuous construction. The largest

5th instar: the hibernaculum is formed low down within grasses and other vegetation

5th instar: at Wheeler's Bay the hibernacula are able to withstand waves crashing over them [AB]

hibernacula that I have found are about the same size as a tennis ball, taking the surrounding vegetation into account, with numerous larvae packed closely together within the centre.

Larvae emerge in the spring when they sun themselves in the vicinity of the hibernaculum, before sprawling over the vegetation, putting down silk webbing as they go. The larvae soon commence feeding on plantains, but only feed in earnest during periods of sun, remaining motionless in dull weather. Larvae are conspicuous since they spend a good proportion of their time basking in groups on top of these webs so that their black bodies absorb the sun's rays, resulting in an elevated body temperature that aids digestion. In *The Butterflies of Britain and Ireland*, Jeremy Thomas quantifies the importance of communal basking: "*Once the temperature in their niches exceeds a threshold of 13°C, they can warm their bodies by a further 20°C through huddling together in a black mass that absorbs heat from the sun*". That being said, the larvae do appear to be somewhat hardy, with an ability to withstand heavy rain, harsh frosts and even snow (Andy Butler, pers. comm.). When fully grown, the fifth instar is around 9.5 mm long.

5th instar: larvae emerge from their hibernacula in the spring and sprawl over nearby vegetation

5th instar: a fully grown larva

Depending on the weather, larvae will start to moult into the **sixth instar** in the second half of March, although this is also very much dependent on the site and even the larval web. Larvae in the same web may also exhibit different rates of development and it is not unusual to find webs containing a mix of fifth and sixth instar larvae of varying sizes. Fifth and sixth instar larvae are almost identical in appearance, apart from their size, so seeing them together is sometimes of benefit if you want to be sure of the instar being observed. When fully grown, the sixth instar is just under 16 mm long.

After a couple of weeks, the larva moults into the final and **seventh instar**. The larva is almost identical to the sixth instar but can be distinguished by its size—when fully grown, the final instar is around 25 mm long. A much better distinction is behaviour—a mature final instar larva is no longer a wholly gregarious creature, and can found wandering over the ground as it seeks out fresh growth of plantain. Anyone who has visited the favoured Isle of Wight chines in the second half of April will almost certainly have come across final instar larvae wandering over footpaths and have to mind their step.

Larvae seek out a suitable location to pupate from late April, which may be within a grass tussock, on old leaves of the foodplant, within some other vegetation, or under stones. Here the larva spins a loose web and suspends itself upside-down from a silk pad before ultimately pupating, the larval stage lasting around 10 months in total. Pupae are occasionally found in the company of others at the best sites. An interesting study in Finland showed that post-hibernation larvae that were under-developed (based on their body mass) would occasionally produce an extra instar, resulting in an incredible eight instars in total.

Pupa

The shiny 14-mm long pupa is formed head-down, attached by the cremaster, and varies in colour from light brown to dark grey, with a series of orange spots running down its back. The adult butterfly emerges after around three weeks, when it can sometimes be found expanding its wings at the base of vegetation before it takes flight.

6th instar: larvae continue to exhibit gregarious behaviour

6th instar: the larva is almost identical to the 5th instar, apart from its size

7th instar: the larvae are now more solitary in their behaviour

A pupa attached to the underside of a stone [AB]

The pupa is often formed within loose webbing

Life Cycles of British & Irish Butterflies

Heath Fritillary
Melitaea athalia

❖ HEATH FRITILLARY IS SUPPORTED BY DR DAN DANAHAR, ALFRED GAY & MIGUEL SANJURJO GARCÍA ❖

	January	February	March	April	May	June	July	August	September	October	November	December
IMAGO												
OVUM												
LARVA												
PUPA												

The Heath Fritillary is one of our rarest butterflies and was on the brink of extinction in the late 1970s. The extinction of the Large Blue in 1979 resulted in a concerted effort to save the Heath Fritillary which was heading the same way—in a 1981 conservation report to the Joint Committee for the Conservation of British Insects (JCCBI), Martin Warren, Chris Thomas and Jeremy Thomas wrote: "*It is unlikely that the Heath Fritillary will survive in the UK for many more years unless deliberate efforts are made to conserve it*". Subsequent research by Martin Warren between 1980 and 1985 identified two characteristics that allowed appropriate habitat management plans to be put in place—adults have limited mobility, and the butterfly requires habitat that is at an early stage of succession where abundant larval foodplants grow in relatively warm ground temperatures.

The butterfly suffered greatly at woodland sites where the cessation of coppicing resulted in the loss of new and sunny clearings that the butterfly needs to thrive, and it is now confined to a few sites where coppicing is still practiced in the Blean Woods complex in Kent, and at reintroduction sites in Essex. Here the butterfly lives up to its moniker of the 'woodman's follower' as it occupies newly coppiced areas as they are created. The butterfly also maintains a foothold in a few unimproved grassland sites in Cornwall and Devon, and in heathland combes on Exmoor in west Somerset where the butterfly moves into regenerating areas of heathland that have recently been burned. On the continent the butterfly is less fussy and, in some areas with unimproved hay meadows, is not only the commonest fritillary, but the commonest butterfly.

The ♀ has a wingspan of between 42 and 47 mm [IL]

The ♂ has a wingspan of between 39 and 44 mm [IL]

The name we use today was provided by Benjamin Wilkes in his *One Hundred and Twenty Copper-Plates of English Moths and Butterflies* of 1749, and was presumably based on his assumption that the larvae fed on a species of heath (from the *Erica* genus that includes Bell Heather), although this is not a known foodplant and this assertion was most likely made in error: "*I found the Caterpillars of this Fly feeding on common Heath in Tottenham-Wood, about the Middle of May, 1745. Six or seven of them were feeding near each other, I observed their Manner of eating,*

♂ and ♀ have similar undersides [IL]

which was extremely quick, and when they moved it was at a great Rate. I fed them on common Heath for three or four Days; at the End of which some of them changed into the Chrysalis, in which State they remained about fourteen Days, and then the Flies came forth".

Male and female are similar in appearance, although the colour and pattern of the wings is somewhat variable in both sexes. Some names given to this butterfly are based on appearance, including the name 'Pearl Border Likeness' given by Moses Harris in his *The Aurelian* of 1766 that is a reference to its similarity to the Pearl-bordered Fritillary.

DISTRIBUTION

From west to east, this butterfly is confined to a predominantly grassland habitat at two sites in the Tamar valley of Cornwall and Devon and at one site in the Lydford valley on the edge of Dartmoor in Devon; heathland combes on Exmoor in west Somerset; and woodland sites in Essex and the Blean Woods complex in Kent. It is absent from Wales, Scotland, Ireland, the Isle of Man and the Channel Islands although, historically, the butterfly was known from Cornwall, Devon, Gloucestershire, Surrey, Sussex, Essex and Kent. Edwin Birchall also reported that it was abundant in Killarney in Ireland in 1866. On a worldwide basis, the butterfly is found across Europe and through temperate Asia as far as Japan.

The butterfly forms discrete colonies in breeding habitat that is usually less than 0.5 ha (70 m × 70 m) in area and rarely strays far—even a strip of unsuitable vegetation can act as a barrier to dispersal, constraining any intermingling of colonies. Conversely, open rides between connecting habitat allow the butterfly to find new suitable patches should its current breeding grounds become overgrown, although this sedentary species needs these new areas to be within 300 m of the current colony if they are to be reached. It can therefore be seen just how fragile this butterfly is in our increasingly fragmented landscape. Although very local in its distribution, this butterfly can be seen in large numbers at some sites in good years, and abundance is usually proportional to the availability of suitable habitat, with changes in woodland acting more rapidly than changes made in heathland and grassland habitats.

HABITAT

Irrespective of the habitat type, there are some characteristics that hold true for all of them. The butterfly is dependent on early successional habitat and new areas must support larval foodplants growing in a relatively open and hot environment. The butterfly is lost when woodland clearings are shaded out, heathland becomes overgrown or scrubbed over, or grasslands are completely abandoned. Numbers peak two or three years after a new area has been created, but rapidly tail off thereafter.

Deciduous woodland sites in Kent and Essex are managed through coppicing (often of Sweet Chestnut or Hornbeam), the felling of groups of trees and ride management, which provide a succession of new clearings and interconnected rides that the butterfly moves into in as little as three years, as its current habitat becomes overgrown. At these sites the larvae feed on Common Cow-wheat that grows abundantly in sparse vegetation. This plant is semi-parasitic on certain deciduous trees and grasses and will appear following winter clearance work.

The yellow flowers of Common Cow-wheat showing at Blean Woods in Kent [BC]

A view from Grabbist Hill, near Minehead, on Exmoor [CB]

A view overlooking the Tamar Valley at Greenscombe Wood [CB]

Heathland combes (valleys) on Exmoor in west Somerset are in an intermediate zone between upland and lowland heathland at 200 to 400 m above sea level, with vegetation that is dominated by Bilberry, heathers and Bracken. These sites are managed through grazing by cattle and ponies, and also burning. Bilberry is a known host for Common Cow-wheat, although Foxglove is used as a secondary foodplant by larvae after hibernation.

Unimproved herb-rich grassland sites in the Tamar valley (that defines the border between Cornwall to the west and Devon to the east), and those in the Lydford valley in Devon, have traditionally been managed through rotational cutting every two to five years, although the habitat management has been amended more recently to include coppicing as these sites have become more wooded. The primary foodplant here has been Ribwort Plantain that grows in short vegetation on stony soils, although Common Cow-wheat has become more prevalent and is also used, with Germander Speedwell and Foxglove occasionally used by post-hibernation larvae.

STATUS

While the butterfly can be abundant in some years, as was the case in 2018, this remains one of our most endangered species and continues to be a priority for conservation efforts. The long-term trend shows a significant decline in both distribution and abundance, with the more recent trend over the last decade showing a moderate decline in distribution and a significant decline in abundance. Using the IUCN criteria, the Heath Fritillary is categorised as Endangered (EN) in the UK.

LIFE CYCLE

At sites in the West Country, this butterfly emerges in mid-May, peaks in early June, and flies until early July. The butterfly emerges at the end of May elsewhere, peaks in mid-June, and flies until late July. A second brood (which is the norm on the continent) may emerge in August at south-eastern sites in some years and, while this brood is normally considered 'partial', some observers noted a 'virtually complete second brood' at Hockley Woods in south-east Essex in 1999.

Imago

Adults feed primarily on brambles, although Bugle, buttercups, heathers, knapweeds and Tormentil are also used. The butterfly is somewhat gregarious and, when basking on shrubs, both sexes can often be found in the company of others. The male is the more conspicuous of the two sexes and spends much of his time patrolling close to the ground in search of a newly emerged female, when he alternates a few flicks of his wings with a short glide. The egg-laden female, on the other hand, has a much more laboured flight.

When a female is found, then the male will flutter around her and, if she is receptive, mating follows. After a few days of waiting for her eggs to ripen, the female will make slow flights just above the ground in search of suitable plants, which are those growing in sunny open areas that provide the warm microclimate required by the immature stages. After landing, she will crawl among the vegetation before laying an untidy batch of up to 150 eggs on a surface close to the foodplant, such as the underside of a bramble leaf or on dead vegetation. Each female will lay up to 600 eggs in total.

A group of basking ♂s [IL]

A mating pair with ♂ below [BE]

Ovum

The 0.5-mm high, almost spherical, egg has around 26 ribs that run from top to bottom. It is pale green when first laid, but soon turns pale yellow, with the head of the larva showing clearly through the shell at the crown just prior to hatching.

Batches may contain up to 150 eggs

Eggs turn pale yellow a few days after being laid

The head of the larva shows through the shell just prior to hatching

Larva

Eggs hatch after between two and three weeks and the newly emerged 1.3-mm long **first instar** larvae eat their eggshells before moving together to the foodplant, where they spin a loose silk web on which they bask and under which they feed on the leaf cuticle. The larva has a shiny brown-black head with a ground colour that is initially pale yellowish-brown but turns greenish-grey as food is ingested. The body is covered in 10 rows of long orange-brown hairs that run the length of the body and each hair emanates from a dark base. The fully grown first instar is 2.5 mm long and develops pale patches just prior to moulting that are a hint of the tubercles of the second instar.

After two weeks, the larva moults into the **second instar** and is now in possession of a shiny black head and seven rows of translucent whitish tubercles, each bearing a mix of dark and pale bristly hairs, along the length of the body. The newly moulted larva is relatively pale, but it soon develops a purplish-brown ground colour. When fully grown, the second instar is 4 mm long, and each row of tubercles exhibits a particular hue (thanks to a shade of brown that develops at the base of each tubercle), giving the larva a striped appearance.

After 10 days, the larva moults into the **third instar** and now exhibits the two-tone colouring that characterises the Heath Fritillary larva, thanks to the amber bases of the tubercles found on each side just below the back. The larva has a purplish-brown ground colour and is covered in numerous grey speckles, the largest of which run along the base of each side where they blend to give the appearance of a whitish stripe. The larva, which has been highly gregarious up until this point, has an increasing tendency to wander away from its siblings as it matures, and this is most noticeable in this instar when smaller groups form. When fully grown, the third instar is 5.5 mm long.

After two weeks, the larva moults into the **fourth instar** and is similar in appearance to the previous instar, although all of the markings are more intense. Toward the end of August, the larva will find a place to overwinter, occasionally in the company of others, such as within a dead rolled up

4th instar: a pre-hibernation larva

4th instar: the hibernating larva has a dense covering of bristles

leaf, where it spins an amount of silk webbing that draws the leaf together. The pre-hibernation larva can attain a length of 6 mm, but the larva shrinks considerably when hibernating to a length just over 3 mm long, when the proximity of the tubercles results in a dense covering of bristles.

The fourth moult into the **fifth instar** takes place within the hibernaculum just before the larvae reappear in March and April. They then recommence feeding and can be seen basking in full sun in order to raise their body temperature, which aids digestion. The larva is similar in appearance to the previous instar although the tubercles are larger and the row of smaller tubercles that run along the middle of the back now take on the same amber colouring as the larger tubercles just below them on each side. Larvae feed intermittently and only when the weather is warm, and it takes several weeks for them to attain their full length of 6.5 mm.

5th instar: newly moulted larvae post-hibernation

5th instar: a fully grown larva

Around the end of April, the larva moults into the **sixth instar** and now has significant grey speckles against its purplish-black ground colour. The amber tubercles are also more significant and are now more conical in shape. The fully grown sixth instar is 11 mm long and almost twice the length of the previous instar.

After two weeks, the larva moults for the last time into the **seventh instar** and is similar in appearance to the previous instar, but has more prominent markings and tubercles. In his *An Illustrated Natural History of British Butterflies* of 1871, Edward Newman writes: "*I kept my specimens on a plant of the narrow-leaved plantain, and covered with a bell-glass; in the middle of the day I always found they*

6th instar: the conical tubercles are much more significant

crawled up the flowering-stems of the plantain, and I was particularly struck with the resemblance of the caterpillars to the flowers of this plant, a resemblance which perhaps serves as a protection against the birds, which at this period of the year are constantly on the look-out for caterpillars wherewith to feed their young". The larva feeds rapidly and will spend long periods basking in order to raise its body temperature so that it can digest its meal, when it can be quite easy to find at good locations. After two weeks, the fully grown larva is between 22 and 25 mm long and prepares for pupation by creating a silk pad under a dead leaf or other surface among leaf litter where it hangs upside-down in a J-shape.

7th instar: the colours are now particularly vibrant

7th instar: a pre-pupation larva

Pupa

The 13-mm long pupa is quite beautiful, with a pearl-white ground colour and numerous striations all over its surface. Curiously, there are amber-based conical protrusions along its back that correspond to the amber tubercles of the larva. The wings of the adult are clearly visible through the pupal shell just prior to emergence and, after two or three weeks in this stage, this most fragile of our butterflies emerges. In his detailed study 'The ecology and conservation of the heath fritillary butterfly, *Mellicta athalia*', published in 1987 in the *Journal of Applied Ecology*, Martin Warren shows that pupae are particularly susceptible to predation by small mammals, which accounted for around 40% of losses in one colony in Cornwall.

A pupa formed under a dead leaf

A pupa just hours before the adult emerged

Life Cycles of British & Irish Butterflies

Riodinidae

The **Riodinidae** is primarily a South American group, where just under 1,350 of the 1,400 or so species in this family are to be found. There is, however, some representation in tropical Asia and Africa and a single species, the Duke of Burgundy, in Europe. The striking metallic colouring of many of the species in this family has resulted in the family name of 'metalmarks' which, unfortunately, does not apply to our sole member. All species in this family are small or medium in size when compared with other families.

Family	Subfamily	Genus	Species	Vernacular name
Riodinidae	Riodininae	*Hamearis*	*lucina*	Duke of Burgundy

Duke of Burgundy ♂ with vestigial forelegs

Duke of Burgundy ♀ with fully formed forelegs

IMAGO

The family is closely related to the Lycaenidae (the coppers, hairstreaks, blues and arguses) and some taxonomic classifications include the Riodininae as a subfamily of the Lycaenidae. However, the Riodinidae are distinguishable in one key respect—the forelegs are greatly reduced and useless for walking in the male, but are fully formed with claws in the female, whereas, in the Lycaenidae, both male and female have six fully functional legs. Another difference is that, unlike many Lycaenids, the male does not possess any androconial scales on his wings. In all species of the Riodinidae, the eye is surrounded by a narrow band of white scales.

The white scales around the eye of a ♀ Duke of Burgundy

OVUM
The relatively small eggs vary in shape and may be spherical or flattened (as in many species of the Lycaenidae) and may be laid singly or in small groups.

LARVA
Larvae are superficially similar to those of the Lycaenidae but usually have more hairs covering their body. Also, while our sole representative of this family does not possess any ant-attracting (or ant-appeasing) devices, such as a honey gland and tentacle organs, many other members of this family do, as in many Lycaenidae. Given the close association with ants that has developed in many species in this family, it is not surprising that there has been much debate over whether the Riodinidae should be accepted as a family in its own right, or treated as a subfamily of the Lycaenidae, with which it shares so many similarities. However, in 'Evolutionary and ecological patterns in myrmecophilous riodinid butterflies', published in *Ant-Plant Interactions* in 1991, P.J. DeVries shows that the location of these devices in the Riodinidae differs from those found in the Lycaenidae, suggesting that the organs in these two families have a different evolutionary origin, and that the status of the Riodinidae as a separate family is justified.

Duke of Burgundy eggs

A final instar Duke of Burgundy larva

PUPA
Pupae are stout, often hairy, and are typically attached to vegetation by a silk girdle around the waist and by the cremaster to a silk pad.

Duke of Burgundy pupa

Life Cycles of British & Irish Butterflies

Duke of Burgundy
Hamearis lucina

❖ DUKE OF BURGUNDY IS SUPPORTED BY NEIL HULME, THE MAKIN FAMILY & MARK TUTTON ❖

	January	February	March	April	May	June	July	August	September	October	November	December
IMAGO												
OVUM												
LARVA												
PUPA												

The diminutive **Duke of Burgundy** is a delightful butterfly that has become much scarcer over several decades although, thanks to the work of several dedicated organisations and individuals, its recent outlook looks more favourable. Some authors referred to this butterfly as the 'Duke of Burgundy Fritillary' since it superficially resembles this group, although it is not a true fritillary. The species does in fact stand apart from all other species and, like the Swallowtail, is the sole representative of its family in Britain and Ireland—in this case, the Riodinidae.

The two sexes are similar in appearance, although the female has more orange on her wings, which are more rounded. As discussed in the Riodinidae family summary, another difference between the sexes is that the forelegs are greatly reduced and useless for walking in the male, but are fully formed, with claws, in the female. A more visible difference is behaviour—the highly territorial male has been described as pugnacious, perky, violent, fierce and a brute, whereas the female is simply described as elusive.

♂ and ♀ (shown, on Cowslip) have similar undersides

A ♂ with a wingspan of 29 to 32 mm, perching on a Primrose

The ♀ is slightly larger, with a wingspan of 31 to 34 mm

DISTRIBUTION

This butterfly is found mainly in central southern England, although scattered colonies are found in the north of England in Cumbria and Yorkshire. This species is not found in Wales, Scotland, Ireland, the Isle of Man or the Channel Islands. Although relatively large colonies of more than 1,000 individuals exist, most contain fewer than a few hundred individuals over the flight period, with a maximum of around a dozen individuals seen on any given day. Some colonies are even smaller than this and seem to be limited by the availability of suitable larval foodplant that is growing in the right conditions. On a worldwide basis, this butterfly is found in the western Palearctic, from Sweden in the north to Spain in the south, and eastwards to the Balkans. It is also found in parts of temperate Asia.

HABITAT

The Duke of Burgundy is known to colonise two very different types of habitat. Historically, this is a butterfly of deciduous and coniferous woodland that has been coppiced or that has extensive rides with large open areas, and where Primrose growing in partial shade is used as the larval foodplant. With the decline of appropriate woodland management, the butterfly has largely shifted to lightly grazed chalk and limestone grassland that contains plenty of shelter, such as chalk pits with surrounding scrub, and where Cowslip is used as the larval foodplant. Scrub edges may also support the butterfly in other habitat types, such as road verges or railway embankments. False Oxlip, the hybrid of Cowslip and Primrose, is occasionally used at some sites.

In all habitats the butterfly requires larval foodplants growing among other vegetation such as grass tussocks or scrub. At woodland sites the ideal habitat seems to develop two to three years after a coppicing cycle, after which the canopy starts to close in, preventing the lush growth of foodplant that this butterfly needs. On calcareous grassland sites the butterfly appears to prefer north- or west-facing slopes which are generally more humid and that encourage the lush growth of foodplant required, rather than the south-facing slopes that are more prone to drought and desiccation of the foodplant.

A woodland site in Hampshire where Primrose is used as the larval foodplant [TB]

Plants that meet these specific requirements are normally a small proportion of those available and knowing which plants to examine pays dividends when trying to locate the immature stages. A case in point is at Noar Hill in Hampshire, a site that puts on one of the finest displays of Cowslips you're ever likely to see, where the plants are dotted all over the site and often in very open and sun-drenched areas. However, you need to find those plants growing in the scrubby edges if you are to stand any chance of finding eggs or larvae.

Duke of Burgundy can be found in the chalk pits at Noar Hill in Hampshire

STATUS

With a decline in appropriate management at woodland sites, such as the cessation of coppicing that creates the partial shade required by the butterfly, many colonies have been lost as their habitat has become overgrown. According to Matthew Oates, a leading expert on this species, the decline of woodland colonies is estimated at 98% between 1950 and 1990. Consequently, most colonies are now found on calcareous grassland sites and these have their own challenges. Based on data from Noar Hill, Oates showed that there is a direct correlation between rabbit numbers, turf height and the size of the butterfly's population. In essence, whatever the habitat type, any management needs to maintain a delicate balance between a sward that is too short and one that is too long so that the foodplants grow in conditions that are optimal for the development of the larva.

The long-term trend shows a significant decline in distribution and a moderate decline in abundance. Thankfully, the butterfly's fortunes have taken a turn for the better with a more recent trend over the last decade showing a stable distribution and a significant increase in abundance. Using the IUCN criteria, the Duke of Burgundy is categorised as endangered (EN) in the UK.

One of the best examples of conservation in action comes from Sussex where many organisations, including the Sussex branch of Butterfly Conservation, have collaborated to turn the fortunes of this butterfly around. In 2003 there were only eight records of the Duke from the entire county whereas total counts now regularly exceed 1,000 each year. The implementation of appropriate habitat management here, based on a deep understanding of this butterfly's needs, has really paid dividends.

LIFE CYCLE

The Duke of Burgundy has one brood each year, with the adults emerging in late April at southern sites and peaking in the middle of May. The butterfly emerges slightly later in the middle of May in northern England, with a peak at the end of the month. A partial second brood may appear in some years, but this is the exception rather than the rule and only occurs at certain sites in southern England, although this is the norm in southern Europe and will perhaps become a more frequent occurrence with climate change.

A territorial ♂ on a favourite perch [MC]

Imago

While the two sexes can be told apart based on their appearance, it is the differences in behaviour that really stand out. Unlike the female, the fast-flying male is highly territorial and will sit on a favourite perch, usually in a sheltered area that catches the morning sun, and will dart out to inspect anything that might be a passing female. Even when disturbed, the male soon returns to his territory, and often the same perch. Favourite areas are reused year after year, possibly by more than one male at a time if the area is large enough, and this makes the males much easier to find when revisiting a site.

Charles Barrett describes this behaviour in his *The Lepidoptera of the British Islands* of 1893, where he also indicates how widespread this species was in woodland: "*A very pretty little butterfly, widely distributed, occurring only in woods, and loving sheltered sunny spots at the bottom of hollows and valleys, and broad, sheltered wood paths. Here it flits about over the lower underwood, constantly returning to the same spot, and settling on the same favourite bush, or even on the same twig. So strong is the liking of the insect for some particular resting-place, such as a little oak sapling, three feet high, in a flowery corner, that when one specimen is captured another will take its place in a few hours, and this may be continued day after day*".

When a male encounters another, then there is a spiralling flight where the pair often disappear from view as they head upwards, although both soon return to ground level and their respective perches. This pattern is often repeated by males with adjoining territories, although there is no evidence of a male being displaced from his territory after such a skirmish.

Most females emerge mid-morning and this is the best time to find a mating pair, although any courtship is minimal. I can still recall my first experience of such an encounter during a visit to Noar Hill in Hampshire, one of the strongholds of this species, where I was watching a female drying her wings after emerging that morning. As soon as she took to the air, a male pursued her to a nearby bush and, by the time I caught up with them, they were already mated. Mating pairs are typically found on a bush and less than a metre from the ground. There are also several records of a female mating more than once—this is based on sightings of very worn females that are paired, and the assumption that all females mate for the first time soon after emerging.

The female does not fly as rapidly as the male, especially when moving from plant to plant as she looks for a suitable plant on which to lay. She will eventually land on the edge of a

A mating pair with the ♀ above [JB]

leaf, curl her abdomen and lay on the leaf underside. Unlike males, who tend to establish territories and stay there, females are also much more mobile, and in one study over half of the females were recorded 250 metres from their original location, indicating that they disperse from the male territories after they have mated. There are also records of females travelling much further—an example is a private site in Hampshire that was colonised, presumably from the nearest known colony, which was 5 km away.

Adults rarely nectar but, when they do, will typically feed from the flowers of buttercups, hawthorns, Tormentil and Wood Spurge, especially in warmer weather. In late afternoon, both sexes roost in tall scrub or trees, with males going to roost before the females.

Ovum

The 0.6-mm diameter spherical eggs are usually laid singly or in small batches of two to five eggs, although there is one record of 23 eggs laid in a single batch in the wild (Tim Bernhard, pers. comm.). Eggs are laid on the underside of a leaf of the foodplant, slightly away from the edge. Large, lush, green-leaved plants are typically used that, as mentioned earlier, grow among other vegetation such as grass tussocks and scrub, and snails are known to cause heavy losses of eggs since they feed on *Primula* leaves during the spring. Eggs are light creamy yellow and give the illusion of being translucent thanks to their shiny surface. As they mature the head and long hairs of the first instar larva become visible through the eggshell, creating a quite beautiful criss-cross pattern when viewed close up.

A batch of four eggs laid on the underside of a Cowslip leaf

The head and long hairs of the larva become visible prior to hatching

Larva

After one to three weeks, depending on the weather, the 1.6-mm long **first instar** larva hatches from the egg by eating away a circular hole in the crown. It may then eat more of the eggshell before moving toward the midrib of the plant, where it rests by day, feeding only at night on the underside of the leaf. The larval feeding damage is quite distinctive and especially noticeable in later instars, when the leaves of the foodplant become peppered with holes while the midrib and main veins are left intact, ultimately leaving just a skeleton of the leaf. This pattern is very different to damage from snails, which feed not only on the leaf edges but also on the main veins.

1st instar: a larva emerging from its egg

Once emerged from the egg, the long black hairs of the first instar that were seen through the eggshell are now on parade. These hairs emerge from prominent black bases and are organised in two rows that run along the back of the larva. The colour of the larva changes as it grows—it initially starts as a creamy yellow, but eventually develops into one of two colour forms—either a greyish-green or a light yellow-brown. The larva also has a distinctive orange head that remains this colour through all larval instars. When fully grown, the first instar is 3.5 mm long.

1st instar: a greyish-green larva
1st instar: a pre-moult light yellow-brown larva
1st instar: larval feeding damage is limited to small holes in the leaves [TB]

After a week or so, the larva moults into the **second instar**. The larva now has a more consistent translucent grey-green ground colour, a more prominent dark broken line running down its back, and the hairs on its back are much shorter in relation to the length of the body. If close-up photos are available, then another difference is that each hair on the back of a first instar shares a root with another hair, whereas those on the second instar larva do not. Larvae emerge at dusk when they can be found feeding, usually on the upperside of the leaves. When fully grown, the second instar is around 7 mm long.

2nd instar: the larva is now a consistent grey-green in colour

3rd instar: the hairy larva is now a dark olive-green

4th instar: the larva is now a rich brown

After around 10 days, the larva moults into the **third instar**. While its distinctive orange head remains, along with the dark broken line along its back, the larva is now a different beast than earlier instars. It is now a dark olive-green in colour with lighter blotches on its sides and is much hairier. It spends daylight hours resting on the underside of the leaf, tucked away at the base of the plant, feeding at night, usually on the upperside of a leaf. When fully grown, the third instar is around 10 mm long.

After two weeks or so, the larva moults into the final and **fourth instar**. The larva is now a rich brown in colour, with paler blotches along its sides and, like the previous instar, is very hairy. The fully grown fourth instar is around 16 mm long, with the larval stage lasting seven weeks in total. According to Oates, larvae are also known to wander several metres from plant to plant, and *Primula* seedlings hidden beneath the sward are an important food source for wandering larvae.

4th instar: a pair of larvae feeding on a *Primula* leaf

Pupa

The larva pupates after a couple of weeks, although their location is somewhat of a mystery since very few have been found in the wild. In captivity, the pupa is always formed in a shaded spot and out of direct sunlight, sometimes on the foodplant and sometimes on other vegetation, attached by a silk girdle and by the cremaster to a silk pad. Various authors have suggested that the pupa is usually formed away from the foodplant in leaf litter or a grass tussock.

A pupa formed within a dead leaf

A coloured-up pupa nearing the time for emergence

The speckled pupa is 11 mm long, a very pale straw in colour and is very hairy when compared with the pupae of other species. The butterfly overwinters in this stage and emerges the following spring.

Life Cycles of British & Irish Butterflies

Lycaenidae

The **Lycaenidae**—the hairstreaks, coppers, blues and arguses—is a family of 4,500 or so species, the majority of which are small in size and rapid in flight. The species found in Britain and Ireland are grouped into three subfamilies—the Lycaeninae (the coppers), the Theclinae (the hairstreaks) and the Polyommatinae (the blues and arguses).

Family	Subfamily	Genus	Species	Vernacular name
Lycaenidae	Lycaeninae	Lycaena	phlaeas	Small Copper
	Theclinae	Thecla	betulae	Brown Hairstreak
		Favonia	quercus	Purple Hairstreak
		Callophrys	rubi	Green Hairstreak
		Satyrium	w-album	White-letter Hairstreak
			pruni	Black Hairstreak
	Polyommatinae	Cupido	minimus	Small Blue
		Celastrina	argiolus	Holly Blue
		Phengaris	arion	Large Blue
		Plebejus	argus	Silver-studded Blue
		Aricia	agestis	Brown Argus
			artaxerxes	Northern Brown Argus
		Polyommatus	icarus	Common Blue
			bellargus	Adonis Blue
			coridon	Chalk Hill Blue

Many species of butterfly have a close association with ants—a phenomenon known as myrmecophily —and this association is most developed in this family. In particular, the larvae of many species produce secretions that provide a food source for ants and, in return, the larvae receive a level of protection from potential predators. When looking for the larvae of Lycaenids, it is sometimes easier to first look for the ants as they scurry about the larvae they are tending.

IMAGO

Many species are brilliantly coloured, and this is often the result of interference effects as light hits the microstructure of the wing scales. Also, as in the Riodinidae, the eyes of the adult are often bordered with dense white scales. All legs are fully functional in both sexes, although the forelegs of the male are slightly less well developed and smaller than those of the female. All legs have a pair of claws, with the exception of the male foreleg that has a single claw.

The undersides of the wings typically differ significantly from the uppersides, although this can be difficult to determine in those species that always land with their wings closed (the Black, White-letter and Green Hairstreaks). Sexual dimorphism

Chalk Hill Blue ♂

Chalk Hill Blue ♀

varies from very subtle, such as in the Green Hairstreak, to blindingly obvious, such as in many of the blues of the Polyommatinae subfamily, where the males are blue and the females brown. Androconial scales are usually present on the uppersides of the male forewings but can be hard to discern.

OVUM

Eggs are typically bun-shaped and covered in a sculptured network of ribs and hollows, giving the appearance of a miniature sea urchin. The egg of the White-letter Hairstreak, however, is the exception to this rule, with some observers making comparisons with a comic book flying saucer.

A Large Blue egg

A White-letter Hairstreak egg

LARVA

A Holly Blue larva scooping out the contents of an Ivy flower bud

The larva is typically shaped like a woodlouse and is covered in short hairs. It has a retractable neck that, in some species, is pushed forward as the larva scoops out the content of a seed pod or leaf on which it is feeding. In early instars, the larva may even enter a seed pod or developing leaf bud to feed.

Lycaenid larvae may have several devices that attract or appease ants. The first device is a number of small, single-celled glands scattered over the larva's skin, known as pore cupolas, which exude secretions. The most visible device, however, is often a pair of retractable 'tentacle' organs on the 11th segment of the larva. These organs take the form of long cylindrical tubes that can be everted (turned inside out) and are topped with a ring of spikes when extended. The third device is a honey gland that is found in the middle of the 10th segment of the larva. This gland exudes a sweet substance that is packed with sugars and amino acids and that is attractive to ants.

A final instar Silver-studded Blue larva with tentacle organs extended at rear

A final instar Adonis Blue larva attended by red ants [CK]

A final instar Small Blue larva attended by a black ant [IC]

PUPA

Pupae are stout and, in some cases, such as in the Black Hairstreak and the White-letter Hairstreak, are attached to the foodplant or other vegetation by a rudimentary silk girdle and the cremaster.

Pupae may equally be found lying free among vegetation or even below ground, as is the case in the Silver-studded Blue and Purple Hairstreak. Both larvae and pupae are known to produce sounds on occasion, which is also thought to be a mechanism to attract or appease ants.

A Black Hairstreak pupa attached to a Blackthorn leaf

The pupa of the Common Blue is formed just below the soil surface in a lightly silked chamber

Small Copper
Lycaena phlaeas

❖ SMALL COPPER IS SUPPORTED BY CHRIS DRIVER, PAUL TATNER & JEAN & ANDY YOUNG ❖

There are only a few butterflies in Britain and Ireland to which the word 'gem' truly applies, and the Small Copper is surely such a species. The metallic copper sheen of the forewing uppersides is quite beautiful, especially in fresh specimens that have yet to lose any scales. This is a fast-flying and widespread butterfly and a familiar and welcome sight for many naturalists throughout the summer months.

In his *Musei Petiveriani* of 1695, James Petiver must have had a brainstorm of adjectives floating around in his head when he named this butterfly 'The small golden black-spotted Meadow Butterfly', which is a bit of a mouthful. I'm not sure if his subsequent naming was an attempt to be more succinct, but he did use the name 'Small Tortoise Shell' in his *Papilionum Britanniae* in 1717 although this is, of course, the name we use for a completely different butterfly today. It is certainly not as blunt as the name 'Copper' used by Moses Harris in 1766 in *The Aurelian*, a name that was extended to 'Small Copper' by James Lewin in his *The Papilios of Great Britain* of 1795, presumably to distinguish it from the now-extinct Large Copper *Lycaena dispar*.

The ♂ has a wingspan of between 26 and 36 mm

The ♀ is slightly larger than the ♂ with a wingspan of between 30 and 40 mm

The two sexes have similar undersides

We have two subspecies in Britain and Ireland—the subspecies *eleus* is found in England, Wales and Scotland and the subspecies *hibernica* is found in Ireland. The differences between these two subspecies are subtle to say the least, where *hibernica* supposedly has a broader copper band on the hindwing upperside, a greyer ground colour on the underside and a more conspicuous and brighter red marginal band on the hindwing underside.

DISTRIBUTION

This butterfly occurs in small discrete colonies and, even during the peak of the flight period, only a few adults will be found on the wing at any one time. It can be observed throughout Britain and Ireland but is absent from mountainous areas and the far north-west of Scotland, the Outer Hebrides and Shetland. On a worldwide basis, this butterfly is found throughout the Holarctic, where the butterfly is distributed throughout Europe, Asia as far as Japan, North Africa and the eastern parts of North America.

Adults may be seen some way from their breeding sites, suggesting that this butterfly is relatively mobile and has an ability to colonise new areas when the opportunity arises. In *The Butterflies of Britain and Ireland*, Jeremy Thomas notes that "One is even recorded to have reached the Royal Sovereign light vessel, more than 12 km off the Sussex coast".

HABITAT

The Small Copper is found in a wide variety of habitats, including woodland clearings, rides and glades; mixed scrub; lowland and upland heathland; calcareous and acid grassland; disused quarries; sand dunes; coastal vegetated shingle; undercliffs; arable field margins and set aside; gardens, parks, churchyards; and disused railway lines. The common factors at all of these sites is that they are relatively open, nectar sources and larval foodplants are present, and they are generally warm

Small Coppers are found in this sheltered and sorrel-rich field margin in Thatcham, Berkshire

and dry. This warmth-loving butterfly suffers during extensive periods that are cool and wet, with predictably negative consequences for those colonies that already inhabit shaded sites, such as those found in woodland rides and glades.

STATUS

It would seem that the status of this butterfly is somewhat stable, although a series of poor years could have a severe impact on its fortunes. The long-term trend shows a moderate decline in both distribution and abundance, with the more recent trend over the last decade showing a stable distribution and a moderate decline in abundance. Using the IUCN criteria, the Small Copper is categorised as Least Concern (LC) in both the UK and Ireland.

LIFE CYCLE

The number of broods varies based on location and weather conditions. In good years the butterfly will typically go through two or three broods in the southern part of its range, with the first adults emerging in early May, occasionally at the end of April, and the last adults seen around the middle of October. In exceptional years there may be a fourth brood in southern England. Two broods are the norm in northern England and Scotland, where the emergence is slightly later, with adults emerging in the second half of May in the most northern part of the butterfly's range.

Imago

The Small Copper is a sun-loving butterfly and can be found basking on flowers, vegetation and even bare earth where it soaks up the sun, and both sexes roost head-down on grass stems. Adults are avid nectar feeders and will use a variety of flowers including Common Fleabane (a particular favourite), buttercups, Daisy, dandelions, hawkweeds, heathers, ragworts, Red Clover, thistles and Yarrow.

Males establish small territories and are the most conspicuous of the two sexes, since they fly up to intercept any passing insect in the hope of finding a virgin female. If another male is

A mating pair with ♀ on the right [NH]

encountered, then the pair enter into what can only be described as a high-speed dogfight before they return to their respective territories. If a mated female is intercepted, then she will land on the ground and rapidly beat her wings, while simultaneously crawling away from her suitor, who eventually loses interest and flies off. However, an encounter with a virgin female results in a rapid courtship that was described by H.J. Henriksen in *The Butterflies of Scandinavia in Nature* in 1982: *"Mating flight takes place at 1-2 m. height, in a rapid, zig-zag flight, which is quicker than the human eye, and with ♂ and ♀ no more than a few cm. from one another. They rest momentarily on a leaf, stone or the gravel between flights and then quickly repeating their agitated ritual. They then settle down in the vegetation, the ♀ crawling deeper with the ♂ in pursuit and mating takes place directly on the ground"*. Male and female remain coupled for around three hours.

Two larvae were found on this lush patch of Common Sorrel in May 2018

Three eggs were found on this regenerating growth of Common Sorrel in September 2017

Egg-laying females are easy to spot since they fly low across the ground searching out suitable foodplants on which to lay. The female will typically land on the ground and then crawl onto the foodplant, tapping the vegetation with her antennae as she goes, ultimately laying an egg before flying off and repeating the process. Egg-laying is typically undertaken in the afternoon and only in bright conditions— as soon as the sun is obscured by cloud the female will rest with wings closed. The primary larval foodplants are Common Sorrel and Sheep's Sorrel, although Broad-leaved Dock is also used when suitable sorrels are in short supply.

Relatively large plants growing among grasses are used in the spring, but later broods choose smaller plants growing in more open positions where the vegetation is sparse. In *The Butterflies of Britain and Ireland*, Jeremy Thomas suggests that this may be because the female will always select those plants that have the highest concentration of nitrogen and nutrients, which are lush tall plants in the spring, and the regenerating growth found on seemingly unsuitable plants during the summer months.

Ovum

Eggs are laid singly toward the base of the upperside of the leaf but may also be laid on the underside and I have even seen females lay on sorrel flowerheads on occasion, but this seems to be the exception. As is often the case, a small subset of available foodplant is chosen by egg-laying females and it is not uncommon to find several eggs on the same plant.

The egg, measuring 0.6 mm wide by 0.3 mm high, is bright white when first laid and is particularly conspicuous on the upperside of a green sorrel leaf where it easy to find. The egg has a curious shape and always reminds me of a deflated football since it not only has the appearance of a compressed sphere, but also an intricate pattern on its surface that resembles an irregular honeycomb with a mix of shapes, including triangles, squares, pentagons and hexagons.

The egg resembles a deflated football with an irregular honeycomb surface

Larva

After a week or two, depending on temperature, the 1-mm long and yellow-brown **first instar** larva emerges from the egg by eating away the crown and immediately moves to the underside of the leaf, close to the midrib, where it starts to feed without breaking through the upper epidermis. This results

1st instar: a young larva next to its eggshell and on its first 'groove' on the underside of the leaf

1st instar: larval feeding damage—a close-up of the transparent windows when viewed from above [WL]

1st instar: a fully grown larva with two rows of black hairs running along its back

in a characteristic groove on the underside of the leaf that is the width of the larva's body, where the larva rests when not feeding. The larva will move to another spot when it hits an obstacle while feeding, such as the midrib of the leaf. The resulting grooves appear as transparent 'windows' that are quite visible to the trained eye when viewed from above, and I have found many Small Copper larvae over the years thanks to these clues that the larvae leave behind.

The first instar is of the typical Lycaenid form, shaped like a woodlouse, and can be distinguished from the second instar by the two prominent rows of black hairs that run along the length of its back. The fully grown first instar is a very pale brown with a green tinge and is around 2.5 mm long.

After around 10 days, the larva moults into the **second instar** but does not eat its old skin, which is the case after each skin change. While similar in shape to the first instar, the second instar is now covered in several rows of long and curved dark brown hairs that run along the length of its body. The second instar continues to feed on the epidermis of the leaf underside and, when fully grown, is 3.5 mm long.

2nd instar: the larva is covered in long dark brown hairs

The colour of the second instar is quite variable—some larvae have a uniform light yellow-green ground colour with no markings, while others have a deeper green ground colour with deep pink markings and, of course, others have a ground colour that is somewhere in between, with subdued pink markings. The colour pattern of a given second instar will subsequently apply to all its remaining instars—if it has a uniform ground colour, it will always have a uniform ground colour, and if it has pink markings, it will always have pink markings. The pink markings consist of a stripe along the back of the larva, and a band that runs continuously around the base of the body.

After 10 days or so, the larva moults into the **third instar**. The larva is similar in appearance to the previous instar, with the same variation in colour, although the hairs on its body are shorter relative to the length of the larva and are no longer curved but stand erect. A behavioural difference between this instar and the second instar is that the larva, while continuing to feed on the underside of the leaf, will perforate it, leaving tell-tale holes, and it will also feed on the leaf edges. When fully grown, the third instar is just over 5 mm long.

3rd instar: the larva will now perforate the leaf when feeding

3rd instar: the hairs on the larva now stand erect and are not curved

Towards the middle and end of October, larvae will start to enter hibernation, although there appears to be no discernible pattern of the preferred instar. In his *Natural History of British Butterflies* of 1924, F.W. Frohawk observed that most of the larvae that he reared in captivity overwintered in their third instar on the underside of a leaf, emerging to feed when the weather was warm enough, which correlates with my own findings in the wild.

If the larva does not overwinter then, after a couple of weeks, the larva moults into the **fourth instar** when its body is now sprinkled with minute white 'warts'. The larva is now a more vivid green and continues to exhibit the variation in colour found in earlier instars. When fully grown, the fourth instar is around 10 mm long, which is twice the length of the third instar and, unfortunately, this is the most reliable diagnostic when trying to distinguish these two instars in the field.

4th instar: a larva with a consistent green ground colour

4th instar: a larva with significant pink markings

5th instar: a larva with a consistent green ground colour

5th instar: a larva with significant pink markings

After two to three weeks, the larva moults into the final and **fifth instar**. The larva is a deeper green than previous instars with hairs that are proportionally shorter relative to the length of the body and a more extensive sprinkling of white warts than in the fourth instar. When fully grown, the larva is around 16 mm long and, when preparing to pupate, fixes itself with a silk girdle and a silk pad at its tail end to a suitable surface, such as the foodplant or, more often, other vegetation such as a dead leaf.

5th instar: the body has an extensive covering of minute white warts

Pupa

The pupa is just under 11 mm long and is attached to its chosen surface by the silk girdle and the cremaster, which is attached to the silk pad. It is thought that the pupa is tended by ants although this is merely a theory that warrants further study. After three or four weeks, depending on temperature, this stunning little butterfly emerges.

A pupa attached to a dead leaf by a silk girdle and the cremaster [WL]

A pupa just prior to emergence, showing the beautiful markings of the adult butterfly

Brown Hairstreak
Thecla betulae

❖ BROWN HAIRSTREAK IS SUPPORTED BY DAVID COOK, DAVID MOORE & MIKE WILLIAMS ❖

The **Brown Hairstreak** is the largest of our hairstreaks and is one of the latest butterflies to emerge each year. It is also a very elusive butterfly, spending much of its time resting, basking and feeding high up in tall shrubs and trees. However, patient observers that wait for the adult to come down to nectar on flowers such as Hemp-agrimony are rewarded with the sight of one of our most beautiful butterflies. The female has particularly striking large orange patches on her forewings that are much reduced (if they are present at all) in the male, while the undersides of both sexes are a mix of browns and oranges that is in perfect harmony with the autumn colours that abound when this butterfly is on the wing.

The different appearance of male and female led early entomologists to think that they were different species. In his *Gazophylacii naturae et artis* of 1703, James Petiver named the male 'The brown double Streak' and the female 'The Golden brown double Streak', with the two sexes eventually identified as one species by Benjamin Wilkes in his *The English Moths and Butterflies* of 1749 where he converged on the name 'The Brown Hair-streak Butterfly'. This is also a butterfly whose specific name is misleading—*betulae* is derived from *Betula*, the genus of birch, despite the butterfly using species of *Prunus*, primarily Blackthorn, as the larval foodplant.

Renowned butterfly expert Jeremy Thomas studied this butterfly, along with the Black Hairstreak, for his 1975 PhD thesis entitled 'Factors influencing the numbers and distribution of the Brown Hairstreak, *Thecla betulae*, and the Black Hairstreak, *Strymonidia pruni*', and much of what we know about this butterfly comes from his work.

♂ with a wingspan of between 36 and 41 mm

The ♂ underside is a mixture of light browns and oranges

The ♀ is larger with a wingspan of between 39 and 45 mm

The ♀ underside has richer colouring than the ♂

Life Cycles of British & Irish Butterflies

DISTRIBUTION

This butterfly has a scattered distribution with several strongholds, including the heavily wooded clays of the west Weald in Surrey and West Sussex, low-lying valleys in north Devon and south-west Somerset, low-lying pastoral areas of south-west Wales and the limestone pavements in and around the Burren in Ireland. In terms of worldwide distribution, the butterfly is found across the Palearctic, with a range that stretches across most of Europe and through temperate Asia as far as Korea.

The butterfly occurs as far north as Worcestershire and Lincolnshire although its former distribution stretched as far as the Morecambe Bay area of south Cumbria and north Lancashire with records from Grange-over-Sands, Arnside Knott and Silverdale in the early 1900s, which is at the northern limit for the butterfly across its range despite its primary larval foodplant of Blackthorn being widespread. There have been recent reports of the butterfly from the Morecambe Bay area, but these are believed to stem from a release.

The butterfly does not form discrete colonies and is typically found in small numbers over a large area with several studies showing that, over its flight period, an inhabited area supports just a handful of adults per kilometre of hedgerow. The butterfly tends to remain in the same areas year after year, although females are known to wander occasionally as they lay, and winter egg surveys often turn up eggs in new areas that are adjacent to an inhabited area.

HABITAT

The Brown Hairstreak can be found anywhere in its range where there are abundant patches of established Blackthorn. This common and widespread shrubby tree is not only used in hedgerows, but is often a component of mixed scrub and woodland edges and clearings. In addition to suitable Blackthorn, inhabited sites also contain prominent 'assembly' trees, usually Ash, that are used by both males and females when looking to find the opposite sex.

The female prefers to lay on sheltered, south- and east-facing young Blackthorn growth, and annual flailing is a particular threat to the welfare of this species when the majority of eggs are destroyed—Jeremy Thomas estimated that between 50% and 100% of eggs are typically lost following a severe trim. Habitat management that benefits the Brown Hairstreak therefore involves a rotational cycle, where a third of the available Blackthorn is cut back each year. If wholesale trimming is necessary, then the recommendation is to undertake the work in early August when the majority of larvae will have pupated on the ground among leaf litter and before females have started to lay their eggs. A level of Blackthorn coppicing has also proved beneficial at some sites, where stands of Blackthorn are cut back and the remaining stumps left to regenerate, with females subsequently laying on two- to five-year-old growth.

An Ash 'assembly' tree within a Blackthorn-rich hedgerow at Hollowfield Farm, Worcestershire [MW]

In their paper 'What type of hedgerows do Brown hairstreak (*Thecla betulae*) butterflies prefer? Implications for European agricultural landscape conservation', published in *Insect Conservation and Diversity* in 2010, Thomas Merckx and Koen Berwaerts not only confirmed some of the observations made by Jeremy Thomas, but also found that hedgerows and woodland with scalloped edges contained significantly more eggs than those with straight edges, suggesting that the resulting micro-climatic conditions are more suitable for the development of the immature stages.

STATUS

Historically, the Brown Hairstreak was a widespread species, but has since declined in many areas due to the loss of hedgerows resulting from agricultural intensification—as described in *Hedges* (New Naturalist volume 58), E. Pollard, M.D. Hooper and N.W. Moore estimated that 140,000 miles of hedgerow were lost in England and Wales between 1945 and 1970, with further losses incurred after that period (50% of our hedgerows have been lost since 1945 according to some sources). That being said, there is evidence that the butterfly is expanding its range in several areas.

The long-term trend shows a moderate decline in both distribution and abundance, with the more recent trend over the last decade showing a stable distribution and a significant decline in abundance. Using the IUCN criteria, the Brown Hairstreak is categorised as Vulnerable (VU) in the UK and Least Concern (LC) in Ireland.

LIFE CYCLE

This butterfly has a single generation and is one of the last butterflies to appear each year, emerging in mid-July and flying until late September, and occasionally well into October, peaking at the end of August and start of September. Males always emerge several days before females and the butterfly overwinters in the egg stage.

Imago

This elusive butterfly, and especially the male, spends much of its time hidden away in tall trees and shrubs where it rests, basks, mates and feeds, the latter on aphid honeydew that has been deposited on leaves. With most of its needs catered for, the butterfly's descent from its lofty home is somewhat unpredictable, such as when honeydew is in short supply, when it will take nectar from a variety of sources, including brambles, Devil's-bit Scabious, Common Fleabane, Hemp-agrimony, ragworts, thistles and umbellifers. There are also reports of the butterfly feeding on overripe blackberries and the berries of Wayfaring-tree.

When it does descend, the butterfly can sometimes be found basking with wings open on surrounding vegetation, gradually closing them as it warms up so that its undersides, which are primarily a mix of oranges and browns, can reflect the light to some degree and prevent overheating. On overcast days, however, the butterfly tends to remain hidden and out of sight, only flying when it is sufficiently warm. When it does fly then, like other hairstreaks, the butterfly has a rapid and jerky flight that, in the case of the Brown Hairstreak, is reminiscent of a Gatekeeper or Vapourer moth *Orgyia antiqua*, which also flash different shades of brown as they dance along hedgerows.

The two sexes are known to congregate around assembly trees, a strategy that some consider analogous to the 'hilltopping' strategy of mate location found in other species that also occur in low densities over a broad landscape. A suitable assembly tree is normally a large Ash with a prominent crown that sits proud of the surrounding canopy, whether at a woodland edge or within a hedgerow. Favoured Ash trees tend to be those with an open aspect to the east that catch the early morning sun (Mike Williams, pers. comm.). Within each colony, many trees will be used for assembly and the same trees appear to be used year after year (Mike Williams, pers. comm.).

The male spends most of his life on an assembly tree and is an early riser, with most activity starting as early as 7 am and concluding by 10 am. Mating also takes place on an assembly tree, when the male walks in tight circles around the female while fluttering his wings, before he attempts to pair. In a 2008 article in *The Entomologist's Record and Journal of Variation*, Andrew Middleton and Liz Goodyear documented three courtships between 7.30 am and 9 am, confirming that this is an early morning activity. Unlike the male, the female only remains on an assembly tree for the week it takes for her eggs mature, before she leaves to search out places to lay her eggs and any Brown Hairstreak found on Blackthorn is almost certainly going to be a female going about her business. The larger-bodied females are only active on the warmest days, with a peak of activity between 10 am and 4 pm.

When egg-laying, which usually occurs between 11 am and 3 pm, the female selects a sheltered plant that is exposed to the sun and with young one- or two-year-old growth that projects from the main body of the plant, although suckering growth is also used. Plants with short shoots due to heavy cutting or deer browsing are generally avoided. Preferred aspects for egg-laying are south- and east-facing, although eggs can even be found on edges that are north-facing, so long as they are sheltered. When a suitable plant has been found, the female will crawl among its branches while simultaneously feeling the surface with her abdomen for a suitable place to lay. Females are quite fastidious when laying and will take their time before laying a single egg, occasionally two or more, on the bark of the foodplant, typically no more than 1.5 metres above the ground and usually at a fork in a branch (especially the junction between new and old growth) or at the base of a thorn.

A ♀ probing a Blackthorn branch with her abdomen [DM]

Although Blackthorn is the most widely used foodplant, the butterfly will also use other species of *Prunus* if Blackthorn is in short supply, and Bullace *Prunus domestica* ssp. *insititia* is often cited as an example. In late 2015, a team from the Upper Thames branch of Butterfly Conservation examined several species of *Prunus* between Buckinghamshire and Oxfordshire, including urban areas, and found eggs on Almond *Prunus dulcis*, Apricot *Prunus armeniaca*, Cherry Plum *Prunus cerasifera* and several subspecies of Wild Plum *Prunus domestica*, namely Plum (ssp. *domestica*), Damson/Bullace (ssp. *insititia*) and Greengage (ssp. × *italica*).

A trio of eggs laid in the fork of a Blackthorn branch

Ovum

The Brown Hairstreak overwinters as a partially developed larva within the egg and, surprisingly, is one of a handful of species that are easier to find as an egg than an adult, so much so that annual winter egg surveys are used as the basis of measuring both distribution and abundance. This is largely because the 0.7-mm wide and 0.6-mm high, urchin-shaped white egg is extremely conspicuous against the dark bark on which it is laid.

Larva

After seven or eight months in the egg, typically in mid-April, the 1.3-mm long **first instar** larva makes its exit by eating a neat circular hole in the top of the egg and, like most species that overwinter as an egg, does not eat the remainder of the eggshell, and the vacated shells can sometimes be found the following year. The larva immediately moves to a developing bud where it crawls into the tender and unfurling leaves to feed while remaining concealed, although larvae can be found on the outside of a bud on occasion. The pale green larva has a shiny black head, greenish-black true legs on the thoracic segments and two rows of long backward-curving hairs along the length of its back, with a series of shorter hairs along each side, with all hairs emanating from a dark base. When fully grown, the first instar is around 2.5 mm long and has a hint of pale and oblique stripes along its sides that are a prominent characteristic of later instars.

A study by H.H. de Vries and colleagues, documented in 'Synchronisation of egg hatching of brown hairstreak (*Thecla betulae*) and budburst of blackthorn (*Prunus spinosa*) in a warmer future', published in the *Journal of Insect Conservation* in 2011, showed that larvae emerged between seven and nine days after Blackthorn budburst, irrespective of temperature, confirming that climate change and changing weather patterns would not be a challenge for this species in terms of access to its larval foodplant.

After 10 days, the larva moults into the **second instar**. The larva is now a brighter green and is superbly camouflaged as it lies flat against the underside of a Blackthorn leaf, with pale markings that mimic the leaf venation, and humps along its back that resemble the leaf edge. The markings along the back comprise two pale lines than run close together for the most part, but these diverge on the first three segments, and less so on the last three segments. Despite this camouflage, which is similar in later instars, Jeremy Thomas found that between 65% and 83% of larvae were predated by invertebrates such as spiders and harvestmen in the first two instars, and by insectivorous birds in later instars, at one of his study sites in Surrey.

The larva now rests by day on a silk pad underneath a Blackthorn leaf and feeds primarily (but not exclusively) at night, moving away from its resting place to feed on a leaf, usually eating from the edge inwards, before returning to the silk pad as dawn approaches. The larva may change its resting leaf every now and again, and a new silk pad is created when it does. When fully grown, the second instar is just under 5 mm long.

After another 10 days, the larva moults into the **third instar** and is identical in appearance to the second instar although, when fully grown and around 10 mm long, it is twice the length. The larva continues to rest on the underside of a leaf, feeding away from its resting place at night.

1st instar: a fully grown larva

2nd instar: the larva is now a brighter green

3rd instar: a larva on the underside of a Blackthorn leaf

After two more weeks, the larva moults into the final and **fourth instar**. Once again, the larva appears to be a larger version of the previous instar with the same shape, colour, markings and behaviour. However, there is a dramatic change in colour as the larva nears pupation, when it turns a dull purple that provides a level of protection as it heads down the dark brown plant stems and finds a suitable pupation site on the ground among the leaf litter. The fully grown fourth instar is around 17 mm long.

4th instar: a fully grown larva under a Blackthorn leaf

4th instar: a larva nearing the time to pupate

4th instar: a pre-pupation larva on a dead leaf [VM]

Pupa

The 12-mm long pupa is formed either in a crevice in the ground, among leaf litter or at the base of a plant. The pupa is also attractive to ants, which are known to bury the pupa in a loose cell of dry earth on occasion, where they continue to tend to the pupa until emergence, although this is a limited defence against the mice and shrews that are believed to eat large numbers of pupae in the wild. The pupa darkens prior to emergence and the adult butterfly hatches out in the morning, after around four weeks in this stage.

The pupa is a mix of browns, with mottling all over its surface [VM]

Life Cycles of British & Irish Butterflies

Purple Hairstreak
Favonius quercus

❖ PURPLE HAIRSTREAK IS SUPPORTED BY KATE ISABELLA AMY DOWDING, PAUL DOYLE & NIGEL KEMP ❖

The **Purple Hairstreak** is the quintessential hairstreak of oak woodland and, despite spending most of its time out of sight in the tree canopy, is our commonest hairstreak and almost certainly the most under-recorded. Adults are occasionally seen basking at lower levels on small trees, shrubs and Bracken and, when the light hits the wings at the right angle, the blue sheen of the male, or the concentrated colours found on the forewings of the female, really do take your breath away—this really is a most beautiful insect.

The ♂ has an all-over blue sheen and a wingspan of 33 to 40 mm

The ♀ has purple patches on her forewings and a wingspan of 31 to 38 mm

Early entomologists struggled to agree on the forewing colour of this species – the names attributed by various authors either use the adjective 'purple' or 'blue', with some authors using both when the two sexes were described as different species. In my opinion, the male has a predominantly blue sheen and the female a purple glow, and so I would avoid colour altogether and simply call this butterfly the 'Oak Hairstreak', which is consistent with its scientific name since the specific epithet, *quercus*, is the genus of oak.

♂ and ♀ have similar undersides

DISTRIBUTION

This butterfly has a widespread distribution across southern England and Wales, with more scattered colonies found in the north of England, Ireland and Scotland. The butterfly is not found in the Isle of Man, Orkney or Shetland. On a worldwide basis, the butterfly is found throughout most of Europe, east as far as the Ural Mountains in Asia, and in parts of North Africa.

The butterfly is not a great wanderer but will disperse and colonise new areas when there has been a build-up of numbers as a result of favourable weather conditions. S.G. Castle Russell writes in *The Entomologist's Record and Journal of Variation* in 1955 where he recalls a decades-old memory that clearly stuck with him: "*At Aldershot in 1887 in a wood nearby there was a large open copse which was full of small birch trees on each of which there was such a swarm of* Thecla quercus *that with two sweeps of the net on one particular tree I could catch fully a hundred of the butterflies*".

HABITAT

This butterfly is associated with woodland that is predominantly made up of oak, the larval foodplant, although it can be found in any location where oaks occur, including lanes, parks, gardens, churchyards and urban areas. A housing estate near to my home has oaks dotted here and there and I suspect that few residents are aware of the spectacle that can be seen on summer evenings as Purple Hairstreaks dance around the tree tops. Large numbers of oaks are not required to support this species and colonies are known to be confined to a single tree at some sites. Larval foodplants include the native species of Sessile Oak and Pedunculate Oak, although the introduced Turkey Oak and Evergreen (Holm) Oak are also used on occasion.

Suitable Pedunculate Oaks line a woodland edge at Pamber Forest in Hampshire

STATUS

Given the elusive nature of this species and the fluctuation in numbers from year to year, then short-term changes in distribution and abundance should be treated with caution. However, when the long-term trend shows a significant decline in both distribution and abundance then it is clear that this species has suffered over the long term, despite a slowing of its decline in the shorter term, with local examples of the butterfly expanding in both distribution and abundance.

The long-term trend shows a moderate decline in distribution and a significant decline in abundance, with the more recent trend over the last decade showing a moderate decline in distribution and a stable abundance. Given the widespread nature of this species then, using the IUCN criteria, the Purple Hairstreak is categorised as Least Concern (LC) in both the UK and Ireland.

LIFE CYCLE

This single-brooded species emerges in the last week of June, peaks in the second half of July and flies well into August. More northern populations, especially those in Scotland, emerge two or three weeks later in the second half of July, peak in the middle of August, and fly until early September.

Imago

The Purple Hairstreak is one of the delights of summer when it can be seen in oak woodlands across its range. It is particularly active in early evening between 5.30 pm and 8 pm, with a peak at around 7 pm as temperatures start to cool, when it puts on a spectacular display with large numbers flitting around the tree canopy, and from tree to tree. Many authors have likened this spectacle to a number of silver coins being tossed into the air as the grey underside of the butterfly contrasts with its darker upperside. 'Vistas' of several individuals chasing one another are not an uncommon sight and I have seen groups of 10 or more embarking in such a display on several occasions at Pamber Forest in Hampshire which is a stronghold of this species. The small number of records of this butterfly turning up in moth traps would seem to confirm that it is, indeed, one that flies late into the evening.

As well as oaks, the butterfly will also inhabit other trees such as Ash and elms although oak is, of course, needed for eggs and larvae. The butterfly feeds primarily on aphid honeydew but will occasionally feed from nectar sources such as bramble, buckthorns, Hemp-agrimony, thistles and a variety of umbellifers when honeydew is in short supply. The butterfly is occasionally found on the ground where it feeds on mineral-rich moisture and, in the hot summer of 2018, adults were seen obtaining moisture from lily pads in ponds, and from leaves in canals. Given the arboreal habits of this species, then seeing it close up is often a case of luck more than judgment since, whenever you visit a suitable site, there is never any guarantee of finding this canopy-dwelling butterfly at a level that offers the opportunity to observe it close up.

In an article published in *British Wildlife* in 2009, David Newland summarised his observations of this butterfly, having spent 23 hours spread over four days at the National Trust site at Sheringham Park in Norfolk, where a wooden gazebo allowed him to spend time in the tree canopy. The observations are, to the best of my knowledge, rather unique and have given us new insights into the hidden world of this elusive species. Newland concluded that the two sexes had little to do with each other outside of their evening forays and spent most of their time basking with open wings, perching with closed wings, or crawling over leaves while feeding on aphid honeydew.

As evening approached, this behaviour altered significantly, with males establishing perches in the higher branches, no doubt on the lookout for virgin females that might fly by. Newland suggests that such a position corresponds to the 'hill-topping' found in other species when the two sexes find one another by visiting high points in their vicinity, although others consider the behaviour to be more akin to a 'lek assembly' where males gather in a prominent position, awaiting a virgin female. Male-to-male interactions led to skirmishes and Newland also records male-female chases taking place and he must be one of a handful of people who has witnessed the minimal courtship of "*a perfunctory face-to-face contact of antennae lasting about 10 seconds*" before mating takes place.

A mating pair [NK]

The ♂ and ♀ face one another during their minimal courtship [MHF]

When egg-laying, the female will fly in and around the boughs of oaks before alighting on a particular sprig, where she then crawls over the plump oak buds and their vicinity, probing with her abdomen as she moves around. Eggs are laid singly (or, less commonly, in groups of two or more) at all heights on the tree but rarely at the top, and always at the base of a plump oak bud or an adjoining twig where they can be found easily during the winter months if you know where to look.

Eggs are always laid on branches that are sheltered and that receive full sun, and these are, therefore, on the south-facing side of the tree. There is also a preference for solitary trees, such as those found at the edges of woods, or those that form part of a hedgerow, rather than trees that are buried deep in woodland. Also, eggs are most often found on the twisted and gnarled branches of relatively mature trees, since these usually have the plumpest buds, rather than relatively young trees or those with small buds. I have searched for Purple Hairstreak eggs every winter for the last couple of decades and knowing these characteristics makes all the

Eggs are laid at the base of an oak bud

Life Cycles of British & Irish Butterflies

difference between success and failure. Despite the elusiveness of the adult butterfly and the ease with which eggs can be found, egg counts are not used when monitoring distribution and abundance trends.

Ovum

The egg is 0.8 mm wide and 0.5 mm high and, like the eggs of most of the Lycaenidae, is bun-shaped. A close examination reveals a beautiful pattern of delicate spines projecting from the surface, the egg resembling a sea urchin. Each egg has a bluish tinge when first laid, but gradually turns white and may become discoloured over the winter months as

When seen close up the egg resembles a sea urchin

The egg becomes discoloured by algae over the winter

exposure to the elements results in algae forming on its surface, making the egg much harder to find since it more closely matches the colour of the surface on which it has been laid. The larva is fully developed within the egg after around three weeks but does not emerge until the following spring, having remained in the egg for around eight months.

Larva

The **first instar** larva emerges from the egg at the same time that the oak buds start to burst, which is as early as the second half of March in favourable years in the southern part of the butterfly's range. The greenish-brown larva takes the best part of a day to create a hole at the top of the crown that is large enough for it to make its exit. On hatching, the 1.5-mm long larva burrows into a developing bud where it initially feeds with just its back end visible although, as the buds unfurl, the larva can be found feeding and resting at the base of a young oak leaf. When it is fully grown, the first instar is light brown and 2.5 mm long. Apart from its size, the first instar is distinguished from all other instars by dark patches found on the first and last segments, which are much lighter in subsequent instars.

1st instar: a newly emerged larva

1st instar: the fully grown larva is light brown

2nd instar: the larva has a series of oblique markings along its sides

Around two weeks after leaving the egg, the larva moults into the **second instar** and now lives under a loosely spun silk web. This web catches all sorts of debris and offers the larva a level of protection as its rests underneath it during the day, moving out to feed on the developing leaves at night. The larva is now covered in a number of fine white hairs and also has a series of oblique markings along its sides that complements the camouflage provided by the silk web. When fully grown, the second instar is just over 4 mm long. The significance of this camouflage is quantified by Jeremy Thomas in *The Butterflies of Britain and Ireland*: "*About half the caterpillars hatching from 104 eggs I once studied in a Surrey wood survived on oak trees, whereas four-fifths of the colony died during the three- to four-week period between leaving their trees and emerging as adults*".

After a week or so, the larva moults into the **third instar**. It is similar in appearance to the second instar larva although the pale patch found on the first segment is now much more prominent and there is a slightly stronger contrast between the various markings. When fully grown, the third instar is around 10 mm long.

After a week or so, the larva moults into the final and **fourth instar**. The colours are now more intense than previous instars and the oblique markings on the sides are a richer brown in colour. The larva spins a few silk strands over the leaf bases and

3rd instar: there is a stronger contrast between the various markings

3rd instar: the larva is well camouflaged when resting among leaf debris

bracts, catching all of the debris that would otherwise fall away as the larva feeds. The larva rests by day underneath this 'tent' and any part of the body that remains exposed is perfectly camouflaged as it lies along an oak stem. This camouflage is so convincing that it has been suggested that the best way to find the larva is not by sight, but by touch. After a couple of weeks and when fully grown, at the end of May or early June, the larva is around 16 mm long.

4th instar: the oblique markings on the sides are a richer brown in colour

4th instar: the larva rests under its silk 'tent' during the day

4th instar: the head of the larva is a reddish-brown

The larva becomes a duller brown as the time for pupation nears, when it descends the tree. Various authors suggest that the pupa may be formed under earth or moss, or in a crevice of the bark, where the larva creates a loose cocoon from a few strands of silk. However, there are few, if any, records of pupae found in these locations. In *The Butterflies of Britain and Ireland*, Jeremy Thomas describes how he once found pupae in ant nests scattered around the periphery of a single oak. He goes on to say that both the larva and pupa are able to 'sing' to ants, confirming that there is, in all likelihood, a strong relationship between this species and ants, as there is in most of the Lycaenidae. Further evidence of this association comes from a separate study that shows that the larva is also covered in 'pore cupolas', which produce substances that attract (or appease) ants.

The mottled brown pupa is formed under soil or moss, or in an ant nest

Pupa

The 10-mm long pupa is usually formed face-down and unattached to the surface on which is it formed. The pupa is mottled with brown markings and the larval skin normally remains attached, despite the lack of hooks on the cremaster. After three to four weeks, the adult butterfly emerges and there have been a few occasions when, on a still and warm summer morning, I have found several newly emerged adults clinging to grass stems, as they expand their wings, below the canopy of a tall oak.

Life Cycles of British & Irish Butterflies

Green Hairstreak
Callophrys rubi

❖ GREEN HAIRSTREAK IS SUPPORTED BY MARTIN DIXON, SCOTT DONALDSON & GREENWINGS WILDLIFE HOLIDAYS (greenwings.co.uk) ❖

	January	February	March	April	May	June	July	August	September	October	November	December
IMAGO												
OVUM												
LARVA												
PUPA												

The **Green Hairstreak** is the most widely distributed hairstreak found in Britain and Ireland and is our only butterfly with green wings, since the green effect on the underside of the Orange-tip, for example, is actually an illusion created through a combination of black and yellow scales. However, the green that we see in the Green Hairstreak is not caused by any pigment but is an iridescent effect resulting from light hitting the lattice-like structure found in the scales, much like the purple we see in the male Purple Emperor, and diligent observers will notice that the 'green' changes depending on the light direction. The resulting effect provides excellent camouflage as the butterfly rests on a favourite perch, such as a Hawthorn branch, when the undersides are all that are on display since both sexes always settle with their wings closed, with the butterfly revealing its brown uppersides only when in flight. This butterfly can also be found positioning its wings at an appropriate angle to the sun in order to regulate its body temperature—the butterfly tilts on its side to catch the sun when warming up, but faces the sun to prevent overheating, when it also casts a minimal shadow and is extremely hard to spot.

The butterfly's current vernacular name is self-explanatory and was provided by William Lewin in his *The Papilios of Great Britain* of 1795 and is slightly more descriptive than the rather bland name of 'Green Butterfly' given to it by Benjamin Wilkes in his *The English Moths and Butterflies* of 1749. The specific name of *rubi* was provided by Linnaeus himself in the 10th edition of his *Systema Naturae* of 1758, in recognition of the *Rubus* genus of bramble which was thought to be the only larval foodplant at the time.

♂ and ♀ have a similar wingspan of 27 to 34 mm

DISTRIBUTION

The butterfly is found throughout Britain and Ireland which is due, in part, to the wide variety of larval foodplants and range of habitats that it uses. It is certainly the commonest hairstreak in northern England, Scotland and Ireland. However, the butterfly is absent from the Isle of Man, Outer Hebrides, Orkney and Shetland. Despite its broad distribution, it is a local species, forming distinct colonies where as few as a dozen individuals fly at any one time during the flight period, although some colonies are much larger where there are large expanses of suitable habitat. In *The Butterflies of Britain and Ireland*, Jeremy Thomas suggests that the butterfly's local nature may be due to both the quality of the foodplant and the density of ants, possibly of a particular species, given the Green Hairstreak's close association with ants when in the pupal stage. On a worldwide basis, the butterfly is found across Europe and temperate Asia, as far as China, and in parts of North Africa.

HABITAT

This butterfly can be found in a wide variety of habitats, including calcareous grassland; woodland clearings, rides and glades; mixed scrub; hedgerows; heathland; bogs; and railway embankments. The strongest colonies are found on calcareous grassland in southern England; moorland in northern England, Wales, Scotland and Ireland; lowland bogs in Ireland; and woodland clearings and their edges in Scotland.

Common features of all these habitats is the presence of scrubby plants and shrubs. The Green Hairstreak is the most 'polyphagous' of all species found in Britain and Ireland in that it

Mixed scrub at Greenham Common in Berkshire

Life Cycles of British & Irish Butterflies

A favoured south-facing hedgerow at Magdalen Hill Down in Hampshire

Isolated shrubs provide shelter at Magdalen Hill Down in Hampshire

uses the widest variety of larval foodplants. These include Common Bird's-foot-trefoil, Common Rock-rose and Dogwood on calcareous grassland; Broom, Cross-leaved Heath, Dyer's Greenweed and Gorse on heathland; brambles and Alder Buckthorn in woodland; and Bilberry in boggy habitats, especially in Scotland.

STATUS
Both long-term and short-term distribution and abundance trends show moderate declines and the conservation status of this butterfly is kept under review as a result. Using the IUCN criteria, the Green Hairstreak is categorised as Least Concern (LC) in the both the UK and Ireland. The greatest threat to this butterfly is habitat loss, as well as inappropriate habitat management that renders sites unsuitable.

LIFE CYCLE
There is one brood each year, with the butterfly appearing in mid-April and flying until the second half of June, although this does vary by location with more northern populations flying slightly later. The Green Hairstreak is also unique among our hairstreaks in that it overwinters as a pupa rather than an egg, remaining as a pupa for about 10 months from the middle of July until the following April. It is therefore our first hairstreak to appear in the year.

Imago

Adults can be found feeding from available nectar sources, including Bluebell, Common Bird's-foot-trefoil, Common Rock-rose, Cowslip, Daisy, dandelions, Gorse, hawthorns, Holly, Horseshoe Vetch, Wild Privet and Wild Strawberry.

However, most sightings of this species are of the territorial males that sit on favourite perches, often at the base of a slope, where they wait for passing females, darting out to investigate any passing insect such as a bee or fly. An encounter with another male results in the pair spiralling around one another as they fly up into the air before they eventually break off from their skirmish, with the resident male returning to his perch, and possibly the same leaf. Both stand-alone shrubs and hedges are used for perching and favoured perches are often reused by different males should the original occupants leave them, even if momentarily, with males using a number of different perches over the course of a day.

One trick I learned while growing up in Cheltenham was that gently tapping a favoured shrub would cause any Green Hairstreak on it to fly up into the air, before it returns back to the same shrub after a few seconds, and I was always amazed at how an apparently barren Hawthorn bush would come to life as it revealed its treasures. It is a shame that the butterfly only settles with its wings closed since the male possesses distinctive pale sex brands on its otherwise dark brown uppersides. These sex brands contain androconial scales that emit pheromones to seduce a female prior to mating.

A ♂ perching on a Hawthorn bush in his territory [MC]

An adult nectaring on Wild Strawberry

A mating pair with ♀ at the top [NH]

The female is more likely to be seen away from the male territories, where she can be found searching out nectar sources and larval foodplants. An egg-laying female will flutter over the chosen foodplant and, after landing, will crawl over the plant while probing it with her abdomen for a suitable spot to lay. Females can be quite easy to find from a distance when egg-laying, and I regularly see several females simultaneously fluttering over the ground at one of my local sites at Greenham Common in Berkshire, where Common Bird's-foot-trefoil is used as one of the larval foodplants.

Eggs are laid singly, and their position is dependent on the larval foodplant selected, but they are always laid on fresh young growth and placed in some nook or cranny that is out of sight, such as between developing leaf shoots on Common Bird's-foot-trefoil and Common Rock-rose, or between flower buds on Buckthorn and Dogwood. The shell of the egg is particularly malleable and is often shaped by the space in which it has been laid.

A ♀ egg-laying on Dogwood [DM]

Life Cycles of British & Irish Butterflies

Ovum

The 0.7-mm wide egg is reminiscent of a doughnut since it is a flattened sphere with a sunken crown, covered in a fine network pattern over its surface. The egg is a very pale blue-green when first laid but turns a pale grey a couple of days before hatching.

The egg is shaped like a flattened sphere

Larva

The 1-mm long **first instar** larva emerges from the egg after a week or so by eating a hole in the crown. It does not eat the remainder of the eggshell, preferring to feed on the tenderest young growth of the foodplant which, depending on the plant in question, may be young developing leaves, a fruit or an unopened flower bud. The larva is of the typical woodlouse shape found in the Lycaenids, is covered in several rows of long black hairs that run the length of the body and has a noticeable dark patch on its first segment, with another small dark patch on the anal segment. The larva also has a dark brown head, a whitish ground colour and several longitudinal chocolate-brown stripes that are most noticeable in the mature first instar. Even at this young age each segment has a noticeable hump (a characteristic shared by all instars that is most visible in the final instar). When fully grown, the first instar is just under 2 mm long.

1st instar: the larva has dark patches on the first and anal segments

1st instar: the larva develops prominent chocolate-brown stripes as it matures

2nd instar: the larva now possesses relatively short and straight hairs

After around 10 days, the larva moults into the **second instar**, partly eating its cast skin before continuing to feed on the tenderest parts of the foodplant when the larva can sometimes be found hunched over the tips of the youngest growth with its head buried inside. After their first moult the larvae also become cannibalistic, with larger larvae devouring smaller larvae, especially those that are preparing to moult and unable to move from the silk pad to which they are attached. While the overall colour of the second instar is similar to that of a mature first instar, it is now covered in numerous straight and relatively short hairs when compared with the curved and long hairs of the first instar. When fully grown, the second instar is around 5 mm long.

3rd instar: the larva now exhibits a beautiful palette of greens

After a week or so, the larva moults into the **third instar** and eats its cast skin before moving to eat the leaves or stems of the foodplant. The larva is now a very different beast, changing colours from the mix of whites and browns of previous instars, to a mix of greens. The resulting palette of greens is quite beautiful—the ground colour is a light green and the larva now has dark green stripes that run the length of its body on its back and sides, along with a whitish stripe along the base of each side. The head is now a shiny black, rather than the dark brown of earlier instars, and the larva is covered in groups of short black hairs. The green colouring provides excellent camouflage and allows the larva to feed more openly on its foodplant. When fully grown, the third instar is around 8 mm long.

After just over a week, the larva moults for the last time into the **fourth instar** and its appearance has changed once again.

4th instar: the oblique lines along each side produce a beautiful pattern when viewed from above

4th instar: the larva continues to sport a pale stripe along the base of each side

4th instar: a larva revealing its shiny reddish-brown head

The larva now has a largely uniform bright green ground colour with yellowish-green lines of varying intensity that run obliquely along each side, producing a beautiful pattern when the larva is viewed from above. The larva is covered in short hairs and its shiny head is now a reddish-brown and is fully retracted into the first segment when the larva is resting or feeding. After around 10 days, the larva is fully grown and around 16 mm long.

Pupa

The 10-mm long pupa is rather stout in appearance, light brown in colour and covered in numerous dark brown speckles that are most visible on the wing cases. The amount of speckling varies considerably and those pupae that have more speckles can appear much darker than those with fewer.

Like many other Lycaenids, this species benefits from an association with ants that provide it a level of protection, although this association is only manifest in the pupal stage in the Green Hairstreak. Pupae have been found in the wild covered in particles of soil believed to have been put there by ants, and a pupa was once found inside an ant nest by Jeremy Thomas. It has also been suggested by some that the pupa is formed at the base of plants among ground litter, occasionally attached to a dead leaf by a silk girdle. However, such observations have been made with regard to individuals reared in captivity and there appears to be little evidence that this is normal behaviour in the wild. In *Butterflies of Surrey Revisited*, Harry E. Clarke shows a direct correspondence between the distributions of Green Hairstreak and *Myrmica sabuleti* in the county which may suggest a dependency on this particular ant species.

Based on individuals reared in captivity, we know that ants are attracted to secretions from pores on the pupa and the butterfly has another trick up its sleeve, which may be used to either attract or appease ants—the pupa (in common with some other Lycaenids) is able to make a squeaking sound, a phenomenon first documented in this species by Christian Kleemann in *Der Naturforscher* in 1774. The sound mimics the squeak of an adult ant and is produced by rubbing together two abdominal segments. Incredibly, it is so loud that it can be heard by the human ear without any amplification.

The pupa is able to make a sound that is attractive to ants

Life Cycles of British & Irish Butterflies 321

White-letter Hairstreak
Satyrium w-album

❖ WHITE-LETTER HAIRSTREAK IS SUPPORTED BY MIKE GIBBONS, KEN HAYDOCK & JILL MILLS ❖

The White-letter Hairstreak is one of our most elusive butterflies, spending much of its time high in its arboreal home where it appears either as a dark speck against the sky as it flits around erratically, or as a small dark triangle when settled on a leaf. Like the Black and Green Hairstreaks, it always settles with its wings closed and regulates its body temperature by angling its wings to the sun. Both sexes have 'tails' on their hindwings, which are shorter in the male, that direct the attention of birds away from the head and it is not uncommon to find individuals with beak-shaped segments missing from their hindwings.

The butterfly is known for its dependence on elms, the larval foodplant, and their decline in the 1970s and early 1980s, as a result of a virulent strain of Dutch Elm Disease that was accidentally introduced in the late 1960s, was a cause of real concern. Spores of the fungus that cause the disease are carried by various species of elm bark beetles, which infect trees when they feed on its bark, or lay their eggs in it. Millions of trees were killed off and the south was especially affected. Our main native species—Wych Elm, English Elm and Small-leaved Elm—are all susceptible to the disease, although it is a small mercy that Wych Elm, which the butterfly prefers, was the least affected.

Surviving White-letter Hairstreak colonies were actively searched out to obtain a better understanding of its remaining distribution and the discovery of several new colonies gave some hope for its future. It had long been thought that the butterfly required flowering elms for its survival since young larvae feed on the flowers and developing seeds, although, in a few cases, the butterfly was found breeding on non-flowering elms and suckers growing near dead trees, with larvae feeding on developing leaf buds. Unfortunately, as trees mature, their bark becomes suitable for the beetle and are reinfected. The Hampshire and Isle of Wight branch of Butterfly Conservation has led a concerted effort to identify elms that are not only disease resistant, but also possess the required qualities to support this butterfly, such as a synchronisation of its flowering with the emergence of larvae in the spring. Started in 2000, Andrew Brookes and other branch members conducted extensive trials and the butterfly was found breeding on one of the cultivars in 2015. The planting of these disease-resistant elms is now treated as a matter of course by many Butterfly Conservation branches.

James Petiver, in his *Gazophylacii naturae et artis* of 1703, named this butterfly, quite simply, the 'Hair-Streak', which was extended to 'Dark Hairstreak' by Moses Harris in his *The English Lepidoptera* of 1775. Confusingly, several authors used the name 'Black Hairstreak' for this butterfly although, in fairness, the Black Hairstreak that we know today had yet to be discovered. Thankfully, John O. Westwood was very specific when he named this butterfly 'The W-Hair-Streak' in his *British Butterflies and their Transformations* of 1841, a name that was expanded to White-letter Hairstreak by William Furneaux in his *Butterflies and Moths* of 1894. This vernacular name and the specific name of *w-album* both make reference to the letter 'W' that is formed from a series of white lines found on the underside of the hindwings.

The slightly larger ♀ has a wingspan of 26 to 36 mm

The ♂ has shorter tails than the ♀ and a wingspan of 25 to 35 mm

DISTRIBUTION

This butterfly is found in lowland areas below 500 m throughout England, south of a line stretching between south Lancashire in the west and the Borders in the east. This species is more local in Wales, and is not found in Ireland or the Isle of Man. In 2017 the butterfly was sighted by Iain Cowe in the Scottish Borders, the first time the butterfly had been seen in Scotland since 1884. A concerted effort was then made to locate eggs and, in February 2018, several were found by Jill Mills and Ken Haydock, including the shell of one that had hatched. This egg will have overwintered in 2016 / 2017 and, therefore, the butterfly must also have been present in Scotland in 2016. On a worldwide basis, the butterfly is found throughout most of Europe and across temperate Asia to Japan.

This butterfly forms discrete colonies which are sometimes very small, containing only a few dozen individuals, although numbers can fluctuate greatly from year to year. Adults are typically found inhabiting a small clump of trees, either in a wood or hedgerow, but a colony may survive on a single isolated tree. The butterfly is not a great wanderer and tends to reuse the same sites year after year, although one mark-release-recapture study showed that individuals regularly moved between trees 300 m apart, and the butterfly has been found several kilometres from known sites on occasion.

HABITAT

This butterfly is found wherever elms, the larval foodplant, grow, especially Wych Elm. The butterfly can therefore be found in broadleaved woodland clearings, rides and glades; mixed scrub; hedgerows; and gardens, parks and churchyards.

Flowering elms were once thought to be essential for successful development of larvae, since the early larval instars feed on the developing seeds and are superbly camouflaged against them although, as mentioned earlier, there is some evidence that non-flowering elms are also used. Since elms produce flowers earlier than their leaves, a concern is that larvae that emerge early will starve if the elm does not produce flowers. However, there is a row of five elms on a main road near to where I live and one thing that I've noticed is that the non-flowering elms come into leaf much earlier than the flowering elms, perhaps because their energy is not put into the production of flowers and seeds.

A mature Wych Elm at Bentley Wood in Wiltshire

STATUS

This species is in serious decline and is therefore a priority species for conservation efforts. Both long-term and short-term trends show a moderate decline in distribution and a significant decline in abundance. Using the IUCN criteria, the White-letter Hairstreak is categorised as Endangered (EN) in the UK.

There is no doubt that the butterfly is under-recorded due to its arboreal habits, with the possibility of it going through its entire life cycle in the tree tops and out of sight. Strange as it may sound, it is probably now easier to find the butterfly than in previous years, since tall mature elms are in short supply, although the butterfly will move to other trees to feed if their leaves are smothered in aphid honeydew, including Ash, Field Maple, oaks, Sycamore and Horse-chestnut.

LIFE CYCLE

This butterfly is single-brooded, with adults starting to emerge in the second half of June, building up to a peak in mid-July, and continuing to fly into the first part of August. The butterfly overwinters as an egg, spending around nine months in this stage.

Imago

Adults are elusive, remaining in the treetops where they feed on aphid honeydew, but will (as a general rule) also come down early in the morning or late in the afternoon to feed from a variety of flowers, especially when honeydew is in short supply. Favourite nectar sources include brambles, Hemp-agrimony, ragworts, thistles, umbellifers, Wild Marjoram and Wild Privet. In the Butterfly Conservation booklet dedicated to this butterfly, authored by Martyn Davies, an analysis of nectar sources is provided with Creeping Thistle coming out as clear favourite at the site studied. Davies also shares observations of males coming down to stony paths where it is thought that they replenish their supply of minerals that are passed to the female when mating.

Clashes between males are often observed when the adults are most active in early morning and late afternoon, when they can be seen spiralling around one another in the tree tops. Mating almost certainly occurs out of sight in the canopy, although mating pairs are occasionally observed at ground level, possibly because females pair soon after emergence near their pupation site.

When egg-laying, the female will flutter around her chosen elm, land, and then crawl slowly along twigs of the foodplant, probing with her abdomen for suitable sites to lay as she goes. The site chosen is usually a rough patch on a sunny and sheltered twig, especially on the girdle scar at the junction between the current and previous years' growth, but may also be at the junction of the main twig and a lateral shoot, or at the base of a bud. Several eggs may be laid close to one another, but this is usually the result of different females laying in the same spot. Wych Elm, English Elm, Small-leaved Elm and their hybrids are all used, although there is a preference for Wych Elm, with eggs laid at all heights on the tree, although most are laid above 2 m from the ground.

A mating pair with ♀ on left [DJ]

A ♀ probing a twig for a suitable place to lay [TP]

Ovum

The 0.8-mm wide and 0.45-mm high egg is a flattened sphere that has been compared to a 'flying saucer', with a whitish rim surrounding a central dome. The egg is green when first laid, but changes to dark brown after 48 hours, with a fully formed larva developing inside the egg before the onset of winter.

Two eggs laid on a girdle scar [IC]

The egg resembles a 'flying saucer'

Life Cycles of British & Irish Butterflies

Larva

Sometime between the start of March and early April, depending on temperature, having spent nine months in the egg, the 1.3-mm long **first instar** larva emerges by eating a hole in the crown. Without eating the remaining eggshell, the larva moves to a nearby bud where it burrows deep within the tissues and, if the bud has opened up sufficiently, will disappear inside. First instar larvae feed on developing seeds and flower buds on flowering elms but must make do with developing leaf buds on non-flowering elms.

1st instar: the newly emerged larva is dark olive

1st instar: the fully grown larva has a purplish-pink ground colour

The black-headed larva has dark discs on the first and last segments, with groups of long hairs that run in two rows along the back, with groups of shorter hairs that run around the base of each side. The larva initially has a dark olive ground colour but, when fully grown and around 3 mm long, develops a purplish-pink hue with a pale patch in the middle of the body—a colour combination that harmonises well with an elm flower.

After around a week, the larva moults into the **second instar** and has more significant humps than in the previous instar, a significant furrow along its back and its head is completely hidden under the first segment when the larva is resting. The larva has more hairs than in the previous instar and those that run along the back are the longest. The colour of the larva is now somewhat variable, ranging from those that are almost green to those that are almost purplish-pink. Most, however, are a beautiful combination of the two colours, resulting in a banded appearance that blends in perfectly with an elm flower, and even those that feed on leaf buds of non-flowering elm are somewhat camouflaged against the brown at the base of the bud and the green of the developing leaves. The larva has a whitish line at the base of each side and a number of oblique pale lines that run along each side. When fully grown, the second instar is 5.5 mm long.

2nd instar: a larva (bottom right) in context against an elm flower

2nd instar: a larva feeding on a developing seed

2nd instar: a larva scooping out the contents of a developing leaf bud

After between two and three weeks, the larva moults into the **third instar** and will now feed on expanding leaves as well as buds and seeds. The larva has a greener appearance and matches the colour of the developing leaves, although it continues to exhibit purplish-pink markings. The colouring, in combination with the humps on the back that resemble the outline of the leaf, provide excellent camouflage. When fully grown, the third instar is around 7 mm long.

3rd instar: a larva with a few purplish-pink markings

3rd instar: a larva that is wholly green

Life Cycles of British & Irish Butterflies

After between one and two weeks, the larva moults for the last time into the **fourth instar** and is similar in appearance to the previous instar, although most larvae usually exhibit only hues of green and are, of course, larger—when the larva is fully grown in early June it is 16 mm long and twice the length of the previous instar. The camouflage is extraordinary, and the larva is almost impossible to discern when resting alongside the edge of an elm leaf, where its profile and colouring perfectly match the leaf outline and venation. The larva also possesses the ant-attracting organs found in other Lycaenids, and wood ants have occasionally been observed feeding from secretions on the body and from the honey gland.

Larvae are most easily found by standing under a tree on a sunny day and looking up, where they are silhouetted against a leaf. A combination of feeding patterns can also indicate their presence—the larva will take chunks out of each side of the leaf, resulting in a diamond-shaped leaf tip. The larva will also create holes in the leaf and even eat panels out of the leaf while leaving the veins intact, although it is thought that the feeding pattern may depend on the species of elm and the toughness of the leaves (Paul Harfield, pers. comm.).

4th instar: a larva in profile against an elm leaf

Typical feeding damage [JBu]

As the time to pupate draws near, the larva gradually exhibits purplish-brown hues, and finds a suitable spot to pupate, which may be on the underside of a leaf, on a stem, in a fork or in a crevice in the bark. Larvae are most commonly found pupating underneath a terminal leaf, with surrounding leaves lightly silked together. The larva is held in position by a silk girdle around its waist, and by its anal prolegs that clasp onto a silk pad. Unusually among our butterflies, the larva takes over a week to finally pupate, having spent almost three weeks in total in this instar.

4th instar: a larva preparing to pupate

A side-on view of the pupa

Pupa

The 9-mm long and hairy pupa is attached to its chosen surface by the cremaster that hooks into the silk pad, and by the silk girdle around its waist. It is speckled in a variety of browns that give the pupa the appearance of an elm bud. Like mature larvae, pupae that are formed on leaves can be found by looking up through the leaves when the sun is shining, when they are in silhouette. This delightful 'butterfly of the elm' hatches after three or four weeks.

Black Hairstreak
Satyrium pruni

❖ BLACK HAIRSTREAK IS SUPPORTED BY ISACE MONMART, JONATHAN PHILPOT & KIRSTY PHILPOT ❖

The Black Hairstreak is one of our rarest butterflies and also a relatively recent discovery, since it was overlooked due to its similarity with its close cousin, the White-letter Hairstreak. It was first discovered in the British Isles in 1828 when a Mr Seaman, an entomological dealer, collected specimens from one of the most famous sites for this butterfly, Monks Wood in Cambridgeshire, describing them as White-letter Hairstreak. When quizzed about the locality of these specimens, having been told that they were not White-letter Hairstreak by Edward Newman, a Victorian entomologist of note, he claimed that he took them in Yorkshire (where they do not exist) in order to keep the location to himself and capitalise on his find. His ruse was finally discovered as this species was subsequently found at several sites in the East Midlands by other entomologists including, as it happens, at Monks Wood.

Both male and female of this elusive insect settle with their wings closed and regulate their temperature not by opening their wings, but by positioning themselves at an appropriate angle to the sun. This is a real shame since the uppersides have a beautiful spread of orange patches along the edges of the wings that are especially prominent in the female. The two sexes can, however, be distinguished by subtle differences in the length of their tails—those of the female are slightly longer than the stubby tails found in the male.

The butterfly was given its name by John Curtis in his *British Entomology* in 1829, and the specific name of *pruni* recognises the larval foodplants, which are various species of the *Prunus* genus, such as Blackthorn and Wild Plum.

The ♂ has stubby tails and a 34 to 39 mm wingspan

The ♀ has slightly longer tails and a 35 to 40 mm wingspan

DISTRIBUTION

This butterfly has a very restricted distribution in the East Midlands of England that follows a line of clays from Oxford in the south-west to Peterborough in the north-east, where the butterfly can be found in dense stands of Blackthorn, the primary larval foodplant, although Wild Plum and other species of *Prunus* are used on occasion. The butterfly also flies at several sites outside of this range which are the result of unofficial introductions.

It comes as some surprise that this butterfly has such a confined distribution when its foodplants are so widespread and it was thought that the root cause was the general lack of the mature Blackthorn stands that this butterfly needs. However, a more plausible explanation is the butterfly's inability to colonise new areas, for this butterfly is not a great wanderer—an entire colony will often confine itself to just a few areas within a wood, with little interchange between them. This lack of movement was quantified by Jeremy Thomas in *The Butterflies of Britain and Ireland* who calculated that, at one site with continuous habitat, the species moved at a rate of just 1 km per decade (100 m per year).

On a worldwide basis, the Black Hairstreak is found throughout Europe and across the Palearctic to Japan, but is absent from southern Spain and some of the Mediterranean islands. Wherever it is found, it is always a local species.

HABITAT

Black Hairstreak colonies are found along woodland edges, rides and glades—wherever dense and mature 3-m to 4-m high Blackthorn stands are found in sheltered and sunny positions—although smaller colonies are occasionally found where the Blackthorn receives sun through a gap in the canopy of mature woodland.

Adults are also found away from woodland, typically in a nearby hedgerow, and one of my most vivid memories was parking in a small layby near Quainton in Buckinghamshire (that is very close to Finemere Wood, a known stronghold of this species), looking out of the window at an adjacent Blackthorn hedge and seeing a Black Hairstreak looking straight back at me. I went on to find several more nectaring on Wild Privet that was growing through the same hedgerow. In some regions, the use of hedgerow Blackthorn is becoming increasingly important, possibly as a result of changes in woodland management that results in Blackthorn getting shaded out (Stuart Hodges, pers. comm.).

Whatever the habitat, management focuses on the creation and maintenance of small patches of the mature Blackthorn that the butterfly requires, with rotational cutting of less than 25% of available habitat applied over a long cycle of between 20 and 40 years, with these areas allowed to regenerate.

Hedgerow with mature stands of Blackthorn, Asham Meads, Oxfordshire [JA]

STATUS

Given the inability of this species to colonise new areas, and sites that are fragmented and isolated, this butterfly is one of our most threatened species and its future depends on appropriate protection and habitat management. Previous losses have been put down to extensive afforestation and coniferisation in the 1950s and 1960s that restricted the 20- to 40-year coppice cycles that result in the dense and mature Blackthorn stands that this butterfly needs.

The long-term trend shows a significant decline in both distribution and abundance, with the more recent trend over the last decade showing a moderate decline in distribution and a significant decline in abundance. Using the IUCN criteria, the Black Hairstreak is categorised as Endangered (EN) in the UK.

LIFE CYCLE

The Black Hairstreak is single-brooded and has one of the shortest flight periods of all our butterflies, emerging at the start of June in good years, more typically in the second week of June, peaking in the latter part of June and flying until the first week of July.

Imago

Adults spend much of their time resting high up on Field Maple, Ash or the primary larval foodplant, Blackthorn, where they crawl over leaves and twigs in search of aphid honeydew on which to feed, making it extremely frustrating to get a sighting of this species. However, the butterfly will come down to feed on various nectar sources, such as Wild Privet, Dog-rose and brambles, especially when honeydew is in short supply.

Adults are most active between midday and mid-afternoon and, like all hairstreaks, are extremely difficult to follow when in flight. To make matters worse, the Black Hairstreak is sometimes found in the company of both White-letter and Purple Hairstreaks and distinguishing these three species in flight is almost impossible.

Females are mated almost as soon as they emerge, although courtship is rarely observed. One account by Vince Lea, and published in the *The Entomologist's Record and Journal of Variation* in 2006, describes a pair that undertook a small number of short spiralling flights, landing on a Blackthorn bush after each flight. The pair eventually flew deep into a bush and mated, remaining coupled for 25 minutes (around an hour is more typical). Given this behaviour, it is no wonder that mating pairs are rarely encountered and I must have had luck on my side when I first went in search of this species at Monks Wood in Cambridgeshire—a Magpie *Pica pica* landed atop a tall Blackthorn bush and displacing a mating pair of Black Hairstreaks that then flew down and landed on a Hazel right in front of me, where I was able to get my first photos of this species.

When egg-laying, the female will crawl around the inside of a Blackthorn bush while probing various nooks and crannies with her abdomen for a suitable site, before laying a single egg on the bark, sometimes in the fork of a small twig. Eggs are deposited on one- to four-year-old growth and at all heights on the foodplant, although most eggs are laid above 1.5 m from the ground.

A mating pair [NH]

A ♀ laying inside of a Blackthorn bush [PW]

Ovum

The egg is blue-green when first laid, but gradually turns dark brown, ultimately turning a pale grey with the onset of winter and often becoming discoloured due to algae that form on it. Eggs are just under 1 mm in width and resemble a flattened sphere with a sunken crown and micropyle, like those of most Lycaenids, but are unusual in that their entire surface is covered in bumps, much like the seed capsule of a Horse-chestnut, with the larva playing the role of the conker.

The egg of the Black Hairstreak

On rare occasions, eggs may be laid in twos and threes

The egg does, however, have the appearance of a small bud and, as a consequence, is extremely difficult to locate in the wild, unlike the easily located egg of the Brown Hairstreak, a species which also lays on Blackthorn but whose white egg stands out against the dark bark. The larva is fully formed within the egg as it passes the winter and the first instar larva emerges the following spring in late March or early April, having remained in the egg for around nine months.

Larva

The **first instar** larva emerges from its egg by eating away the crown but does not eat the remainder of the shell. The tiny 1.3-mm long larva immediately makes its way to the end of its branch where it feeds by day, using its extensible neck to bury its head deep into a flower bud, or into the unfurling leaves if available, where it scoops out the insides, much like the behaviour of a Holly Blue larva as it feeds on the unopened flower buds of Holly, Ivy and other plants.

The newly emerged first instar is a dark wine red with several longitudinal lines of relatively long white hairs. The larva develops a lighter band around its middle as it matures, and the resulting colour combination renders the larva almost invisible when it is resting on a woody stem. In fact, the larva is extremely well camouflaged in all of its four instars, always blending in with the youngest leaf growth that it prefers to feed upon. Until I made an effort to study the immature stages of this species in detail, I had always considered, incorrectly, that Blackthorn was predominantly green leaves on brown woody stems. However, unfurling young Blackthorn leaves reveal a subtle mix of yellow, green and pink, and it becomes clear as to why the larvae have the colouring they do.

This camouflage makes it difficult to find larvae and early collectors resorted to 'beating' for larvae as this was the simplest way of obtaining this species for their collections, although this method was not always met with positively. An entry by George B. Dixon in the 1898 volume of *The Entomologist's Record and Journal of Variation* expresses concern at this approach: "*I feel constrained to add yet another growl to the columns of the* Entomologist's Record. *I went, last season, to the happy hunting-ground of* Thecla pruni. *On arrival I was filled with disgust and contempt. Someone, who evidently is well "in the know" of at least one good locality for this insect had beaten the bushes most unmercifully, for not only were the terminal and lateral shoots of the stems broken off, but the bark and part of the stout stems themselves were cut off with the stroke … Someone may say 'But you cannot obtain* T. pruni *without beating'. Persons cannot, of course, if they are blind, and if they are, they ought not to be entomologists*".

After two to three weeks, when just under 2 mm long, the larva prepares to moult by attaching itself to a dark leaf bud. This is in contrast with every other instar when the larva attaches itself to a leaf prior to moulting.

The **second instar** larva is similar in appearance to the previous instar, although it now has a very obvious two-tone colouring, with dark red markings at front and rear, and green across its middle. Also, the hairs on the body are now black, rather than the white of the previous instar. When fully grown, the second instar is just over 4 mm long.

After another two weeks or so, the larva moults into its **third instar**. The larva is now extremely variable in colour, ranging from almost green through to the colouring found in earlier instars (front and rear a dark red with a green band across the middle). The body also has more obvious humps, with a central furrow running along the back. The fully grown third instar is around 9 mm long.

After two weeks or so, the larva moults into the final and **fourth instar** where, again, individuals vary in appearance, ranging from almost complete green through to a mix of yellow, green and pink. The distribution of colours is somewhat different in this instar, with any red colouring now running

down the back of the larva from the second segment to the anal segment. However, the key difference between this instar and previous instars is that the colour of the head, which up to this point has been a shiny black, is now a pearly white. Another key difference is that the humps found on the back of the larva are confined to the fifth to ninth segments.

When fully grown, the final instar is just under 16 mm long and may move up to two metres before finding a suitable pupation site, such as a leaf or a stem. It then spins a silk pad to which it clings, and also attaches itself with a girdle around its waist. The larva becomes increasingly pale and, having spent 10 days in this instar, moults for the last time into the pupa.

4th instar: a primarily green larva showing its characteristic white head

4th instar: a larva exhibiting a mix of yellows, greens and pinks

4th instar: a pre-pupation larva

Pupa

The pupa of the Black Hairstreak exhibits a similar strategy of avoiding predation as an early instar Swallowtail larva—it mimics a bird dropping. This mimicry is so perfect that the larva chooses to position itself in plain sight on the upper surface of a Blackthorn leaf or stem before the resulting pupa is revealed. The pupa itself may be brown or black but is always adorned with white markings that contrast with the ground colour. Despite this mimicry, it is actually quite visible to the trained eye and does

When shown in context, the pupa perfectly mimics a bird dropping

not always escape the attention of birds—in *The Butterflies of Britain and Ireland*, Jeremy Thomas says that pupal mortality is high (around 80%), mainly from bird predation.

The pupa is just under 10 mm long and is attached to a leaf or stem by the silk girdle and by the cremaster that attaches to the silk pad. The pupa is a pale green just after pupation but gradually darkens, reaching its final colour after 48 hours. The extremely local but delightful adult butterfly emerges after around three weeks.

The brown form of pupa

The black form of pupa

A newly emerged ♂ with its pupal case

Life Cycles of British & Irish Butterflies

Small Blue
Cupido minimus

❖ SMALL BLUE IS SUPPORTED BY JAYNE CHAPMAN, THE MAKIN FAMILY & JAMIE MARLAND ❖

Both the vernacular and scientific names of this species recognise the smallest of our butterflies, whose wingspan varies between 18 and 27 mm in both sexes. This butterfly is, however, a giant when compared with the world's smallest, the Western Pygmy Blue *Brephidium exilis* from North America, whose wingspan is a mere 12 mm. Male and female Small Blues are similar in appearance, although the male upperside is almost black with a dusting of blue scales, whereas the female upperside is dark brown with no blue scales. Both sexes have silvery grey undersides that are not unlike those of the Holly Blue, although the two species live in very different habitats, with the Small Blue confined to areas where its sole larval foodplant, Kidney Vetch, is found. This foodplant, which is also known as 'Woundwort' (not to be confused with woundworts in the dead-nettle family), was used by traditional herbalists to relieve a whole host of maladies including, as we might guess, kidney problems.

The ♂ is almost black with a variable amount of blue scaling [NF]

The ♀ is mainly brown and has no blue scales [CK]

Both sexes have a silvery grey underside

DISTRIBUTION

While the Small Blue is found from the north of Scotland to the south of England, and from the west of Ireland to the south-east of England, it has a surprisingly sparse distribution outside of its strongholds on the chalk and limestone grasslands of southern England, especially those found in the Cotswolds and on Salisbury Plain. Other colonies are often found in isolated pockets in coastal locations that includes the north coast of Cumbria in England, the Moray and Angus coasts in Scotland, the south-west coast of Wales, and the west and south-east coasts of Ireland, especially in the counties of Donegal, Sligo, Clare, Limerick and Kerry in the west, and the counties of Wexford, Wicklow and Dublin in the east. The butterfly is absent from the western and northern Scottish isles, the Isle of Man and the Channel Islands. It has also largely disappeared from northern England and there have been very few sightings from Northern Ireland in recent years. On a worldwide basis, the butterfly has a Palearctic distribution and is found throughout Europe and across temperate Asia.

Colonies tend to vary in size from year to year, which may be linked to the quality of Kidney Vetch, and while most colonies are very small, consisting of less than 30 adults at the peak of the flight period, a few colonies at the best sites can reach over a thousand adults. Most adults fly less than 40 m during their life and studies show that males are, surprisingly, more sedentary than females. That being said, based on mark-release-recapture data, adults have been recorded 1 km from a neighbouring site and the butterfly has also been found several kilometres from any known colony on occasion. It is therefore not surprising that the butterfly is able to colonise newly created sites, especially when they are close to existing colonies.

HABITAT

The ideal Small Blue habitat is sheltered, warm and free draining, with a good amount of Kidney Vetch, along with grasses and shrubs that are used for perching and roosting. A wide variety of habitats is used, including unimproved chalk and limestone grassland, abandoned quarries, sand dunes, undercliffs, railway embankments and even steep-sided road verges. All sites used by the

Life Cycles of British & Irish Butterflies

The Small Blue thrives in chalk scrapes that are rich in Kidney Vetch at Magdalen Hill Down in Hampshire

butterfly have sparse vegetation with some disturbance (caused by erosion, or the activities of rabbits and ants) that allows Kidney Vetch to become established, and plenty of nectar sources.

STATUS

The Small Blue has suffered a significant long-term decline in distribution, especially in the northern half of England, and is now confined to a handful of colonies in many regions. Due to the isolation and the somewhat-limited mobility of this species, many remaining colonies are also prone to extinction when any local losses cannot be balanced by recolonisation. Appropriate habitat management of existing sites that encourages growth of Kidney Vetch while maintaining surrounding shrubs and shelter is therefore essential.

Fortunately, it is easy to create the right conditions for this butterfly and there are numerous examples of the butterfly colonising newly created areas once Kidney Vetch has established itself, with females able to circumvent any obstacle they face en route. Designed by Dan Danahar, the Liz Williams Butterfly Haven at the Dorothy Stringer School in Brighton, East Sussex, is an excellent case in point. The site was colonised by the Small Blue two years after its creation and, according to Neil Hulme in *The Butterflies of Sussex*, the nearest colony is 3 km away, and required a gravid female Small Blue to make its way across a city landscape to reach its new home.

The long-term trend shows a moderate decline in distribution and a stable abundance, with the more recent trend over the last decade showing a stable distribution and a moderate decline in abundance. Using the IUCN criteria, the Small Blue is categorised as Near Threatened (NT) in the UK and Endangered (EN) in Ireland.

LIFE CYCLE

Adults start to appear in mid-May at southern sites, reach a peak at the end of May and start of June, and produce a partial second generation at the end of July and start of August. Adults may appear in the first half of May in good years. Emergence is later further north where there is no second generation, with adults emerging in the first part of June in northern Scotland.

Imago

This delightful little butterfly spends a good amount of time basking and resting, when small groups of twos and threes can be found together, with individuals periodically making sorties in search of their favourite nectar sources of Common Bird's-foot-trefoil, Kidney Vetch and Horseshoe Vetch. Males can sometimes be found feeding together while taking on salts and minerals from damp mud, animal droppings and carrion. Both sexes may be found from late afternoon onwards in communal roosts, facing head-down in long grass where they open their wings to catch the last of the sun's rays.

When on the lookout for a mate, males perch between 30 and 120 cm above the ground on small shrubs or grass stems in sheltered pockets, often at the base of a south-facing slope that is devoid of Kidney Vetch, where they will fly out to investigate passing insects in the hope of finding a virgin

A ♂ attempting to mate with a newly emerged ♀

A mating pair

A ♀ egg-laying on Kidney Vetch, the sole larval foodplant

female. Newly emerged females will deliberately fly into these areas where they are quickly mated without any elaborate courtship, the pair remaining coupled for around an hour.

Once mated, the female moves away from the male perching areas and spends most of her time searching out suitable plants of Kidney Vetch on which to lay, which are normally those whose inflorescences are still expanding and relatively immature, although those whose flowers are starting to wither may be used if Kidney Vetch is in short supply. After laying a single egg on a floret (or between two florets), she then rubs her abdomen over the flowerhead, leaving a scent that is thought to deter other females from laying on the same plant, since the larvae are cannibalistic, especially in the first three instars.

Several eggs are often laid on a single plant at good sites

Ovum

Eggs are of the typical Lycaenid bun shape and, given the size of the butterfly, are very small, being only 0.4 mm wide and 0.2 mm high. The ground colour is a pale blue-green, although a fine network of white ridges that cover the surface of the egg gives it a white appearance. Several eggs are occasionally found on the same inflorescence, although this is usually the result of visits from different females, or the same female visiting at different times. Eggs are quite easy to find at suitable sites, even when the gleaming white egg is laid on the white downy calyx of a florct, although the eggs do darken before the larva emerges when they become more visible. Locating eggs is, in fact, often the easiest way to determine if a site has been colonised.

The white egg is laid on the white downy calyx of a Kidney Vetch floret

The egg turns dark grey just before hatching

Larva

After between one and three weeks, depending on temperature, the 0.8-mm long **first instar** larva emerges from the egg by eating an untidy exit hole in the top of the egg. The larva does not eat the remainder of the eggshell and this can give away the presence of a larva to the keen observer. The larva has a pale yellowish-brown ground colour, a shiny bronze-black head and an olive-brown patch on the first and last segments. The larva is also covered in several rows of hairs that run the length of its body.

On hatching, the larva immediately burrows into a floret by eating its way through the downy calyx and corolla, ultimately living inside the floret where it feeds on the ovary and developing seed. Surprisingly, it is sometimes possible to make out the silhouette of the larva as it sits on the inside wall of the floret. Given the limited resources available on a single Kidney Vetch inflorescence it is perhaps not surprising that the larvae are highly cannibalistic, with any younger larva eaten by its elder. When fully grown, the first instar is just under 1.5 mm long.

After only a few days, and while inside the floret, the larva moults into the **second instar**. The larva continues to exhibit a pale yellowish-brown ground colour, but this is overwhelmed by several rows of reddish-brown markings that give this instar a lilac appearance. The larva now has a black head, is covered in numerous and relatively short hairs, and retains the olive-brown patch on the first segment, although it no longer possesses a patch on its last segment. The larva also exhibits a well-developed honey gland on the 10th segment that is a characteristic of many larvae of the Lycaenidae.

The young second instar continues to feed inside a floret but starts to feed from outside a floret as it matures, when it can be found with its rear end exposed while its front end is buried deep inside the floret to reach the developing seed. As the contents of each floret are eaten, the larva moves on to another, leaving tell-tale holes at the base of each floret as it goes, along with frass that gets trapped within the floret hairs. When looking for these diagnostic characteristics, it should be remembered that the base of each floret is very fragile and can easily break off, and any search should therefore be undertaken with care. When fully grown, the second instar is around 3.5 mm long.

After around a week, the larva moults into the **third instar** and is now just over 4 mm long. The overall appearance is similar to that of the second instar although, apart from an obvious difference in size, the ground colour is now pale brown, and the larva now has a more prominent dark lilac stripe running along its back from the second segment, with several less prominent stripes of the same colour along its sides. It also has an almost-white line along each side just below the spiracles that is very faint in the second instar, if it is visible at all. While the larva retains its shiny black head, the patch on its first segment is now very pale when compared to that of the second instar. The larva is now of a size where it spends all of its time outside of a floret and it often takes up position head-down along the side of a floret where it continues to get access to the developing seed by eating a hole in the floret wall. When fully grown, the third instar is just under 6 mm long.

After around a week or so, the larva moults for the last time into the **fourth instar**. Like earlier instars, the larva withdraws its shiny black head under the first segment when at rest, but can extend it forward when feeding, thanks to its extensible neck. The larva now has a pale brown ground colour with a smattering of subtle pink markings, including a pink stripe that runs along its back. It is also covered in a series of fine bristles of varying length, each of which is set in a dark bulbous base and which, in combination with numerous small dark spots, result in a speckled appearance when the body is viewed close up.

The larva possesses the ant-attracting devices of pore cupolas and a honey gland but lacks the tentacle organs found in other Lycaenids. Ants are only rarely found in attendance, possibly because the ant species present in Britain and Ireland do not forage as high as Kidney Vetch flowers, although ants are frequent visitors to larvae on the continent.

4th instar: a newly moulted larva in context of Kidney Vetch flowerheads

When fully grown, the larva is around 9 mm long and it will either overwinter or pupate and go on to produce a second brood. Those that overwinter are believed to do so at ground level, either under moss, within some other vegetation, or in a crevice in the soil. In the *Sussex Butterfly Report* for 2014, Dave Harris writes of two larvae that had been found inside a nest of the black ant *Lasius niger*. The larva re-emerges in the spring and, without feeding further, wanders off to find a suitable pupation site. When preparing to pupate, the larva attaches itself to a grass blade, leaf or some other vegetation by a silk girdle and a silk pad at their back end. In captivity, the larva will occasionally attach itself to a Kidney Vetch floret.

4th instar: the larva is adorned with subtle pink markings

4th instar: a rare sighting of a larva attended by a black ant [IC]

4th instar: a larva preparing to pupate

Pupa

The 8-mm long pupa is attached to its chosen surface by a silk girdle and by the cremaster, which hooks into the silk pad created by the larva. The pupa has a beautiful pattern of varying intensity on its wing cases and relatively long hairs on its head and along the back of its body. This diminutive but delightful butterfly emerges in around two weeks, depending on temperature.

The pupa is extremely hairy

The pupa is attached by a silk girdle and the cremaster

A coloured-up pupa just hours before the adult emerged

Holly Blue
Celastrina argiolus

❖ HOLLY BLUE IS SUPPORTED BY DAVE MILLER, JAN SCHUBERT & MIKE CREIGHTON & BILL SEAGER ❖

	January	February	March	April	May	June	July	August	September	October	November	December
IMAGO												
OVUM												
LARVA												
PUPA												

The **Holly Blue** is a familiar sight in gardens, where two of its primary larval foodplants, Holly and Ivy, are often grown. This double-brooded species is unique amongst the butterflies of Britain and Ireland in that its different broods use different foodplants, and for it to be named after one of the primary foodplants used by the spring brood could be seen as rather misleading—the butterfly might equally be called the 'Ivy Blue' after the primary foodplant used by the summer brood.

James Petiver thought that the male and female were different species when he first named them in his *Papilionum Britanniae* of 1717, giving the male the name 'Blue Speckt Butterfly', which recognises the pattern of spots found on the underside, and the female the name 'Blue Speckt Butterfly, with black tipps', which acknowledges the dark band found on the apex of the forewings of the female. William Lewin also recognised a key habitat type for this species when he named this butterfly the 'Wood Blue' in his *The Papilios of Great Britain* in 1795. It was not until 1853 before the butterfly was given the name we use today by the Reverend F.O. Morris in his *A History of British Butterflies*.

Males of the spring and summer broods are similar in appearance, but females of the summer brood have much broader dark bands on their forewing uppersides than females of the spring brood. Despite being one of our 'blues', both male and female have a lilac tinge when seen at certain angles to the light, although the butterfly never fully opens its wings.

♂ and ♀ have similar undersides and a wingspan of 26 to 34 mm

♂s of the two broods are similar in appearance

The ♀ of the spring brood has reduced bands on its forewings [NF]

The ♀ of the summer brood has prominent bands on its forewings

The Holly Blue is often confused with a male Common Blue, although the spotting on their undersides is quite different and, behaviourally, the Holly Blue normally flies much higher than the Common Blue, behaving much like a hairstreak, whereas the Common Blue usually flies just a few feet off the ground. The location of the individual may also provide a clue—a butterfly seen in suburbia or in woodland is almost certainly going to be a Holly Blue.

A *Listrodomus nycthemerus* wasp that has recently emerged from a Holly Blue pupa

This species is known to fluctuate wildly in numbers, forming a predictable cycle over four to six years, which is thought to be driven by the level of parasitism from the ichneumon wasp *Listrodomus nycthemerus* whose sole host is the Holly Blue. The wasp lays its eggs in early instar Holly Blue larvae, one egg per larva, with the adult wasp eventually emerging from the Holly Blue pupa. As the level of parasitism increases, Holly Blue numbers decline, to the point that numbers of the parasite decline due to the unavailability of its host, allowing the butterfly to increase in numbers. And so the cycle continues.

DISTRIBUTION

The butterfly is widespread and common across most of England and Wales, but has a more scattered distribution in Ireland. It is also found on the Isle of Man and has been recorded in Scotland on occasion, where it has been found as far north as Edinburgh. This is a widespread species and its range extends throughout Europe and northern Asia to China and Japan. It is also found in North Africa, and in North and Central America. While the butterfly will be found in the same locality year after year, it does not form discrete colonies and will wander across the landscape.

HABITAT

This butterfly is found in many different types of habitat, including gardens, parks, churchyards; woodland; mixed scrub; hedgerows; and arable field margins and set-aside—anywhere its foodplants and nectar sources can be found—I remember once watching several Holly Blues flying around a clump of Ivy in central London.

A clump of Ivy growing over a World War II pill box on the Kennett and Avon canal path

STATUS

The long-term trend of this species indicates a moderate increase in both distribution and abundance, with a spread northward, showing that this species is doing well in the long run. The short-term trend is, however, less positive, with a moderate decline in distribution and a significant decline in abundance. Using the IUCN criteria, the Holly Blue is categorised as Least Concern (LC) in both the UK and Ireland.

LIFE CYCLE

The Holly Blue is the first of our 'blues' to show itself during the year, with the spring brood emerging from overwintering pupae at the start of April in good years, peaking in early May, and giving rise to a summer brood that emerges in the second half of July and that flies until early September. While the butterfly is normally double-brooded, this does vary by region—the most northern populations produce a single brood and, conversely, some southern regions manage to produce a third brood in favourable years, with sightings in October and, on the rare occasion, in November.

Imago

Like many species, the adults have a predisposition toward aphid honeydew as a primary source of nourishment, although both sexes also visit a variety of nectar sources including brambles, Bugle, buttercups, forget-me-nots, Holly, Ivy, thistles, Water Mint and Wild Privet. Males will also come down to the ground to take salts and minerals from damp mud and animal waste.

F.W. Frohawk used an unconventional light source when observing the roosting habits of this species, as described in his *Natural History of British Butterflies* of 1924: "*This little butterfly passes dull weather settled among the foliage of trees and bushes, where it also rests for the night ... A female argiolus settled on a laurel leaf in the sunshine ... It at first sat with closed wings facing the sun, so only cast a thin streak of shadow .... At 4.48 it opened its wings and listed to the left to get the full rays of the sun. At 4.50 it flew off straight to a privet hedge six feet away, and immediately settled on a leaf two or three inches inside, to roost for the night ... It did not move when the writer held a lighted match near to examine it*". I can recall my own childhood experiences of the Holly Blue roosting within the branches of a conifer that grew in the front of the family home, when the odd butterfly would fly very late into the evening, with activity still noticeable at 10 pm in the height of summer, before the butterflies settled down to roost. I am unsure if the street lights or house lights influenced this behaviour, but I was amazed that I was still able to partake in my favourite hobby so late into the evening.

Of the two sexes, the male is the most often encountered when launching itself from its perch at a passing butterfly, or when fluttering around each plant in turn as it searches for a virgin female. In *Butterflies of Surrey Revisited*, Ken Willmott describes how the species frequently uses the same 'master trees' from year to year on which to perch, which are normally at a high point in the landscape—whenever I visit Noar Hill in Hampshire to see the Duke of Burgundy in spring or Brown Hairstreak in late summer, I invariably find Holly Blues fluttering around the tall conifers at the top of the reserve.

If the male encounters an already-mated female, then she will rapidly beat her wings, and even drop into the undergrowth, until the male loses interest and flies off. Any courtship and subsequent mating is a rarely observed event. In the Butterfly Conservation booklet on the Holly Blue, Ken Willmott writes: "*A resting female was discovered by an actively searching male and she immediately led him from her resting place onto a suitably sited 'platform', in this case a large ivy leaf, part of an extensive growth covering a large expanse of brick wall. They paired at 17.13 hrs. and were found still 'in cop' at 18.45 hrs. that evening. The pairing site was re-visited early next morning and the couple were found parted, in the shade, on separate ivy leaves, in close proximity*".

A mating pair [RC]

An egg-laying female will flutter around a foodplant that is growing in a sunny and sheltered situation, before settling on a flowerhead, curving her abdomen underneath an unopened bud and depositing a single white egg. She then flies off to repeat the process, often on the same bush. A surprisingly large variety of plants is used, including Holly, Spindle, dogwoods and Gorse that are chosen by the spring brood and Ivy, buckthorns, brambles and snowberries that are chosen by the summer brood. Holly is dioecious (separate male and female plants) and only a female plant produces the fruit whose development provides sustenance for any larvae. Eggs are occasionally laid on male trees that do not produce fruit and, in this case, larvae reach maturity by feeding on the terminal and tenderest leaves of the plant. As documented in a 1994 issue of *British Wildlife*, Richard Revels, an expert in this species, observed that the butterfly seemed to prefer dogwoods over Holly for the spring brood, when both were available.

A ♀ of the summer brood laying on Ivy [TP]

Unopened Ivy flower buds are used by the summer brood

Ovum

The egg is 0.6 mm wide and 0.3 mm high, shaped like a flattened sphere. It is quite beautiful when seen close up and resembles the shell of a sea urchin, with a delicate pattern of peaks and hollows that cover its surface. The egg has a pale greenish-blue ground colour, although the peaks are pure white, and the overall white appearance makes it easy to find eggs on favoured plants, especially in good years when the number of eggs laid is relatively high.

Larva

After a week or so, the tiny 1-mm long **first instar** larva emerges from the egg by eating away the crown. The remainder of the eggshell is not eaten and empty shells are quite easy to find, especially once a larva has been located. The larva starts to feed by burying its head deep into the bud on which the egg was laid, before scooping out the contents. Much like the calm persona exhibited by a duck that is frantically paddling under the water, the Holly Blue larva appears completely still while it feeds within the flower bud. The first instar has a series of long hairs along its back that curve backwards, with smaller hairs slightly below these, and also along its sides. The larva has a shiny black head in all instars, but always keeps it hidden under the first segment when not feeding. The larva feeds by both day and night and leaves tell-tale feeding damage on the flower buds.

After around seven days, the fully grown 2-mm long first instar larva moults into the **second instar** and proceeds to eat its old skin, which it does after each moult. The larva is now pale green and has a dense covering of long white hairs. When fully grown, the second instar is 3 mm long and is now a brighter green, with a pale line running along each side.

After seven days, larva moults into the **third instar**. Structurally, the larva is similar to the previous instar, although the hairs are shorter, and the pale line running along each side is much more prominent. One visible change is that the third instar now exhibits a honey gland, which appears as a scar on the 10th segment. When fully grown, the third instar is around 6 mm long.

Third and fourth instar larvae have different forms and F.W. Frohawk describes three in his *Natural History of British Butterflies* in 1924. The first form has a clear green ground colour with subtle pink markings and with a white line running along each side. The second form is a pale greenish-brown with rich pink markings and with a yellow line running along each side. The third form, which is the most common by far, is light green without any pink markings and with a pale yellow line along each side. William Buckler, in his *The Larvae of the British Butterflies and Moths* of 1886, that precedes Frohawk's work, describes five forms in total. Suffice to say, we can simply conclude that the larva is quite variable in terms of its ground colour, its pink markings, and the colour of the line on each side of the body.

1st instar: a close up of the larva

2nd instar: a newly moulted larva next to its eggshell

2nd instar: the fully grown larva is a brighter green and has a pale line along its sides

3rd instar: the larva has shorter hairs than previous instars

After around a week, the larva moults into the final and **fourth instar**. This is the easiest instar to find in the field, not simply because it is larger than previous instars, but because the feeding damage on developing flower buds is quite noticeable—the flower buds that have been eaten become discoloured and stand out from the others, making it very easy to scan a patch of foodplant in a short amount of time (and there may also be frass in the vicinity of the feeding damage). The buds that have been eaten also have a distinctive hole in one side where the larva has bored into the bud before scooping out the contents.

In his *magnus opus* of 1906, *A Natural History of the British Lepidoptera*, James William Tutt writes: "*In some few cases where the buds showed that they had been tenanted but the larva was absent, it was usually found on the back of one of the nearest ivy-leaves, and, apparently, in each case, in the act of moulting*". Therefore, if obvious feeding damage is found on a plant, but no larva can be located, then it is either because it has moved off to pupate, or it is in the process of moulting and the undersides of nearby leaves should be searched. This may explain any mysterious disappearance and reappearance of any larva being monitored.

4th instar: a larva with rich pink markings along its back and sides [NH]

4th instar: a larva revealing its shiny black head

4th instar: a green form of the larva, and with no pink markings

4th instar: A larva being tended by the black garden ant *Lasius niger*

4th instar: the larva turns a dull purple prior to pupation

4th instar: a pre-pupation larva attached to the underside of a leaf

The larva continues to exhibit different forms in this instar, although the colours are now much more brilliant than in the third instar. One interesting observation made by Frohawk is that a larva of a particular form in the third instar does not necessarily retain the same colouring when it moults into the fourth instar. When fully grown the fourth instar is between 13 mm and 15 mm long.

I am lucky to have a patch of flowering Ivy in my garden that plays host to the Holly Blue every year and, in some years, the larvae are tended by the black garden ant *Lasius niger* (red ants have been recorded as tending Holly Blue larvae on grassland sites), where the ants can be seen feeding from the honey gland on the 7th abdominal segment (the 10th segment), that is most developed in this instar. The larva also has a pair of tentacle organs on its 8th abdominal segment (the 11th segment).

Prior to pupation, the larva gradually turns a dull purple and starts to wander away from the flower buds to find a location to pupate, when its new colouring possibly offers better camouflage as it crawls over woody stems and branches. There are reports of pupae being found at the back of dense Ivy that is growing up a wall as well as among Ivy roots, although I suspect that many larvae pupate at ground level, and possibly within an ant nest. Wherever it pupates, the larva attaches itself to its chosen surface with a very fine silk girdle.

Pupa

The larva pupates after a few days, revealing a 9-mm long stumpy pupa that is typical of the Lycaenidae, and that is attached by the silk girdle and by the cremaster to a silk pad. The pupa is a mix of browns with some mottling on its surface, resulting in an overall 'burnt' appearance. Offspring from the spring brood emerge in two to three weeks, whereas offspring of the summer brood overwinter and spend around 210 days as a pupa, before emerging the following spring.

The pupa is attached to its chosen surface with a fine silk girdle

Life Cycles of British & Irish Butterflies

Large Blue
Phengaris arion

❖ *LARGE BLUE IS SUPPORTED BY HABITAT DESIGNS LTD, HELEN STERNE & MARK TUTTON* ❖

	January	February	March	April	May	June	July	August	September	October	November	December
IMAGO												
OVUM												
LARVA												
PUPA												

Of all the butterflies of Britain and Ireland, the history of the Large Blue is surely the most fascinating, with our emerging understanding of its complex life history intertwined with a tale of decline, extinction, reintroduction and expansion. I also feel a personal connection with this butterfly since I grew up in one of its previous locations on the south side of Cheltenham on the edge of the Cotswolds, with an 1858 record of the butterfly just 100 m from my childhood home in Leckhampton.

While this is the largest of our blues, the Large Blue is only slightly larger than a Chalk Hill Blue and is most easily identified from its upperside, with streaked spots and dark borders on the forewings that are unmistakable, although these markings are highly variable and much reduced in some individuals. Both sexes are superficially similar in appearance, although the upperside markings are more prominent in the female. The best way to distinguish the two sexes is to look at the abdomen—that of the male is long and thin, whereas that of the female is shorter and wider since her abdomen is filled with eggs.

♂ with a wingspan of 38 to 48 mm [NH]

♀ with a slightly larger wingspan of 42 to 52 mm

♂ and ♀ have similar undersides

The complete life cycle of the Large Blue was a mystery to early entomologists—while they were aware that eggs and young larvae could be found on the flowerheads of Wild Thyme, they were unable to determine what the larvae ate in their last instar. William Buckler wrote the following in his *The Larvae of the British Butterflies and Moths* of 1886: "*I had wild thyme ready for them, planted in a large flower-pot, and full of bloom; on this some of the larvae lived till July 28th, when they were seen to be restless as if in search of something I had not given them, and after that I could note no more ... All these failures puzzled us much; after three weeks' feeding on the thyme flowers the larvae seemed to want something else – what could it be?*". Three names figure in addressing this gap in our knowledge—F.W. Frohawk, T.A. Chapman and Capt. E.B. Purefoy.

In 1903 Frohawk wrote of his experiences in the field in *The Entomologist*: "*From observations I made last year concerning the deposition of the eggs in a natural state, I felt convinced that some connection existed between arion larvae and the common yellow ant (Formica flava) by the preference shown by the butterfly in selecting the thyme growing on ant-hills for oviposition ... On Aug. 11th, many having passed their third moult, when they cease feeding on thyme, I started investigating what relation there might be between the larvae and ants, thinking in all probability that they might feed either on the larvae or pupae of the latter ... I supplied this larva with an ant's cocoon with one end removed; it at once began eating it*".

It was a few more years before Frohawk was able to reapply his detective skills and, in 1906, he once again shared his thoughts in *The Entomologist* following a thorough examination of an area where Large Blues had been seen laying: "*...we then determined on searching all the most likely-looking ants' nests ... upon shaking part of the crown of the nest over a cloth, a goodly-sized, plump, cream-coloured, grub-like larva fell out, which I instantly identified as a full-grown arion larva ... its head was so disproportionately small for the size of its body, that I at once concluded it had not passed through another moult*". Frohawk had therefore confirmed that Large Blue larvae remained in an ant nest for the duration of their final and fourth instar.

The next major discovery was made by Chapman who, while in the company of Frohawk, found a Large Blue larva that he subsequently analysed, as summarised in the *Transactions of the Entomological Society of London* in 1915: "Mr. Frohawk's discovery … compelled us to regard … the question of how this minute larva, one-eighth of an inch long, grew to its mature dimensions of well over half an inch long and correspondingly thick; in fact, a larva suitable to produce a butterfly as large as L. arion. What was its food? On May 14, 1915, on pulling up plants over a nest of Myrmica scabrinodis var. sabuleti [now classified as *Myrmica sabuleti*] … a larva of L. arion *was found. Unfortunately in the rough process necessary in disturbing plants and soil the larva suffered an injury … there remained, however, the possibility of learning what it had eaten by examining the contents of the alimentary canal … the hairs in the arion agreed precisely with those of the full-grown larva of* M. scabrinodis, *and that the chitinous triangles agreed exactly with the mandibles of the same larva. Nothing of a vegetable character was found amongst these contents, and it could not be doubted that the L. arion* larva had eaten many larvae of scabrinodis and nothing else for a long time". Chapman had therefore confirmed Frohawk's suspicions that Large Blue larvae did, indeed, feed on ant grubs in their final instar.

Later in the same year, Chapman determined that the Large Blue larva was carried into the ant nest, with the adopting ant no doubt being fooled into thinking that one of its grubs had somehow left the nest: "*The ant examines it and proceeds much as ants do when milking Lycaenid larvae. At length … the larva assumes a most extraordinary form, swelling up the thoracic segments at the expense of the others … The ant then seizes it behind the thorax and carries it into her nest*".

Meanwhile, Purefoy was attempting to rear the Large Blue in captivity, following the transplanting of ant nests to his garden at East Farleigh in Kent. Frohawk, who was recording the life history of the Large Blue in detail, summarised their collective observations in a 1915 article in the *Transactions of the Entomological Society of London* where he says: "*Capt. Purefoy tells me that in every case … the ant which first finds the larva is the one that carries it away*". Incredibly, after much trial and error, Purefoy eventually succeeded in rearing Large Blues from the egg, with adults emerging in 1917.

DISTRIBUTION

In addition to the discovery of its elaborate life cycle, the decline, extinction, reintroduction and expansion of this butterfly is a key part of its history. Even early entomologists considered the Large Blue to be a rare and local butterfly—"*This species of butterfly is but rarely met with in England*" writes William Lewin in his *Papilios of Great Britain* of 1795—and they would make pilgrimages to known sites each season to take specimens for their collections. Core areas included the Poldens in Somerset (first recorded in 1833 and found at 3 sites), Northamptonshire (1837, 7 sites), the Cotswolds in Gloucestershire (1850, 33 sites), the South Devon coast (1856, 6 sites), Dartmoor in Devon (1870, 6 sites) and the Atlantic coast of Devon and Cornwall (1891, 34 sites). Collectors also noted that specimens from different regions also had subtly different hues that no doubt added to their appeal.

The demise of the butterfly at seemingly suitable sites was a concern even in the late 1800s and theories for the decline were offered by several entomologists, including Herbert Marsden who shared his thoughts in the *Entomologist's Monthly Magazine* in 1884: "*The year 1870, however, is the one to be marked with a white stone by the lovers of Lycaenidae. It would, I am sure, have been possible for an active collector, of the greedy school, to have caught over 1000 "large blues" during the season, for in a few visits I secured about 150, not netting half of those seen, and turning many loose again. Now come the dark days. Part of June, 1877, was damp and broken … In 1878 the weather was worse … 1879 was yet worse … In 1878, not over a dozen were seen, mostly worn and weather-beaten, for there were hardly ever two consecutive fine days. In 1879 they were yet scarcer, while in 1880 only two were obtained, and two or three more seen. For the four years, 1881-84, not one has been seen in the Gloucestershire district that I have been able to trace*".

Unfortunately, the situation was more complex than anyone could have imagined, with the butterfly continuing to decline over the following decades. Several causes were suggested, including over-collecting, a cooling climate and habitat loss due to development and agriculture. Inappropriate grazing regimes, abandonment and the effect of myxomatosis on rabbit numbers in the 1950s all contributed to sites becoming overgrown, resulting in cooler ground temperatures that caused the ants to move out and the butterfly to decline. However, none of these suggestions provided an adequate explanation for losses at sites that remained in suitable condition, with plenty of Wild Thyme and numerous colonies of red ants. Despite many attempts to save this butterfly over the years, it ultimately became extinct in each of its known regions—the Poldens in the late 1840s,

Northamptonshire in around 1860, the South Devon coast in 1906, the Cotswolds in the early 1960s, and the Atlantic coast of Devon and Cornwall in 1973. At this point the Large Blue was confined to a single colony on the edge of Dartmoor that contained around 250 adults.

Despite the insights provided by Frohawk, Chapman and Purefoy some 60 years earlier, there was clearly some unknown factor that had been overlooked that might explain the terminal decline of this species. Thanks to the pioneering work of Jeremy Thomas, it was determined that Large Blues require sites that have a high density of one particular species of ant—*Myrmica sabuleti*—which requires particularly high temperatures and is best-suited to south-facing sites where the turf height is less than 3 cm (other *Myrmica* species will adopt larvae, but survival rates are very poor). The conclusion was that declines in *Myrmica sabuleti* had gone unnoticed at Large Blue sites, leading to the loss of the butterfly. Unfortunately, Thomas' discovery came just too late, with the last colony on Dartmoor going extinct in 1979.

Interestingly, in their paper 'Host specificity among *Maculinea* butterflies in *Myrmica* ant nests', published in *Oecologia* in 1989, Jeremy Thomas, Graham Elmes, Judith Wardlaw, and Michal Woyciechowski have since shown that each European *Phengaris* species depends on a single, and different, host species of *Myrmica* ant for its survival—*Myrmica sabuleti* for the Large Blue, *Myrmica scabrinodis* for the Scarce Large Blue (*Phengaris teleius*), *Myrmica rubra* for Dusky Large Blue (*Phengaris nausithous*), *Myrmica ruginodis* for Alcon Blue (*Phengaris alcon*) and *Myrmica schencki* for Mountain Alcon Blue (*Phengaris rebeli*)—and this dependency may explain losses of all of these 'Large Blue' species in apparently suitable sites.

Following its extinction in Britain, the Large Blue became the subject of a highly organised reintroduction programme. A period of habitat restoration followed, based on the refined understanding of the butterfly's ecological requirements which had been documented in detail by Jeremy Thomas. David Simcox subsequently undertook a search across a number of different sites in France, Belgium, Denmark and Sweden to find a potential donor population, with Thomas and Simcox converging on a population in Öland, an island off the east coast of Sweden. Large Blue eggs were brought back and early instar larvae reared through on Wild Thyme, with the larvae put down at sites on Dartmoor in 1983. The success of this initial release provided the encouragement needed to take things further.

Former sites, and those with potential, were surveyed to determine their suitability based on a number of factors, including the density of *Myrmica sabuleti* ant nests, the abundance of Wild Thyme and the ability to graze the site. In 1992 the Large Blue was introduced onto the Somerset Wildlife Trust reserve at Green Down in Somerset, again from livestock obtained from Öland. The butterfly flourished and produced sufficiently large egg counts to support further reintroductions and every Large Blue that now flies in Britain originates from Green Down. Reintroductions continue to this day through the combined efforts of the 'three amigos' (as I like to call them)—Jeremy Thomas, David Simcox and Sarah Meredith—and many partner organisations. Reintroduction sites are all in the south-west of England currently, with notable colonies in the Polden Hills in Somerset (such as Green

Jeremy Thomas, David Simcox and Sarah Meredith collecting Large Blue eggs for a reintroduction

Down and Collard Hill) and the Cotswolds in Gloucestershire (such as Daneway Banks). 2018 was a good year for many butterflies, not least the Large Blue, with an estimated 250,000 eggs laid at Green Down and 166,000 eggs laid at Daneway Banks.

While Large Blues live in close-knit colonies with a limited amount of dispersal, the butterfly has managed to colonise several new areas close to the original reintroduction sites in the Poldens, demonstrating that the butterfly will expand its range so long as there is suitable habitat available within striking distance of an existing colony. On a worldwide basis, the Large Blue is found across Europe and through temperate Asia as far as China.

HABITAT

The Large Blue inhabits both alkaline limestone grassland and neutral-to-acid grassland where Wild Thyme is found in abundance and where *Myrmica sabuleti* flourishes. Chalk downland is considered too dry to support the ant in sufficient numbers and, therefore, the Large Blue. Typical sites are also closely grazed and on a south-facing slope that receives the full effect of the sun. Overgrown conditions, on the other hand, result in a cooler ground temperature and a smaller population of *Myrmica sabuleti*, making the site unsuitable for the Large Blue and, specifically, the development of final instar larvae.

In Sweden, the butterfly flies around one month later than the endemic British subspecies, and there was a concern that the Swedish stock would not be suitable in terms of the crucial synchronicity between the phenologies of the butterfly and Wild Thyme (such as the peak emergence of females synchronising with the peak development of suitable Wild Thyme flower buds). Happily, it turned out that there was an 86% overlap in Devon and a 51% overlap in the Poldens in Somerset. Unfortunately, there was only a 13% overlap in the Cotswolds and the original reintroductions there failed. Twelve years after the reintroduction, however, adult emergence in Somerset had moved forward 10 days, resulting in perfect synchronicity. This provided the opportunity to reintroduce the Large Blue to the Cotswolds using home-grown Large Blues that had adjusted to their new home. Suffice to say, these reintroductions are proving to be much more successful.

In 1995 another incredible discovery was made. At a much warmer site in the Poldens, near to Green Down, *Myrmica sabuleti* is able to inhabit a tall sward thanks to the increased ground temperature. Incredibly, the Large Blue was found to be laying its eggs here on Wild Marjoram, that flowers around two weeks later than Wild Thyme, with the resulting offspring successfully completing their development in ant nests below the plants. The butterfly was also found to be using Wild Marjoram on a nearby site that David Simcox and Jeremy Thomas had designed specifically for this butterfly.

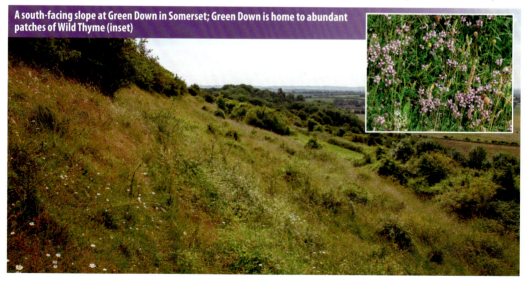

A south-facing slope at Green Down in Somerset; Green Down is home to abundant patches of Wild Thyme (inset)

In the space of 22 years the butterfly has therefore moved its phenology forward to synchronise with the development of Wild Thyme flower buds in Britain, before moving its phenology back by two weeks to synchronise with the development of Wild Marjoram flower buds. In more-or-less the same region the butterfly therefore exhibits two phenologies and exploits two larval foodplants. This opens up incredible possibilities for the future of this highly adaptable butterfly and, at Daneway Banks in Gloucestershire, both 'races' of the butterfly have been released which are able to breed over a larger area, utilising two different larval foodplants, resulting in a larger overall population. What is more exciting is that the late-emerging 'race' will be better-suited to potential reintroductions at sites in Devon and Cornwall where the butterfly historically emerged later than other populations.

STATUS

Thanks to the extraordinary research by Jeremy Thomas, the subsequent field work by David Simcox and Sarah Meredith, and the funding and other support provided by partner organisations, the future of the Large Blue in Britain looks to be secure, with the reintroduction of this species held up as a 'poster child' for conservation in action that is recognised worldwide, and rightly so. Despite the overwhelming success of the Large Blue reintroduction programme, the Large Blue is categorised as Critically Endangered (CR) in the UK using the IUCN criteria.

LIFE CYCLE

This butterfly has one generation each year and a relatively short flight period, with adults typically seen on the wing from the first week of June until early July, with a peak in mid-June. Of course, these dates may vary in years that are considered early or late—in 2018, for example, the first sighting was reported in late May.

Imago

This is a warmth- and sun-loving butterfly that spends much of its day resting on a shrub and often out of sight. However, the adults are quite conspicuous when flying and can often be found feeding on a variety of plants, especially Wild Marjoram, Wild Thyme and Selfheal. The butterfly has a slow and fluttering flight, but is surprisingly nimble, flying effortlessly from the bottom to the top of the steepest of slopes and outpacing even the most athletic of observers. In bright sunlight the adults keep their wings closed, and this is one of the few times when intermittent sunshine or overcast skies are welcomed by photographers, when the adults will bask with their wings held open, revealing the characteristic pattern on the forewings. During dull weather, and when roosting for the night, the butterfly will head for various shrubs and tall grasses where it rests head-down.

A male looking for a mate will fly to the lowest parts of a site, such as the bottom of a slope, from where he will patrol up and down, dipping down every now and again in search of a virgin female. Once her wings have dried, a virgin female will also make her way to the bottom of the slope and, once she has been intercepted by a male, the pair mate following a brief aerial courtship, and remain together for around an hour, after which the female hides in vegetation while her eggs ripen. While some females will lay on the day of their emergence, most lay the bulk of their eggs (up to 100) the following day.

A mating pair

An egg-laying female is easy to spot as she flutters low over the grassland in search of a suitable flowerhead of Wild Thyme or Wild Marjoram, before landing and immediately probing the unopened flowerheads with her abdomen, preferring plants whose flowerheads are still relatively compact where she deposits a single egg before flying off. Two or three eggs are often found on the same flowerhead at good sites, but these have almost certainly been laid by different females since the larvae are cannibalistic in their early instars, although a plant may be revisited by the same female if flowering thyme is in short supply. Jeremy Thomas writes in *The Butterflies of Britain and Ireland* that he has found over 100 eggs on the same plant on occasion and, in an earlier paper, Thomas says that each female may lay between 200 and 300 eggs in total, although most will lay only 60 or so due to predation and unsuitable weather curtailing opportunities to lay. Females may also probe a flowerhead but not lay any eggs, a process that my friend Guy Padfield has usefully termed 'oviposturing'.

A ♀ egg-laying on Wild Thyme

The different shades of these two eggs indicate that these were laid at different times

Ovum
The 0.5-mm wide egg, like those of most Lycaenids, is shaped like a compressed sphere with a sunken crown. It has a greenish-blue ground colour but appears white due to a network of white ridges that covers its surface. Eggs are easy to find since they stand out against the dark pink and unopened flower buds to which they are attached, although they are not usually visible unless the buds are gently prised apart. The crown of the egg darkens slightly before hatching, providing a hint of the black head of the first instar larva.

Larva
In 2018 I was able to rear, under license, a number of Large Blues from egg to their fourth and final instar (after which the larvae were released on the site from which the eggs were obtained), and I am indebted to David Simcox and Sarah Meredith for their support. Suffice to say, my personal observations and those of David and Sarah are included in the descriptions below.

After between five and 10 days, depending on temperature, the tiny 0.8-mm long **first instar** larva eats an untidy hole in the eggshell, usually towards the crown, before crawling out. This is a slow process and it may be several hours before the larva emerges. The larva initially has a pale grey ground colour, a shiny black head and is covered in hairs of different lengths. Its most distinctive features, however, are a dark disc on its first segment and another on its anal segment. The larva immediately wanders to a suitable flower bud where it eats a hole in the base before crawling inside, where it goes on to feed on the tender tissues inside the bud, when it develops a yellowish-brown ground colour. As the larva grows it will move to other flower buds but becomes too large to completely enter the bud, and larvae can be found feeding with the front part of their body inside the bud, but with their rear end exposed. When fully grown and 2 mm long, the larva has a pale pink tinge, when it perfectly matches the flower buds of the foodplant.

1st instar: a larva next to its eggshell, and another about to emerge from its egg

1st instar: a larva making a hole in the base of a flower bud

1st instar: a fully grown larva preparing to moult

Larvae are cannibalistic in their first three instars, which was first explained by Frohawk in an article in *The Entomologist* in 1903: *"On July 29th I found two of the larvae rolling about together under the thyme blossom; upon close examination I found the smaller one had seized the larger with its jaws, which were buried into its side, apparently sucking it. Upon pulling them apart I placed*

2nd instar: a larva feeding on a flower bud

3rd instar: the larva has longer hairs on its body

3rd instar: a fully grown larva

the victim under the microscope, and found a deep hole in its side, with the surrounding surface shrunken, and liquid exuding from the wound. This conclusively proves the cannibalistic habits of these larvae, which I had always suspected, as on previous occasions large numbers of larvae had disappeared in a mysterious manner". Frohawk provided additional observations in his *Natural History of British Butterflies* of 1924: "During the first three stages the larva readily devour each other, and so strongly developed are cannibal habits that even the smaller larva will seize and feed on the larger ones. Immediately after the completion of the third and last moult the cannibalism disappears … which appears all the more curious owing to the fact that it is solely carnivorous during its last stage". The absence of any cannibalism in the final instar is surprising since more than one larva may be adopted (up to 40 larvae have been recorded in the same ant nest), and most nests are only able to support a single Large Blue larva, with starvation the main cause of mortality. It is thought that overcrowding within an ant nest results in the dwarf adults that are occasionally encountered.

After around a week, the larva moults into the **second instar**. It now has a milky white ground colour and is covered in light pink markings that run in rows along the length of the body and that give the larva a level of camouflage within the flower buds. The hairs of the first instar are now replaced with short white bristles that cover the body, and the discs on the first and last segments are still present but are more subdued than in the first instar. Behaviourally, the larva now withdraws its head under the first segment while at rest and, when fully grown, the second instar is 3 mm long.

After another week, the larva moults into the **third instar**. It is similar to the second instar in most respects and the discs on the first and last segments are still present, although it is slightly brighter in appearance, has longer hairs on its body and the honey gland on the tenth segment is now clearly visible. The fully grown third instar is still tiny and only 4 mm long.

After a week or so, the larva moults for the last time into the **fourth instar**. The larva is duller than previous instars, with a milky white ground colour and more uniform pink markings that run longitudinally along the body. In this instar the larva has a prominent furrow that runs along its back (although this diminishes as it grows) and it is covered in four rows of long hairs. The larva retains the disc found on the first segment and its honey gland is particularly conspicuous, looking like a small incision on the tenth segment. In profile, this instar is unlike any other—the first segment slopes to the front, the last three segments slope to the back, and the remaining segments are humped.

4th instar: a newly moulted larva

The majority of larvae remain on the plant until between 5 pm and 7 pm—a peak time for foraging *Myrmica sabuleti*—when each will drop to the ground and crawl under a leaf or soil particle, waiting (sometimes for several hours) to be found by a red ant. An ant that comes across the larva will tap it, causing the larva to secrete a droplet from its honey gland that the ant drinks, when the ant is said to be 'milking' the larva. Although other ants may be in attendance, they eventually disperse, leaving the original ant alone with the larva. Eventually, after a period of between 25 minutes and four hours, the larva distorts its body by rearing up on its prolegs (its 'claspers'), mimicking an ant grub that has somehow left the nest. The ant immediately picks the larva up by clasping its thorax in its jaws and carries it back to its nest where it lives alongside the ant grubs that will form its future diet. I had the pleasure of experiencing an adoption in 2018 and was surprised that, after a lengthy 27 minutes of watching the larva being 'milked', the adoption happened in just a few seconds before the ant hurried away with the larva. Despite the brevity of the grand finale, it is an experience that will not be forgotten.

4th instar: a larva awaiting adoption by an ant

4th instar: a larva being tended by a *Myrmica sabuleti* ant [MS]

4th instar: a larva rearing up to mimic an ant grub [JT]

4th instar: an ant adopting a larva before taking it into its nest [MS]

4th instar: a fully grown larva inside an ant nest [MS]

4th instar: a larva feeding on an ant grub [MS]

When in the ant nest, the larva spins a silk pad on which it rests, in an empty cell near the main chambers, which it leaves to feed on ant grubs. The larva grows quickly as it eats the largest ant grubs, and it soon becomes much larger than its hosts and its grubs, with its colour changing from a dull red to a pale white. Written by Jeremy Thomas, the species descriptions in *The Butterflies of Britain and Ireland* are a tour de force, but the Large Blue species description is very special, where he writes: *"Preying on an ant brood is a dangerous activity which the caterpillar reduces to a minimum. It rests on its silk pad for days at a time before gliding into the brood chambers to binge-feed on grubs, then returns to digest them in its haven for perhaps another week"*. Thomas goes on to explain the many hazards that the Large Blue larva must contend with when in the ant nest, including the potential for ants to turn on it if they are looking to eliminate large grubs that could develop into queen ants that rival the current incumbent.

The larva hibernates deep within the nest, returning to the upper chambers that are closer to the surface in the spring. Early entomologists found it hard to believe that the larva did not go through another instar to become fully grown, when it is around 15 mm long and almost five times its original length in this instar and 100 times heavier than when it was taken into the ant nest. In one of his 1915 articles, T.A. Chapman writes: *"The small size of the head for so large a larva is almost ridiculous"*. An explanation for this was provided by J.W. Tutt in his *A Natural History of the British Lepidoptera* of 1906: *"This small size of head implies that the feeding of the larva in this stage … does not require greater power and energy on the part of the larva …. It lends, therefore, some support to the idea that throughout this stage the larva is in some way supported by, or at the expense of the ants and not in the ordinary way of eating vegetable tissues"*.

Pupa

The 13-mm long pupa is formed in the chamber originally inhabited by the larva but has no cremastral hooks and invariably lies loose in the chamber without any attachment. It is pale yellowish-orange when first formed but darkens to an orange-brown after a few days. The pupa is constantly attended by ants and it has been shown that it is able to make a sound that is similar to a queen ant and gets the same amount of attention as a result. After around three weeks, the adult is fully formed and can be seen quite clearly through the pupal casing. The adult emerges from the pupa early in the morning, usually between 7 am and 10 am, crawls

Pupae within an ant nest [MS]

through the ant chamber to the surface, then ascends a grass stem or other vegetation where it inflates its wings. Finding a newly emerged adult drying its wings is made all the more special when remembering that this butterfly has one of the most elaborate life cycles of all of our butterflies.

A coloured-up pupa [DS]

A newly emerged adult expanding its wings [MS]

Silver-studded Blue
Plebejus argus

❖ SILVER-STUDDED BLUE IS SUPPORTED BY HARRY E. CLARKE, JAMES FOWLER & GEOFF ROGERS ❖

	January	February	March	April	May	June	July	August	September	October	November	December
IMAGO												
OVUM												
LARVA												
PUPA												

The **Silver-studded Blue** gets its name from the light blue scales found on the underside of the hindwings of most adults, which really stand out when light reflects off them. This name was provided by Moses Harris in *The English Lepidoptera* of 1775, albeit with the spelling 'Silverstuded Blue'. The specific name of *argus*, on the other hand, is a reference to the many-eyed shepherd of Greek mythology, in recognition of the spotting found on the undersides. While the male has a blue upperside, that of the female is a less-conspicuous brown, dotted with a varying number of orange lunules around the wing edges. The two sexes are similar in size, with a wingspan of between 26 and 32 mm.

A particular characteristic of this butterfly is its variable appearance across its range, with notable differences in the amount of blue found on the female upperside, the brightness of the male and wingspan. This has led to the naming of four subspecies on our shores, namely the nominate subspecies *Plebejus argus argus* that is found throughout most of its range; the extinct subspecies *cretaceus* (that was slightly larger with a brighter blue in the male) that was formerly found on chalk downland of Dorset, Hampshire, Surrey and Kent; the extinct subspecies *masseyi* (in which the female had an amount of blue on the hindwings and base of the forewings) that was formerly found on the mosses of south Cumbria and Lancashire; and the subspecies *caernensis* (that is smaller, with a wingspan of 25 to 28 mm, and in which the female has a large amount of blue on both hindwings and forewings) that is found on limestone on the Great Orme in North Wales.

A ♂ from Prees Heath in Shropshire

A ♀ from Stedham Common in Sussex [MC]

A ♂ from Silchester Common in Berkshire

A ♀ from Prees Heath with prominent light blue 'studs' on her hindwings

Several other notable populations exist, although these have not formally been given subspecific status—on mossland at Hafod Garregog NNR in North Wales; on limestone on the Isle of Portland in Dorset; at Holy Island off Anglesey; at Prees Heath in Shropshire; and on the sand dunes of south-west England. In the article 'Population differentiation and conservation of endemic races: the butterfly, *Plebejus argus*', published in *Animal Conservation* in 1999, Chris Thomas and colleagues showed that these 'races' differ in their

♂ and ♀ of subsp. *caernensis* from the Great Orme in North Wales

Life Cycles of British & Irish Butterflies

appearance, habitat choice, larval foodplant, performance on different larval foodplants, and their chosen host ant species from which the butterfly receives protection against predators. Most notably, Thomas and colleagues found that habitat preference, attractiveness to a particular ant species and larval performance on particular foodplants were maintained, even if the butterfly (or larva) was placed in a different habitat. This suggests that there is some evolutionary divergence between these races beyond their appearance, even though they have not been formally recognised as genetically distinct and worthy of subspecific status.

DISTRIBUTION

The butterfly's strongholds are undoubtedly on the heathlands of southern England (especially those found in Dorset and Hampshire), on various sand dunes in Cornwall and on limestone at the Great Orme in North Wales. In addition to the populations already mentioned, the butterfly can also be found in often coastal locations in Devon and Cornwall, Pembrokeshire in South Wales, Norfolk and the Suffolk Sandlings. The butterfly is also found on Sark in the Channel Islands, although the butterfly is not found in Scotland, Ireland or the Isle of Man. The butterfly was formerly more widespread and was to be found in much of central, eastern and northern England and Scotland. On a worldwide basis, the butterfly is found throughout most of Europe and across temperate Asia as far as Japan.

The largest populations run into tens of thousands and seeing adults in such large numbers is one of the must-see spectacles in butterfly watching on our shores. Watching adults going to roost head-down in large groups, or opening their wings in unison in the morning, are sights that will never be forgotten. Most colonies, however, have between a few dozen and a few hundred adults flying at any one time, although this diminutive butterfly is always a delight to behold, whatever the numbers.

Adults are highly colonial and, even when there is an expanse of seemingly suitable habitat, will confine themselves to a few specific spots. Adults are also highly sedentary, and rarely move more than 50 m during their lifetime, with most moving less than 20 m. This limited dispersal ability is a concern when sites become isolated, especially where the butterfly resides within a metapopulation structure, requiring interchange between adjacent patches as habitat suitability changes from year to year.

HABITAT

Our remaining populations are found on heathland, limestone grassland and sand-dune systems. This is a warmth-loving butterfly and all sites provide sheltered areas that are in full sun, as well as warm microclimatic conditions for the development of the larva and, in particular, the conditions that support the colonies of black *Lasius* ants that offer the butterfly a level of protection in all of

Lowland heathland at Silchester Common in Berkshire

its stages. Many sites, especially those further north, are on south-facing slopes that heat up more quickly than surrounding areas.

Suitable heathlands may be those that are relatively dry and dominated by Bell Heather, or those that are relatively wet and humid and dominated by Cross-leaved Heath. The soil may also vary from being very sandy to very peaty. The butterfly prefers short and sparse vegetation that can be maintained by periodic burning and through appropriate grazing, either by rabbits or by cattle and ponies. A variation on this habitat type is the last known mossland population at Hafod Garregog NNR, whose wet and peaty heathland is maintained using cattle. In this habitat type the butterfly uses heathers and gorses as larval foodplants.

Limestone grassland populations are found on the Great Orme in North Wales (and the Dulas Valley, 13 km to the east, where the butterfly was introduced in the 1950s), and on the Isle of Portland. On the Great Orme, grazing by sheep and goats maintains the short sward required and here the butterfly uses Common Bird's-foot-trefoil, Common Rock-rose, Black Medick and Hoary Rock-rose as larval foodplants whereas, on the Isle of Portland, the butterfly uses Common Bird's-foot-trefoil, Common Rock-rose and Horseshoe Vetch. In the paper 'When is a habitat not a habitat? Dramatic resource use changes under differing weather conditions for the butterfly *Plebejus argus*', published in 2006 in *Biological Conservation*, Roger Dennis and Tim Sparks show that shrubs are also an essential component of the habitat on the Great Orme, with adults using them for roosting, resting, basking and mate location, and for shelter when the weather is cool, cloudy and windy. In warm and calmer conditions, the butterfly moves into more open areas that are usually dominated by larval foodplants and nectar sources. Their conclusion is that the bounds of the habitat change with conditions and that habitat management should accommodate all of the butterfly's requirements.

Limestone grassland at Happy Valley on the Great Orme in North Wales

Limestone grassland at Tout Quarry on the Isle of Portland

A group of adults going to roost amongst shrubs on the Great Orme

Holywell Dunes in North Cornwall [PB]

The sand-dune systems of south-west England are less well-studied, although it is likely that larvae here feed on Common Bird's-foot-trefoil, and this is certainly my experience from studying this species at Penhale Sands in Cornwall.

STATUS
The Silver-studded Blue has suffered serious declines over the last century, that some have estimated as large as 80%, with the butterfly becoming extinct in Scotland, northern England, and throughout most of central, eastern and south-eastern England.

The root causes of this demise are the loss of habitat to housing development, industry (for mineral extraction, for example), agriculture and forestry. Accidental or overly extensive burning of heathland can also have a negative impact. Inappropriate habitat management can also have a negative effect—for example, the abandonment of heathland may lead to it becoming overgrown and unsuitable for this species, the first indication of which is the loss of the ant colonies on which the butterfly depends. Ants may also be lost due to other factors—in her aptly titled MRes thesis 'Is it all about the ants? What are the factors influencing the presence of *Plebejus argus* (the Silver-studded Blue butterfly) on Studland Peninsula?', Lorraine Munns shows that, on the Studland Peninsula, the two ant species *Lasius niger* and *Formica rufa* out compete each other and do not coexist, with the butterfly largely absent from areas devoid of *Lasius niger*. The loss of chalkland colonies was accelerated due to the decline in the rabbit population as a result of the introduction of myxomatosis in the 1950s, and grazing by rabbits continues to play an important role at some sites. Such losses are compounded by the butterfly's limited powers of dispersal, when it is unable to colonise nearby suitable habitat in a timely fashion.

The long-term trend shows a significant decline in distribution and a moderate increase in abundance, with the more recent trend over the last decade showing a moderate increase in distribution and a stable abundance. Using the IUCN criteria, the Silver-studded Blue is categorised as Vulnerable (VU) in the UK.

LIFE CYCLE
The butterfly is single-brooded in Britain and, although the flight period is staggered based on location, most colonies start to emerge in early June, peak at the end of the month, and fly to the end of July. The butterfly emerges at the end of May in early years and at predictably early sites such as the Great Orme. Conversely, the butterfly will fly into August at some sites and in late years.

Imago
Adults are active on all but the windiest and most inclement days and will even fly in dull weather so long as it is warm enough. They are most active, of course, on sunny and windless days, when their undersides allow them to sparkle like so many jewels as they dance around their habitat, just above the vegetation. Neither sex spends much time visiting flowers to nectar, but adults will occasionally feed on Bell Heather, brambles and Common Bird's-foot-trefoil, depending on habitat.

Males are impossible to miss when patrolling their sites, as they flutter low above the vegetation looking for newly emerged females. When a female is found then the male will land beside her, vibrate his wings for a couple of seconds and, if she is responsive, mate with her with the pair remaining 'in cop' for around an hour.

Females have some very specific requirements when egg-laying. The first is that eggs are usually laid in a relatively warm microclimate, close to the ground at a height of less than 10 cm, where they overwinter. Eggs are attached either to a woody stem or twig of the foodplant, or some other nearby surface such as leaf litter, moss or lichen, usually at the edge of short vegetation and bare ground and where new growth of the foodplant is likely to be found.

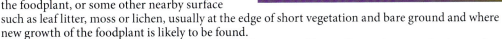
A mating pair with ♂ on left [NKi]

In their paper 'The distribution and density of a Lycaenid butterfly in relation to *Lasius* ants', published in *Oecologia* in 1992, Diego Jordano, José Sánchez Rodríguez, Chris Thomas and Juan Fernandez Haeger also showed that there is a direct correlation between the location of Silver-studded Blue eggs, larvae and pupae and the density of *Lasius* ants, suggesting that females detect the presence of ants when finding a suitable place to lay. A study of a South African Lycaenid, the Roodepoort Copper *Aloeidas dentatis*, showed that the butterfly laid in the presence of pheromones left on ant trails, and this may explain the findings documented in 'Observations on the life-history of the Silver-studded Blue *Plebejus argus* L.', published in *Transactions of the Suffolk Natural History Society* in 1987, in which Howard Mendel and Eric Parsons tell of females egg-laying on the underside of Bracken fronds on the Sandlings of East Suffolk, with ants known to be attracted to the floral nectaries of the plant.

The white egg has a fine network pattern on its surface

Ovum
The egg has the shape of a compressed sphere with a sunken crown and is 0.6 mm in diameter and 0.3 mm high. It is covered in a fine network pattern that is incredibly beautiful when seen close up, the white projections on the surface giving the egg a white appearance, despite having a greenish-white ground colour. The larva is fully developed inside the egg as it overwinters.

Larva
At the end of February and in early March, the 1.1-mm long **first instar** larva emerges from the egg by eating away the crown. It is of the typical Lycaenid shape, resembling a woodlouse, with humps along its back and with flattened sides. It has a shiny black head that is held under the first segment when not feeding and two rows of long white hairs that run along the back, with each hair emerging from a dark base, with another row of white hairs at the base of each side of the body. The larva is a pale brown when newly emerged, but colours up as it feeds and, when fully grown and 2.1 mm long, has several dark and pale lines that run the length of its body and that are reminiscent of later instars. The larva also has an olive-brown patch on

1st instar: a larva eating away at the crown | 1st instar: a newly emerged larva | 1st instar: a larva eating the cuticle of a leaf | 1st instar: the mature larva has a striped appearance

Life Cycles of British & Irish Butterflies

its first segment, with a similar but smaller patch on the anal segment. The larva feeds on the tenderest parts of its foodplant, such as young shoots, buds and flowers, and, when feeding on leaves, will only eat the cuticle, without breaking through the leaf.

Even at this young age the larva is tended by ants, although the exact nature of this relationship is unknown. Experiments in both the laboratory and in the wild show that placing a newly emerged larva near an ant nest will result in an ant carrying the larva into the nest, and both first and second instar have been found in ant nests in the field. In *The Butterflies of Britain and Ireland*, Jeremy Thomas, writes: "*We do not yet know how the hatchling Silver-studded Blue caterpillar interacts with ants in the wild. In captivity, they eat tender plant tissue, and it is possible that this is their food inside ant colonies, for I have often found the roots and blanched shoots of gorses, trefoils and other foodplants twisting through the internal passageways of their* Lasius *nests. But if they do feed above the ground at this stage, I suspect they are carried there by the ants, in the same way that workers transport their herds of domesticated aphids from one suitable foodplant to another*".

2nd instar: the larva has more prominent markings

After between two and four weeks, the larva moults into the **second instar** and, while superficially similar to a mature first instar, now has much more prominent markings. The larva has a pale brown ground colour, with white-bordered dark brown lines along the centre of its back and along the middle of each side, with a whitish line along the base of each side. The body is sprinkled with dark spots and there are several rows of white hairs of varying size that run the length of the body, with many hairs emerging from dark tubercles. A dark patch remains on the first segment and, when fully grown, the second instar is 4.2 mm long.

While the larva does not yet possess all of the ant attracting organs, the honey gland does show as a small incision on the 10th segment and produces large droplets of clear liquid in response to the attention of ants. Larvae now live within the chambers of an ant nest by day and emerge at dusk to feed on the foodplant. In their paper 'Specificity of an ant-Lycaenid interaction', published in *Oecologia* in 1992, Diego Jordano and Chris Thomas show that different races of Silver-studded Blue attract a particular species of black ant—on the limestone grassland of the Great Orme, larvae were tended by *Lasius alienus*, whereas heathland colonies were tended by *Lasius niger*. However, it should be noted that both of these species have been 'split' relatively recently, in the early 1990s, resulting in four *Lasius* species—*Lasius psammophilus* was split from *Lasius alienus*, and *Lasius platythorax* was split from *Lasius niger*—and these splits are not always accounted for in the studies described here. One exception is in *The Butterflies of Sussex*, where Neil Hulme suggests that *Lasius platythorax* is the key ant species used on the heathlands of the Sussex Weald.

After two to three weeks, the larva moults into the **third instar**. The larva now comes in two colour forms—light brown and light green—although intermediates appear too. Irrespective of its predominant ground colour, the larva has a white-bordered dark brown stripe along its back. Both forms also have a dark line of the relevant hue that runs along each side, made up of two pairs of adjacent dark and pale markings that are found on each segment. The larva also has a whitish line that runs along the base of each side and each colour form retains the dark patch on the first segment. The most noticeable difference from the previous instar is the presence of a pair of tentacle organs on the 11th segment that, along with 'pore cupolas' that are sprinkled all over the body, gives the larva the full complement of ant-attracting organs. These tentacle organs are periodically raised as the larva slowly glides along. Mendel and Parsons write: "*Larvae appeared to dislike the attention of ants at their heads and would usually retract them beneath the pro-thorax and erect the retractile 'brushes' in response. The effect would be to displace and attract the ants thereby guiding them to the honey-gland*". When fully grown, the third instar is 6.3 mm long.

3rd instar: the brown form of larva

3rd instar: the green form of larva

4th instar: the brown form of larva with an attendant ant [SO]

4th instar: the green form of larva

4th instar: a larva of an intermediate colour

After another two weeks, the larva moults into the final and **fourth instar** and continues to exhibit two colour forms, although the green form is much brighter than in the previous instar. The markings are much bolder in both forms although, in all other respects, the larva is similar in appearance to the previous instar. A key difference is, of course, size and, after a couple of weeks in this instar and when fully grown, the final instar is twice the length of the previous instar—it is just under 12.5 mm long when at rest and 14 mm long when crawling.

4th instar: a view of the tentacle organs (only one is everted) and the honey gland

Pupa

A newly formed pupa

A ♂ pupa just hours before the adult emerged

The extraordinary relationship that this species has with ants continues into the pupal stage. In *The Butterflies of Britain and Ireland*, Jeremy Thomas writes: "*The Silver-studded Blue's pupa is formed in the same chambers inhabited by the caterpillars. The ant expert, John Pontin, was the first to find them, lined up along the passages of* Lasius niger *nests in the New Forest. More recently, I have found hundreds of chrysalises in this situation with* L. alienus, *generally 1-3 per ant nest but with up to 20 close together in the best colonies, often accompanied by final and penultimate stages of the caterpillar*". A potential benefit of living life below ground, at least on heathland, is avoiding both deliberate and accidental burning.

Pupae have also been found outside of ant nests, under mats of gorse, under stones, and in a variety of crevices and holes, and there is even an historic record of pupae being found under the bark of a pine trunk. Whatever the location, the 8.5-mm long and yellowish-green pupa is always tended by ants, which may even establish a small nest around the pupa.

A newly emerged ♂ with *Lasius* ants in attendance [JB]

After between two and three weeks, the butterfly is clearly visible within the pupal case. As the butterfly starts to emerge, usually before 10 am, ants will scurry around and feed on droplets exuded by the adult. At the best sites, it is not uncommon to find an adult expanding its wings from a suitable platform, such as a sprig of heather or a grass stem, where ants continue to tend the adult. In their study, Jordano and Thomas found that the newly emerged adults were only ever tended by *Lasius* ants, but were attacked by *Myrmica* and *Formica* worker ants, confirming the specialist association that the butterfly has with particular ant species.

Life Cycles of British & Irish Butterflies

# Brown Argus
Aricia agestis

❖ BROWN ARGUS IS SUPPORTED BY PATRICK BARKHAM, DOUGLAS GODDARD & MARTIN & RACHEL PARTRIDGE ❖

The **Brown Argus** is often dismissed as a dark female Common Blue, although there are several clues that can help address any confusion. The first is that the Brown Argus has a consistent chocolate brown upperside with no blue scales, with the most beautiful orange crescents along the wing fringes that are especially prominent in the female. Both sexes also have a prominent black cell spot on each forewing, which occasionally has a white surround. Of course, the long and thin abdomen of a male Brown Argus is another clue that a given individual is not a female Common Blue although, when confronted with a female butterfly, the last resort is to examine the underside spotting—the Brown Argus has two spots that form a 'figure of eight' toward the top of the hindwing underside, whereas these spots are spaced apart in the Common Blue. Despite these clues, a fresh Brown Argus may give off a blue sheen when viewed at certain angles to the light, which adds to the confusion.

The name that we use for this butterfly today originates from the work of James Petiver who, in his *Gazophylacium naturae et artis* of 1702, named this butterfly 'The edg'd brown Argus'. The word 'edg'd' is a compression of 'selveged' (a finished edge of fabric), where the butterfly's orange crescents resemble the stitching, and 'argus' is the many-eyed shepherd of Greek mythology, which is a reference to the numerous spots found on the butterfly's underside. Subsequent authors recognised that the butterfly was closely related to the blues, leading William Lewin to call this butterfly the 'Brown Blue' in his *The Papilios of Great Britain* of 1795. The lack of any blue scales in either sex is specifically recognised in the names of related European species, which are known as 'anomalous' blues.

The ♀ has a shorter and wider abdomen [IL]

The ♂ has a long thin abdomen

♂ and ♀ have similar wingspans of 25–31 mm [IL]

The ♂ may give off a blue sheen when viewed at certain angles to the light

DISTRIBUTION

This butterfly's distribution extends as far north as Yorkshire, although it is absent from the north-west of England and parts of the border between England and Wales. In Devon, Cornwall and North and South Wales, the butterfly has a curious distribution since it is only found in areas close to the coast. The butterfly is absent from Scotland, Ireland, the Isle of Man and central Wales, but is present in the Channel Islands. On a worldwide basis, the butterfly is found throughout most of Europe and temperate Asia, and in parts of North Africa.

The Brown Argus occurs in small and compact colonies of only a few dozen adults, with even the largest sites containing no more than 700 individuals. The butterfly is not a great wanderer and typically travels only a couple of hundred metres from where it emerged, although individuals will travel much further on occasion, up to several kilometres, allowing the butterfly to colonise new areas as they become suitable.

HABITAT

The butterfly's strongholds are on chalk and limestone grassland where its key larval foodplant, Common Rock-rose, is abundant. The butterfly is, however, able to use quite a wide variety of habitats including woodland clearings; heathland; disused quarries; sand dunes and undercliffs; arable field margins and set-aside; brownfield sites; road verges and disused railway lines.

The butterfly thrives on the lower slopes of Magdalen Hill Down in Hampshire

STATUS

The butterfly experienced a 40% decline up to the 1980s before its fortunes changed for the better. In the 1990s, it started a rapid expansion, especially in northern, central and eastern England, and has not only recolonised areas from where it had been absent for over 50 years, but has also spread to new areas. It is thought that this expansion is due to an increased use of Common Stork's-bill, Dove's-foot Crane's-bill and other related species as larval foodplants, which are now on the menu thanks to two recent changes. The first is the greater availability of these foodplants thanks to EU subsidies that encouraged farmers to create set-aside in their fields, allowing these plants to become established. The second is that climate change has allowed the butterfly to use areas that were previously too cold for the successful development of its immature stages, in addition to the sun-drenched and relatively hot sites on calcareous grassland where Common Rock-rose is used as the larval foodplant.

The long-term trend shows a significant increase in distribution and a moderate decline in abundance, with the more recent trend over the last decade showing a stable distribution and a moderate decline in abundance. Using the IUCN criteria, the Brown Argus is categorised as Least Concern (LC) in the UK.

LIFE CYCLE

Adults first emerge in southern and central England in early May, peak at the end of May and beginning of June, and give rise to a second brood that emerges at the end of July and into August.

In North Wales and northern England, the first emergence starts in early June with a partial second brood appearing in mid-August. A partial third brood may appear in southern England in good years, which is the norm in southern Europe.

The butterfly has been observed overwintering as a second, third or fourth instar larva, with most overwintering while in their third instar. Larvae of the first brood are fully grown after six weeks or so, whereas those that overwinter are fully grown in around eight or nine months.

Imago

This is a warmth-loving species that is most active in sunshine, where it can be found in sheltered areas or on south-facing slopes. The butterfly also flies rapidly as described by James Tutt in his *A Natural History of the British Lepidoptera* of 1906: "*This species may be called the 'zigzagger', for it darts swiftly to and fro when in flight, showing first its grey underside and then its black upperside, so that one can hardly follow it with the eye*". The resulting effect when flying has been likened to a spinning silver coin. Adults feed on a variety of nectar sources, including buttercups, Common Bird's-foot-trefoil, Common Rock-rose, Horseshoe Vetch, dandelions, ragworts, White Clover, Wild Marjoram and Wild Thyme.

A group of Brown Argus roosting head-down

The butterfly will roost communally, head-down, on grass stems at night and during inclement weather, and often in the company of the Common Blue that exhibits similar behaviour. Visiting good sites in late evening is an excellent time to see this spectacle when scores of Brown Argus can be seen for quite some distance as they roost on the tallest grass heads and above the sward, when their pale undersides stand out from the surrounding grassland.

Males have two strategies for finding a mate. The first is perching, often on the ground at the base of a slope or in a sheltered hollow, where the male awaits a passing virgin female and will dart up to investigate any passing insect. The second strategy is patrolling, where the male will fly up and down an area looking for a female resting on a grass head. When a female is found, there is a short flight close to the ground before the pair settle among the grass and mate.

A mating pair, with ♀ on the right

Once her eggs have ripened and there is warm sunshine, the female will meander close to the ground in search of suitable foodplants on which to lay. She then lands and crawls over the foodplant, tasting it with her feet as she goes, until she finds a suitable spot. She then curls her abdomen under a leaf and lays a single egg close to the midrib before immediately flying off (although the eggs are laid on the upperside of a leaf on occasion). On chalk and limestone grasslands, the primary larval foodplant is the Common Rock-rose whereas, in other habitats, the butterfly will use annual foodplants that include Common Stork's-bill, Dove's-foot Crane's-bill and other related species.

Preferred plants are those with lush leaves and that have high concentrations of nitrogen and, in the case of Common Rock-rose, plants growing in a short turf that is no taller than 10 cm and

A ♀ egg-laying on Dove's-foot Crane's-bill [DM]

Life Cycles of British & Irish Butterflies

surrounded by bare ground. The suitability of Common Rock-rose was studied by Nigel Bourn and Jeremy Thomas and described in their paper 'The Ecology and Conservation of the Brown Argus Butterfly *Aricia Agestis* in Britain', published in *Biological Conservation* in 1993. They found that the 'Efficiency of Conversion of Ingested Food' (ECI), which is a measure of how much of the food ingested is converted into the recipient's mass, was increased by 10% given a 1% to 3% increase in the nitrogen content of the foodplant. In other words, a larva can feed up 10% more quickly, and be less prone to predation, if the female selects the right plant. They also suggest that the higher nitrogen content may result in better quality secretions from the larva, which may contribute to the relationship that this larva has with ants and the level of protection it receives.

An egg laid on the underside of a leaf of Common Rock-rose

Ovum

The 0.5-mm wide egg resembles a flattened sphere with a sunken crown, that is typical of the Lycaenids. It is covered in a network pattern with a prominent point where two ribs join. It is greenish-white when first laid, becoming a more opaque pearl-white just before the larva emerges.

Larva

The 1-mm long **first instar** larva emerges from the egg in around a week, depending on temperature, by eating away a portion of the top or side of the egg, partially eating the remainder of the eggshell on occasion, before moving to take up position on the underside of the leaf. The larva is pale yellow, but turns pale brown as it grows, and has a shiny black head and several rows of long white hairs that run the length of its body. The larva feeds by day and proceeds to eat the underside of the leaf, but does not break through the upper surface. The resulting 'windows' are visible from above which can give away the presence of a larva underneath the leaf. The fully grown first instar is around 2 mm long.

1st instar: a larva feeding on the underside of a leaf

1st instar: the feeding damage appears as 'windows' in the leaf cuticle

2nd instar: a larva, next to its eggshell, feeding between the upper and lower cuticles of a leaf

3rd instar: a dark larva with particularly prominent markings

After a week or so, the larva moults into the **second instar**. The larva now has a pale green hue and, while there are other subtle distinctions with the first instar, a key difference is that the larva now tunnels between the upper and lower cuticle using its extensible neck to feed, leaving the rear part of its body exposed. This strategy also means that the area being 'mined' is free of frass. The fully grown second instar is just under 3 mm long.

Larvae that overwinter will do so underneath a leaf low down on the foodplant or in some other vegetation at the base of the plant, where they remain fixed until the spring without feeding over the winter, even on relatively mild and sunny days.

If the larva does not overwinter then, after another week or so, it moults into the **third instar**. The ground colour is now a pale green with a dark line along its back, although other markings are quite variable. Darker individuals have several rows of dark green oblique markings along their sides, with an off-white stripe along each side close to the spiracles that is bounded both above and below by pink borders, whereas lighter individuals have the same markings, but these are much paler. The larva remains covered in longish white hairs and continues to feed on the leaf cuticle and, when fully grown, is around 5 mm long.

If the larva does not overwinter while in the third instar then, after around 10 days, it moults into the **fourth instar**. The larva is similar to the previous instar in appearance, but the colours are much more brilliant. Both dark and pale forms of the larva are produced—the former has a noticeable dark pink stripe running along its back that is dark green in pale individuals. At this stage the larva continues to feed primarily on the underside of a leaf but will now eat the leaf edges and will also break through the cuticle, leaving distinctive holes in the leaf surface when seen from above. When fully grown, the fourth instar is around 6.5 mm long while resting.

After another 10 days or so, the larva moults for the last time into the **fifth instar**. It is similar in appearance to the fourth instar with almost identical markings although, unlike the fourth instar whose hairs are relatively long and curved, those on the fifth instar are relatively short and bristle-like. The fully grown larva is around 11 mm long.

4th instar: a dark individual with a dark purple stripe along its back

4th instar: a pale larva with a dark green stripe along its back

Larvae now feed more openly and are often given away by the presence of attendant ants, including the red ant *Myrmica sabuleti* and the black ant *Lasius alienus*, that are attracted by the pore cupolas, tentacle organs and, especially, the honey gland, all of which are the most developed in this instar. While this relationship could be considered symbiotic since the ants feed on sugary secretions from the larva, and the larva is afforded some protection by the ants, it could well be that the butterfly is simply looking to appease any ants to stop them from attacking it, especially since Common Rock-rose is often found growing on top of an ant hill at calcareous grassland sites.

After another week or so, the larva prepares to pupate by moving to the base of the foodplant. Pre-pupation larvae and pupae are thought to be covered in earth by ants, thereby creating a protective 'cell', and may even be carried away by ants on occasion.

5th instar: a fully grown larva

5th instar: a larva being tended by a black ant

Pupa

The 8.5-mm long pupa is surprisingly similar to the larva in colour, with a light green ground colour and a white stripe along the side of its abdomen, in the vicinity of the spiracles, which is bordered by pink above and below. Another characteristic of the Brown Argus pupa is that it has a black crescent over each eye. Although there are no cremastral hooks, the pupa often retains the shed larval skin. In his *Natural History of British Butterflies* of 1924, F.W. Frohawk mentions that some pupae are also held in position by a flimsy silk girdle spun around their waist.

The colouring up of a Brown Argus pupa, as the adult butterfly develops, is fascinating. The pupa initially changes colour to a yellow-brown, the eyes darken and the orange crescents on the forewings of the adult butterfly become visible through the pupal case. The pupa then darkens further until the final form of the adult can be seen quite clearly. After two or three weeks as a pupa, the adult butterfly emerges.

A newly formed pupa

A coloured-up pupa a day before the adult emerged

Northern Brown Argus
Aricia artaxerxes

❖ NORTHERN BROWN ARGUS IS SUPPORTED BY ABBIE MARLAND, ISLA PLUMTREE & TAM STEWART ❖

	January	February	March	April	May	June	July	August	September	October	November	December
IMAGO												
OVUM												
LARVA												
PUPA												

For many years, the Northern Brown Argus was considered to be a subspecies of the Brown Argus and various studies, starting with those of F.V.L. Jarvis and O. Hoegh-Guldberg in the late 1950s, have attempted to determine the true status of the Northern Brown Argus once and for all. These rearing studies and a more recent genetic study by Kaare Aagaard and colleagues, documented in 'Phylogenetic relationships in brown argus butterflies (Lepidoptera: Lycaenidae: *Aricia*) from north-western Europe' and published in *Biological Journal of the Linnean Society* in 2002, confirm that the Northern Brown Argus is, indeed, a separate species.

Early entomologists suspected that there might be two species in the Brown Argus 'group' in Britain—in comparison with the Brown Argus, the Northern Brown Argus has white spots on its forewings (in Scotland), is a deeper chocolate brown, has a different phenology (the timing of the life cycle stages), and has a more northern distribution, with the Northern Brown Argus found only in northern England and Scotland. The Northern Brown Argus also has one brood rather than two each year, although the Brown Argus is also single-brooded in the most northern part of its range. That being said, it is thought that hybrids between Northern Brown Argus and Brown Argus may occur where the two species meet in the Peak District.

Hybridisation aside, the Northern Brown Argus has two subspecies in Britain. The first is the nominate subspecies, *Aricia artaxerxes artaxerxes*, that is found in Scotland and the second is subspecies *salmacis*, that is found in northern England. The nominate subspecies has distinctive white spots on the forewings but, in subspecies *salmacis*, these are black, occasionally with a surrounding white halo. Also, the spots on the underside of the nominate subspecies are often 'blind' in that they lack a 'pupil', whereas those of subspecies *salmacis* often have well-developed pupils. Interestingly, the research by Aagaard and colleagues showed that the white-spotted subspecies *artaxerxes* is not endemic to our shores as was previously thought but is closely related to the Northern Brown Argus found in Scandinavia.

Before the Northern Brown Argus was recognised as a species in its own right, several names were used to describe the two subspecies. The *artaxerxes* subspecies was given names such as the 'Brown

The subsp. *artaxerxes* ♀ has a similar wingspan to the ♂, and more substantial orange lunules

The subsp. *artaxerxes* ♂ has a wingspan of between 25 and 31 mm

Subsp. *artaxerxes* ♂ and ♀ have similar undersides

Subsp. *salmacis* typically lacks the white spots found in subsp. *artaxerxes* [NF]

Some colonies of subsp. *salmacis* have spots with prominent pupils [IL]

Life Cycles of British & Irish Butterflies

Whitespot', provided by William Lewin in his *The Papilios of Great Britain* of 1795, and the 'Scotch Argus', provided by Adrian Haworth in his *Lepidoptera Britannica* of 1803, before this name was assigned to a completely different species. The *salmacis* subspecies was first discovered at Castle Eden Dene in County Durham and is still informally referred to as the 'Castle Eden Argus' or the 'Durham Argus' by some. However, it was left to T.G. Howarth to provide a single unifying moniker that embraced both subspecies, naming this butterfly the 'Northern Brown Argus' in his *South's British Butterflies* of 1973.

DISTRIBUTION

As its name implies, the butterfly is found in the northern parts of Britain. The *salmacis* subspecies is found in England in a band that runs between Lancashire and Cumbria in the west, through parts of Yorkshire to County Durham in the east, and the *artaxerxes* subspecies is found in Scotland, from Dumfries and Galloway in the west to the Borders in the east, with an eastern distribution heading further north where the butterfly's range extends as far as south-east Sutherland.

On a worldwide basis, the butterfly has a patchy distribution in the western Palearctic, and is found across Europe into western Asia, and also in parts of North Africa. In all of these regions the butterfly's appearance is reminiscent of subspecies *salmacis* and it also tends to inhabit mountainous regions such as the Alps and Pyrenees in Europe, and the Atlas Mountains in North Africa, where it is known as the 'Mountain Argus'.

Most Northern Brown Argus colonies occupy less than one hectare (100 m × 100 m) of breeding habitat, and support less than 200 adults in total, of which less than a third fly on the peak day. Sam Ellis, who studied this species in some detail in the 1990s, says that a minimum breeding area is 0.25 hectares (50 m × 50 m), although even smaller areas may be occupied if they lie close to others and where there is some interchange of adults. At the other end of the spectrum, a few sites that spread over a much larger area may support a few thousand adults.

Ellis also found that this is not a particularly mobile species with an average movement of just 30 m per day for both sexes, and less than 20 m per day at small and isolated sites. While the butterfly is known to be capable of crossing large bands of scrub, the general consensus is that many sites are isolated and prone to extinction, with only a limited chance of them being subsequently recolonised unless they are part of a wider metapopulation. Nevertheless, some individuals must disperse since there are occasional records some distance from existing colonies and because some isolated sites have been recolonised (Sam Ellis, pers. comm.).

HABITAT

The butterfly can be found wherever the primary larval foodplant of Common Rock-rose grows on calcium-rich soils. Typical habitats include well-drained and lightly grazed unimproved grassland; coastal valleys; quarries; and limestone pavement and outcrops. The butterfly also likes shelter and

Subsp. *artaxerxes* flies on south-facing slopes at St Abbs Head in Berwickshire

Subsp. *artaxerxes* flies south of Kirkandrews in Kirkcudbrightshire, Scotland [JA]

Subsp. *salmacis* flies at Bishop Middleham Quarry in County Durham

warmth, preferring south-facing sites that contain hollows, or an amount of scrub with patches of bare ground.

Ellis found that females will not lay their eggs at heavily grazed sites, even if the foodplant is abundant, preferring a varied vegetation structure with an uneven sward that is of medium height (between 6 cm and 10 cm) or taller. This structure can be maintained through light grazing by livestock or rabbits at grassland sites, and through appropriate scrub management at ungrazed sites. Iain Cowe, the Borders butterfly recorder for the East Scotland branch of Butterfly Conservation, has seen numbers dwindle at one site in Berwickshire due to overgrazing by rabbits, even though Common Rock-rose remains abundant (Iain Cowe, pers. comm.).

STATUS

Population levels of this butterfly have shown an ongoing decline in some regions due to inappropriate habitat management, such as overgrazing and lack of scrub control, and loss of suitable habitat to tree plantations. The long-term trend shows a moderate decline in distribution and a significant decline in abundance, with the more recent trend over the last decade showing a moderate decline in distribution and a stable abundance. Using the IUCN criteria, the Northern Brown Argus is categorised as Vulnerable (VU) in the UK.

LIFE CYCLE

The butterfly is single-brooded, although the timing of the emergence is dependent on the weather. In England, adults typically emerge in early June, peak in late June and fly until the middle of July. Emergence is later further north and, in northern Scotland, adults may not appear on the wing until the middle of July and fly until the second half of August.

Imago

This is a warmth-loving species and both male and female are very active in warm sunshine when they can be found flying from flower to flower. They feed on a variety of nectar sources, including Bell Heather, brambles, Common Bird's-foot-trefoil, Kidney Vetch, knapweeds, Red Clover, ragworts, Small Scabious, thistles, Wild Thyme and, when it is available, Red Valerian.

Both sexes roost head-down high up on a grass head or stem, or some other suitable surface such as dead knapweeds or plantain flowerheads, often in a sheltered area at the base of a slope, and, like the Brown Argus and Common Blue, will roost communally.

The male will patrol up and down an area when looking for a mate and, when a female is found, will pursue her low over the ground before the pair settle on vegetation or among grass and mate.

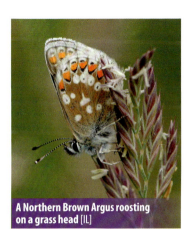

A Northern Brown Argus roosting on a grass head [IL]

Life Cycles of British & Irish Butterflies

The female will fly low over the ground when looking to lay her eggs and eventually lands on a suitable plant before she crawls over it to taste the leaves with her feet. Once a suitable spot has been found, she lays a single egg on the top of the leaf, near the midrib, before flying off. This is unlike the behaviour of the Brown Argus female that typically lays her eggs on the underside of leaves. Ellis also found that eggs are often laid on the second or third pair of leaves from the top of the hostplant. Like the Brown Argus, the Northern Brown Argus selects those plants that are richer in nitrogen since this allows the larva to grow more quickly, thereby reducing the potential for it to be predated. Jim Asher also reported at Butterfly Conservation's Eighth International Symposium in 2018 that suitable egg-laying areas were sheltered by rock or Gorse at his coastal study sites in south-west Scotland. Eggs are extremely easy to find as their white shells are very conspicuous against the green of a Common Rock-rose leaf—I once recorded over 40 eggs at the base of a 20-m stretch of hillside at Burnmouth in less than 30 minutes.

A mating pair [IL]

The egg is laid on the upperside of the leaf

A delicate network pattern is visible when the egg is viewed close up

Ovum

The 0.5-mm wide egg has the typical Lycaenid shape, resembling a flattened sphere, and has a sunken crown. Unsurprisingly, it is identical to the egg of the Brown Argus, and is covered in a network pattern with prominent points wherever two ribs join.

Larva

After around two weeks, depending on temperature, the 1-mm long **first instar** larva emerges from the egg and, without eating the remainder of the eggshell, moves to the underside of the leaf. Here it feeds on the lower epidermis, without breaking through the upper surface, and this leaves tell-tale pale patches that can give away the presence of a larva when seen from above, although some other insect species, such as the Cistus Forester *Adscita geryon* moth, leave similar clues. The larva will drop to the ground at the slightest disturbance and it is thought that this is a defence against grazing animals. The larva has a shiny black head and is covered in several rows of long hairs that run the length of its body. Its ground colour is a very pale yellow, almost white, when it first emerges from the egg, but the larva turns slightly greener as it ingests food. When fully grown, the first instar is around 2 mm long.

After two weeks, the larva moults into the **second instar** and is now around 2.5 mm long when crawling. The larva now has a green hue, many more hairs on its body and a dark furrow along its back, although it is otherwise similar in appearance to the first instar. When fully grown, the second instar is 3 mm long.

After around three weeks, the larva moults into the **third instar** and now has a pale green ground colour, but retains the shiny black head and long hairs of earlier instars. Paler larvae have a whitish stripe that runs along the base of each side and, in darker larvae, this may be accompanied by a pale pink stripe, along with pale and oblique stripes along each side. The larva now continues to feed on the leaf cuticle but, before it is fully grown in this instar, in September and early October, will move low down on the foodplant, or on some other vegetation at the base of the plant, where it hibernates until the following spring. The larva emerges from hibernation on warm days in late February and early March when it resumes feeding and, when fully grown, the third instar is 5 mm long.

A couple of weeks after feeding has resumed, the larva moults into the **fourth instar**. The larva now has more brilliant and contrasting colours than the third instar and the hairs on its body are now more bristle-like. The larva also changes its feeding habits and, while it will continue to feed on the

1st instar: the newly emerged larva is a very pale yellow

2nd instar: a larva feeding underneath a leaf

3rd instar: a pre-hibernation larva

4th instar: the colours are now much more brilliant

5th instar: the larva has a pale plate on the first segment

5th instar: the larva has a shiny black head in all instars

underside of a leaf, will now break through the upper surface, and it will also eat the leaf edge. When fully grown, the fourth instar is 6.5 mm long.

After around 10 days, the larva moults for the last time into the **fifth instar**. It is similar in appearance to the fourth instar although the colours have even more contrast, the pale and oblique stripes along the sides of the body are more prominent, and the larva possesses a distinct dark green line along the middle of its back. The larva also has a pale plate on the first segment. When fully grown, in late May, the final instar is 12 mm long. When preparing to pupate, the larva will descend to the ground and move under loose vegetation or soil particles that it holds together with a few silk threads.

The larva possesses all of the ant-attracting (or appeasing) organs, namely a honey gland, pore cupolas and tentacle organs, and, while ants are often seen tending to Northern Brown Argus larvae on the continent, this association has only occasionally been observed in Britain when the black ant *Formica lemani* was seen in attendance.

Pupa

The 8.5-mm long pupa is both paler and plainer than that of the Brown Argus and the only prominent feature is a black crescent over each eye. The pupa also lacks any cremastral hooks, although it retains the larval skin once it has been shed. The pupa darkens as the butterfly develops, until the adult can be seen clearly through the pupal case and, in the *artaxerxes* subspecies, the white spots on the forewings are quite visible. This delightful little butterfly emerges after three or four weeks as a pupa.

A newly formed pupa

A coloured-up pupa of subsp. *artaxerxes*, with the distinctive white spot visible

As its name suggests, the Common Blue is, indeed, the commonest and most widespread of our blues, and is found throughout Britain and Ireland. However, it is only the male that has a uniform lilac-blue upperside—the female has a brown upperside with a variable amount of blue scaling and orange crescents along the wing edges. The underside of the male is also a mix of grey and light brown, whereas that of the female is slightly darker and a richer brown. Whatever the colour, there is no denying that this diminutive butterfly brightens up our countryside wherever it is found.

The first vernacular name given to this species by James Petiver in his *Musei Petiveriani* of 1699 was 'The Little Blew Argus' although this butterfly is far from the smallest of our blues. Petiver later confused the two sexes, naming the male 'The blue Argus' and the female 'The mixt Argus' in his *Gazophylacii naturae et artis* of 1704, with Moses Harris providing the name we use today in his *The English Lepidoptera* of 1775.

We have two subspecies in Britain and Ireland—the nominate subspecies *icarus*, and the subspecies *mariscolore* ('colour of the sea') where, according to most authors, the male of the latter is a more brilliant blue than the nominate subspecies and the female is more heavily marked with blue, with exceptionally well-developed orange crescents along the outer margin of the wings. However, the distribution of these two subspecies is somewhat confused. There is general agreement between authorities that the nominate subspecies is found throughout England, Scotland, Wales, the Isle of Man and the Channel Islands, and that subspecies *mariscolore* is found in Ireland. However, individuals that conform to the description of the nominate subspecies can be found in Ireland and, conversely, it is possible to find individuals that conform to the description of subspecies *mariscolore* in Britain, including north-west Scotland. Putting these differences aside, coming across a subspecies *mariscolore* individual is a special moment wherever it is found and, after several years of searching around different parts of Ireland, I finally connected with this subspecies at Craigavon Lakes in County Armagh, Northern Ireland. This location holds a special place in my heart since it is also where I first came across the Cryptic Wood White and, as any butterfly aficionado will tell you, the first sighting of any species (or subspecies) will stay with you forever.

The ♂ has a wingspan of between 29 and 36 mm

A typical ♀ with a mainly brown upperside and a similar wingspan to the ♂

A ♀ with an unusual amount of blue on its upperside

A ♀ subsp. *mariscolore* with well-developed orange crescents

DISTRIBUTION
The Common Blue is the most widespread Lycaenid found in Britain and Ireland and can be found almost anywhere, including Orkney and parts of Shetland, although it is absent from mountainous areas of Wales and Scotland. This butterfly lives in discrete colonies that, at the peak of the flight period, are only ever measured in the tens or hundreds, unlike the other members of the *Polyommatus* genus, the Adonis and Chalk Hill Blues, whose colonies are often measured by the thousand. Worldwide, the Common Blue is found across Europe, through temperate Asia into China, and in parts of North Africa. The butterfly is not a great wanderer, although it will turn up away from any known colony on occasion and I get one or two individuals in my suburban garden in most years.

HABITAT
The strongholds of the Common Blue are open grasslands that are rich in Common Bird's-foot-trefoil, the primary larval foodplant. However, given this butterfly's broad distribution, it should come as no surprise that it is found in a wide variety of other habitats, including woodland clearings, rides and glades; meadows and pasture; disused quarries; sand dunes; undercliffs; arable field margins and set-aside; brownfield sites; road verges and disused railway lines.

Despite this being our commonest blue, the butterfly is particularly prone to drought and populations crashed following the hot summer of 1976 when the larval foodplants desiccated in the scorching summer sun, taking several seasons for the species to recover. A site that is local to me demonstrates this fragility all too well in most years—in the spring, the Common Bird's-foot-trefoil is lush and the ground soft, but by mid-summer the ground in some areas is hard and the foodplant so dried out that it crunches under foot.

The yellow flowers of Common Bird's-foot-trefoil at Greenham Common in Berkshire

STATUS
This butterfly has experienced a modest decline over the last several decades, mainly as a result of agricultural intensification that has led to a loss of suitable grassland and the accompanying foodplants. However, despite a general decline in distribution, this butterfly remains widespread and is not currently a species of conservation concern. The long-term trend shows a moderate decline in both distribution and abundance, although the more recent trend over the last decade shows a stable distribution and abundance. Using the IUCN criteria, the Common Blue is categorised as Least Concern (LC) in both the UK and Ireland.

LIFE CYCLE
This species has two broods in the southern half of Britain and Ireland and, in good years, adults may be seen from early May, with a peak at the end of May, giving rise to a second generation that emerges in the second half of July that peaks in the middle of August. There may also be a third

brood in favourable years, especially in the southern English counties. The overlap between broods means that this butterfly can be found on the wing almost continuously from May through to early September and even into October and November in good years. There is a single brood further north in northern England, Scotland and the north of Ireland, which typically emerges in June and reaches its peak in July.

Two ♀s and a ♂ roosting head-down

Imago

This butterfly is most active in bright sunshine and is a frequent visitor to flowers, although overcast conditions or weak sunshine provide the best photo opportunities when the adults will bask, often on a flowerhead, with wings held wide open. The territorial male is the more conspicuous of the two sexes and can be found either perching or patrolling in search of a virgin female. The female, on the other hand, spends most of her time resting when not feeding or egg-laying. Adults are avid nectar feeders and will visit a variety of flowers including those of Common Bird's-foot-trefoil, Bugle, clovers, Common Fleabane, Kidney Vetch, knapweeds, ragworts, thistles, Wild Marjoram, Wild Thyme and vetches.

In very dull weather this species roosts head-down on a tall grass stem and it is not unusual to find several adults in the same vicinity and even the same stem. Watching these groups following a period of inclement weather or in early morning is one of the great joys of butterfly watching as each member of the group slowly adjusts itself to catch the sun before slowly opening its wings when the lilac-blue males, especially, sparkle like jewels against the tall grasses.

Mating is a rapid affair without any preliminary courtship, and couples can often be found sitting on top of a flowerhead where they will remain until they separate, unless disturbed.

A mating pair with ♂ on the left

An egg-laying female will fly low over the ground before landing and then crawling over low-growing plants, while drumming with her feet in order to detect a suitable plant on which to lay. When found, she dips her antennae and curls her abdomen before laying a single pale green egg on the upperside of a young leaflet, although she may occasionally lay on nearby vegetation. The primary larval foodplant is Common Bird's-foot-trefoil although Black Medick, Common Restharrow, Greater Bird's-foot-trefoil and Lesser Trefoil are also used. I have watched females lay eggs many times and have been surprised at some of the locations chosen—on one occasion at Chiddingfold Forest (that spans Surrey and West Sussex), an egg was laid on one of the puniest Common Bird's-foot-trefoil plants I have ever seen, and which was growing in the middle of a footpath.

An egg laid on Common Bird's-foot-trefoil

Ovum

Eggs are 0.5 mm in diameter and 0.25 mm high and are a pale green when first laid, but soon turn a brilliant white as they dry. They are covered in a beautiful network of undulations when viewed close up and have a sunken crown and a greenish micropyle, with a flattened bun-like appearance—as James William Tutt said in his *A Natural History of the British Lepidoptera* of 1906: "*They looked as if treated by a steam-roller*". Eggs are quite conspicuous and are easily found at good sites—I remember one occasion at Greenham Common in Berkshire when I was able to scan a carpet of Common Bird's-foot-trefoil with some close-focusing binoculars and find over a dozen eggs in less than 10 minutes, along with the remainder of several eggs from which the larva had already emerged.

Larva

Ten days or so after the egg has been laid, the 1-mm long **first instar** larva emerges by eating away the centre of the crown, leaving the remainder of the shell. The larva is a pale greenish-yellow and has several rows of long pale hairs that curve backwards over its body. The main distinguishing feature of this instar, however, is that its surface is covered in minute black spots that are absent in later instars although, unfortunately, this particular feature is almost impossible to discern in the field.

Some authors state that the early instars feed solely on the underside of the leaflet but, in my experience, the larva feeds on the epidermis of both sides of the leaflet but will never break through from one side to the other, always leaving the leaflet structurally intact. This feeding behaviour leaves characteristic 'windows' on the leaflet surface and can give away the presence of an early-instar larva. As well as feeding on the leaves, larvae in all instars will also feed on the flowers of the foodplant and, when at rest, the larva conceals its dark brown head under the first segment.

1st instar: a newly emerged larva feeding on the leaf epidermis

1st instar: the fully grown larva is a very light brown

1st instar: typical larval feeding damage

2nd instar: the larva now has prominent humps on its back

3rd instar: the larva has two pale yellowish lines running along its back

After nine days or so, the fully grown first instar is a very light brown and approximately 1.5 mm long, and moults into the **second instar** but does not eat its old skin. The larva now exhibits prominent humps on its back, is slightly hairier than the first instar and has several very faint pale lines running along each side. Otherwise, in terms of both appearance and behaviour, it is very similar to the first instar, although the surface of its body is no longer covered in minute black spots. When fully grown, the second instar is 2 mm long.

After approximately 16 days, the larva moults into the **third instar**. This instar is similar in appearance to the second instar but is now covered in numerous hairs of varying size, with each appearing to emanate from a dark base when viewed close up. The body is of a consistent dark green, although there are two pale yellowish lines running down the back of the larva, with another running along each side below the spiracles. Larvae will now crawl more readily over the plant and will feed on the edges of leaves rather than the epidermis.

The rate of development of the third instar determines whether or not it will go on to contribute to another brood within the same year or hibernate. In his *Natural History of British Butterflies* of 1924, F.W. Frohawk writes: "*As a rule only a small number of the larvae from the first brood feed up and pupate the same summer; the majority feed and grow very slowly, therefore they remain small, and after their second moult enter into hibernation*". Of course, such a statement from the early 1900s must be adjusted to accommodate our changing climate given that three broods are occasionally encountered in the south of England, although the sentiment is the same—some larvae will have an accelerated growth and go on to produce adults in the same year, whereas others will grow more slowly and, if they overwinter, will do so in their third instar, developing into adults the following year. Larvae that hibernate do so either low down on the larval foodplant under a leaf, or within a nearby piece of moss or leaf litter, changing colour from green to olive. Larvae emerge five or six months later in early March when they recommence feeding and restore their green colouring, soon reaching their full length of just over 3 mm. Larvae that overwinter spend five to six months in this instar and those that do not just over three weeks.

4th instar: yellowish-green lines run along the base of each side

The **fourth instar** is similar to the third instar but is a more uniform green, with shorter hairs, although it retains the pale lines running down the back. It also retains the lines that run laterally along each side of the larva below the spiracles, although these are now very distinct and a yellowish-green. After approximately 30 days, and before the fourth and final moult, the larva is 6.5 mm long.

5th instar: a fully grown larva

The final and **fifth instar** is very similar to the fourth instar, with pale hairs all over its body that are similar in length to those of the fourth instar, and with the light yellowish-green lateral stripes along its sides below the spiracles. However, the larva is much longer than the previous instar—the final instar larva is 12 mm long when fully grown and almost twice the length of the fourth instar.

Like most Lycaenids, the larva is attractive to ants and possesses a honey gland on the 10th segment (that is most developed in the final instar) and tentacle organs on the 11th segment, although it is less attractive to ants than its close relatives, the Adonis and Chalk Hill Blues. It has also been shown that the quality and quantity of secretions from the honey gland are directly related to the quality of the foodplant on which the larva has fed and this, in turn, influences the degree to which the larva will attract ants from which it will receive protection. It is therefore no surprise that the female pays a lot of attention to where she lays her eggs.

Pupa

The 10-mm long pupa is formed either on the ground or just under the soil surface within a flimsy cell made up of soil particles that are held together with a few strands of silk, and which may be put in place by ants looking to conceal the larva. The pupa is not attached to its cell by any girdle or the cremaster since, like many closely related species, there are no cremastral hooks, although the cast larval skin often remains attached to the end of the pupa. The pupa darkens as the adult butterfly develops and the wings can be seen quite clearly through the pupal case just before this jewel of a butterfly emerges, having spent two weeks in this stage.

A pupa outside of its earthen cell | A fully coloured-up ♂ pupa | A fully coloured-up ♀ pupa

Life Cycles of British & Irish Butterflies

Adonis Blue
Polyommatus bellargus

❖ ADONIS BLUE IS SUPPORTED BY PATTY GARDNER, COLIN & SUE KNIGHT & LORRAINE MUNNS ❖

	January	February	March	April	May	June	July	August	September	October	November	December
IMAGO												
OVUM												
LARVA												
PUPA												

The aptly named **Adonis Blue** is truly a jewel of chalk and limestone grassland—the electric blue of the male upperside is the most brilliant of all of our Lycaenids. The female, on the other hand, has chocolate brown uppersides that are sprinkled with a variable amount of blue. Two of the early names for this species refer to known localities—the 'Clifden Blue' (after Cliveden, in Buckinghamshire) and the 'Dartford Blue' (in Kent)—but these were superseded by its current vernacular name when W.S. Coleman used 'Adonis Blue' in his *British Butterflies* of 1860. Even the scientific name of this butterfly pays homage to the stunning appearance of the male—*bellargus* is a combination of the Latin words *bellus* (beautiful) and *argus* (the many-eyed shepherd of Greek mythology, which is a reference to the spots found on the butterfly's underside).

The brown ♀ has a similar wingspan to the ♂ [IL]

The brilliantly coloured ♂ has a wingspan of between 30 and 40 mm

The ♂ has a slightly lighter underside than the ♀

♀ roosting on a grass stem

DISTRIBUTION

Like its close relative, the Chalk Hill Blue, the distribution of this species follows that of Horseshoe Vetch which, in turn, follows the distribution of chalk and limestone. However, the Adonis Blue has a more restricted range than the Chalk Hill Blue, indicating that it has more precise habitat requirements. That being said, this butterfly can be found in large numbers where it does occur, especially in its core areas of chalk downland in Dorset, Wiltshire, Sussex and the Isle of Wight. Increases have also been recorded from the Chilterns, the North Downs and in East Kent. This species is absent from central England, northern England, Wales, Scotland, Ireland, the Isle of Man and the Channel Islands. On a worldwide basis, the Adonis Blue is distributed across the western Palearctic, through central and eastern Europe and into Russia, but is absent from Fennoscandia and parts of the Mediterranean.

The Adonis Blue lives in discrete colonies and even small barriers such as a line of scrub or a hedgerow can prevent any significant dispersal, with most adults flying less than 250 metres and, whilst the butterfly has recolonised sites that are 15 km from the nearest inhabited patch, this is

thought to be the result of the butterfly using intermediate areas as 'stepping stones'. Populations fluctuate greatly from year to year and between the spring and summer broods of this double-brooded species, and the butterfly can occasionally be found by the hundred at good sites, and even by the thousand at the best, especially in the larger second brood.

HABITAT

This is a warmth-loving species that prefers sheltered, south-facing grassland slopes where the larval foodplant, Horseshoe Vetch, is abundant. Sheltered hollows such as chalk pits and abandoned quarries are also used, as are coastal cliff tops. The butterfly requires sites where the turf is closely cropped and, in their article 'Seasonal variation in the niche, habitat availability and population fluctuations of a bivoltine thermophilous insect near its range margin', published in *Oecologia* in 2003, David Roy and Jeremy Thomas show that the spring brood female lays her eggs on Horseshoe Vetch growing in vegetation that is between 1 cm and 7 cm tall (in mid-summer, when temperatures are high), whereas the second brood female shifts to laying among vegetation that is less than 3 cm tall (in August/September when temperatures are cooler). Such precise vegetation requirements are typical of a species that is at the northern limit of its range in Britain and a habitat management regime using cattle, ponies and sheep needs to account for any impact by rabbits to prevent overgrazing, and also aim to provide a varied turf height that will cater for both broods.

On the continent, where temperatures are higher, the butterfly is much less fussy and can be found in areas of relatively tall vegetation, so long as the larval foodplant is present. The Adonis Blue also has a strong association with ants and requires ant colonies that are active from early March (when the overwintering larvae become active) until late October (when the overwintering larvae go into hibernation) and this, again, requires sites that have relatively high temperatures.

The yellow flowers of Horseshoe Vetch on the slopes of Mill Hill, East Sussex [NH] and (inset) a ♀ feeding on Horseshoe Vetch

STATUS

Like several other species, the Adonis Blue has suffered from the loss of unimproved calcareous grassland to development and agriculture. Inappropriate habitat management also impacts this butterfly, either through overgrazing (by livestock or rabbits) where precious nectar sources are lost, or undergrazing (as a result of abandonment or a decline in the rabbit population) where a tall sward results in unsuitably cool conditions. The Adonis Blue, along with the Chalk Hill Blue, can also be impacted by a period of drought that causes Horseshoe Vetch to wilt. The long and hot summer of 1976, enjoyed by so many of our species, was disastrous for the Adonis and Chalk Hill Blues, whose numbers plummeted, and it took several generations for these species to recover.

All of these factors took their toll as the butterfly declined steadily throughout the 1950s and 1960s. Thankfully, the butterfly had a resurgence in the 1980s and 1990s when there was a focus on the needs of this butterfly on nature reserves and the targeting of this species within agri-environment schemes, which resulted in increased grazing by livestock. The recovery of rabbit populations following the introduction of myxomatosis also had a major impact as overgrown downs became suitable once again.

The long-term trend of the Adonis Blue shows a stable distribution and a significant increase in abundance, indicative of its recovery, with the more recent trend over the last decade showing a moderate decline in both distribution and abundance. Using the IUCN criteria, the Adonis Blue is categorised as Near Threatened (NT) in the UK.

LIFE CYCLE

The Adonis Blue has two broods each year, with the spring brood flying from mid-May to mid-June, peaking at the end of May and start of June, and the summer brood flying from early August to mid-September, peaking at the end of August and start of September. In good years, the summer brood may start its emergence at the end of July and, in late years, stragglers may still be seen in early October.

The summer brood is invariably larger than the spring brood and it is thought that this is due to the greater availability of suitably warm areas in June and July used by the spring brood females. Females of the resulting summer brood, however, have to be more selective, laying only on those plants growing in short vegetation and in sheltered areas. It is thought that the more precise habitat required by the summer brood females is the primary constraining factor influencing the size of the smaller spring brood, although the much longer time for the development of their overwintering offspring (and the greater potential for predation, for example) may also play a part.

Imago

Adults will only fly when the sun is shining, when they can be found taking nectar from a variety of plants and usually whatever flowers are present. Those of the spring brood feed on the flowers of Horseshoe Vetch, whereas those of the summer brood feed on knapweeds, ragworts, thistles and Wild Marjoram. Males will also congregate on animal droppings where they take on minerals and salts.

This species can be found roosting communally head-down at night, and an early morning visit to a good site often pays dividends. I have a vivid memory of visiting Figsbury Ring in Wiltshire at 7 am and watching butterfly after butterfly open its wings in the hazy morning sunshine. The end result was a chalk earthwork that was bespeckled with electric blue jewels along its length—an unforgettable sight.

The male is the most conspicuous of the two sexes, not only because of its colour, but also because of its behaviour as it flies swiftly in a zig-zag pattern close to the ground, searching out newly emerged females. If a virgin female is found, then the male will land alongside her and mate without any discernible courtship, and it is not unusual to find a mating pair where the female has yet to dry her wings. After pairing, the female will spend long periods resting on the ground in sheltered hollows and depressions whilst her eggs develop.

Females are very particular when selecting egg-laying sites, choosing only those plants growing in sunny and sheltered areas, among closely cropped vegetation, and often in the vicinity of bare ground. The female flutters low to the ground when searching out suitable plants on which to lay. She ultimately lands, crawls over the plant and lays a single white egg on the underside of a terminal leaflet, often crawling or flying to a nearby plant and laying another egg before flying off. Eggs are occasionally laid on the upperside of a leaflet and also on a stalk of the foodplant. Several eggs may be laid on the same plant (Jeremy Thomas has found as many as 40), although these will usually have been deposited by different females.

A group of ♂s around a fox scat [NH]

A ♂ mating with a newly emerged ♀

Ovum

The 0.6-mm diameter and 0.3-mm high greenish-white egg resembles a flattened sphere and has a fine network pattern all over its surface that F.W. Frohawk likens to 'rough frosted glass'.

The egg resembles a flattened sphere

Larva

The **first instar** larva emerges from the egg by eating a hole in the crown, and often to one side, leaving the remainder of the eggshell uneaten. Eggs from the spring brood, laid in May and June, hatch in around two weeks, whereas those of the summer brood, laid in August and September, take three or more weeks to hatch, indicating that the speed of development of the first instar while in the egg is temperature-dependent.

The 1-mm long larva starts feeding on one surface of a leaflet, using its extensible neck to full effect as it scoops out material below the epidermis, but does not break through to the opposite side (although it is perforated on occasion). The shiny black head of the larva (it is this colour in all instars) can be seen, under high magnification, moving backwards and forwards as it feeds, and this pattern of feeding leaves pale discs on the leaflet surface. The larva has a pale yellow-brown ground colour with a pale stripe at the base of each side and is covered in several rows of long hairs. Unlike its close relative, the Chalk Hill Blue, the larvae are diurnal, feeding during the day. When fully grown, the first instar is 2.5 mm long.

According to Frohawk, larvae from the second brood will hibernate while in their first, second or third instar and, as a rough guide, this means that larvae that emerge from the egg in August, September and October will typically overwinter in their third, second and first instar respectively. Whatever the instar, a hibernating larva will spin a silk pad on the underside (occasionally the upperside) of a leaflet or low down on a stem, remaining attached to the pad until the middle of March the following year.

First brood larvae that do not overwinter moult into the **second instar** after a week or so, whereas those that do overwinter moult after around 180 days, and the larva may or may not eat its old skin. The second instar is similar in appearance to the previous instar but now has more hairs over its body and, as it grows, develops two pale yellow stripes that run the length of its back, with another at the base of each side. The larva continues to rest beneath a leaflet or on a stem, and feeds on the epidermis of a leaflet, although the feeding damage is more extensive than that of the first instar. When fully grown, the second instar is around 4 mm long.

The association with ants is well-developed in the Adonis Blue (a relationship that starts in the second instar according to Jeremy Thomas, who has studied this species extensively) and it is sometimes easier to find the well-camouflaged larva by first looking for the attendant ants. The ants feed on secretions from the microscopic pores (the 'pore cupolas') that are scattered all over the body of the larva, and also from the honey gland, which ants stimulate by 'drumming' on the body of the larva with their antennae. The larva is also known to 'sing' to get the attention of (and possibly pacify) ants although the sounds produced are inaudible to the human ear. The larva benefits from this relationship by receiving protection from invertebrate predators and parasitoids. Several species of both red and black ants have been recorded as suitable hosts, especially the red ant *Myrmica sabuleti* and the black ant *Lasius alienus*, and they are known to bury larvae that are resting at night (Adonis Blue larvae only feed during the day), or those that are moulting, in loose earth cells. The cells may contain several larvae as well as several ants that continue to 'milk' the larvae for secretions.

1st instar: a fully grown larva

2nd instar: a larva feeding on the underside of a leaflet

If the larva does not overwinter, then it moults into the **third instar** in about two weeks, depending on temperature. The larva has an olive-green and blotchy ground colour, the yellow stripes on the back and sides are much more prominent, the spiracles are noticeably black, and the body is now covered in numerous long and greyish-white hairs. The larva now feeds on the edges of leaflets and, when not feeding, rests at the base of the foodplant, often on bare soil.

While the relationship with ants starts in the second instar, the honey gland is more easily located in the third instar, when it can be found on the 10th segment, along with a pair of tentacle organs on the 11th segment. The fully grown third instar is 6 mm long.

After around two weeks, the larva moults into the **fourth instar** and is similar in appearance to the third instar, although it now has a greener and more uniform ground colour, with deeper yellow markings, and the hairs on its body are now largely brown, with white at their bases. When fully grown, the fourth instar is 9 mm long.

3rd instar: the larva now rests at the base of the foodplant

4th instar: a newly moulted larva

4th instar: a larva being 'milked' by a black ant

After around 10 days, the larva moults into the final and **fifth instar**. The larva is now covered in relatively short and stiff hairs that are the same colour as those of the fourth instar, and it now has a pale patch on its first segment. While the overall appearance of the larva is similar to the fourth instar, there is an obvious difference in size—when fully grown the final instar is around 16 mm long and almost twice as long as the fourth instar. When ready to pupate, the larva retires to the roots of the foodplant, or a crevice in the ground, where ants cover it in soil thereby forming a cocoon, the larva pupating after four or five days. Pupae have also been found in the warm upper chambers of ant nests, where they are continually tended until they emerge, although it is not known if the larvae crawl into the nests or are carried there.

5th instar: a larva showing its black head and a pale patch on its first segment [CK]

5th instar: the camouflaged larva is most easily found by first looking for attendant ants [CK]

Pupa

The 11-mm long pupa is a translucent pale green when first formed. As the adult butterfly develops, the wings and other body parts become increasingly visible through the pupal case, the adult emerging after around three weeks in this stage. This event usually takes place early in the morning, between 9 am and 10.30 am, and often in the company of ants as the newly emerged adult makes its way to a suitable platform to expand its wings.

The newly formed pupa is a translucent pale green

A coloured-up ♂ pupa

A coloured-up ♀ pupa

Chalk Hill Blue
Polyommatus coridon

❖ CHALK HILL BLUE IS SUPPORTED BY ADA GREENAWAY, BENJAMIN GREENAWAY & NELLY GREENAWAY ❖

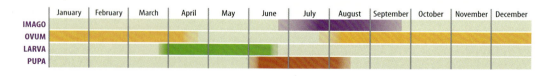

As its name implies, the Chalk Hill Blue is found on chalk and limestone downland where its sole larval foodplant, Horseshoe Vetch, grows in abundance. The adult butterfly is most often seen in bright sunshine and, at the best sites and in good years, the ground may appear to shimmer with the activity of hundreds, if not thousands, of males flying just a few inches off the ground as they search for newly emerged females.

Aside from its habitat, the butterfly is known for the unmistakable colour of the male that was recognised in the first name of 'The pale blue Argus' given to this species by James Petiver in his *Gazophylacii naturae et artis* of 1704, before Moses Harris gave us the name we use today in his *The English Lepidoptera* of 1775 that has been used consistently ever since (albeit with spellings of 'Chalkhill', 'Chalk-Hill' and 'Chalk Hill'). While the two sexes are strongly dimorphic—the males are a pale sky blue and the females a chocolate brown—they do have similar wingspans of between 33 and 40 mm, making them larger than all of our other blues, with the exception of the Large Blue. The butterfly is known for its variation and *A Monograph of the British Aberrations of the Chalk-Hill Blue Butterfly*, authored by P.M. Bright and H.A. Leeds and published in 1938, describes over 400 aberrations.

The pale sky blue ♂ has a wingspan of between 33 and 40 mm

The ♂ underside is a very pale greyish brown

The chocolate brown ♀ has a similar wingspan to the ♂

The ♀ underside is a richer brown than that of the ♂

DISTRIBUTION

The distribution of this species does not extend beyond the distribution of Horseshoe Vetch which, in turn, follows the distribution of chalk and limestone grassland. This species is found only in England, south-east of a line running from the Cotswolds of West Gloucestershire in the west to Cambridgeshire in the east, although the foodplant does occur much further north than the butterfly's current distribution, suggesting that the range of this species is limited by climate rather as well as the availability of foodplant. This species is absent from most of central and northern England, Scotland, Wales, Ireland, the Isle of Man and the Channel Islands. The butterfly's worldwide distribution covers the western Palearctic, from much of Europe in the west (excluding Fennoscandia and the south Mediterranean) to western Asia as far as the Ural Mountains.

The butterfly lives in discrete colonies that vary in size from tens of adults to hundreds of thousands and, in *The Butterflies of Sussex*, Neil Hulme describes a population explosion in 2012 at Friston Forest Gallops in East Sussex, with the number of adults estimated at over 800,000 on 3 August. Given the size of the largest colonies, it is unfortunate that the butterfly has limited dispersal abilities, although both sexes have been recorded at least 30 km from any known colony in good years. These dispersing adults are mostly male, indicating that the females are more sedentary, although the male is the most conspicuous of the two sexes and this may provide an element of recording bias.

HABITAT

This is a species of unimproved calcareous grassland where plentiful Horseshoe Vetch is present. This is also a warmth-loving butterfly, and it is typically found on sheltered, south- and west-facing hillsides with short and sparse vegetation, with patches of bare ground that adults occasionally use to warm themselves. The Chalk Hill Blue, however, is more tolerant of slightly cooler sites than its close relative, the Adonis Blue, and is able to breed in a taller sward, with an ideal vegetation height of between 5 and 10 cm.

Both species fly at sites with a varied sward, which can be maintained through grazing with cattle or sheep although, as with the Adonis Blue, any grazing regime should be monitored to prevent overgrazing, especially given the contribution of a fluctuating rabbit population. In *The Butterflies of Sussex*, Neil Hulme writes: "By 1993 rabbit numbers on Mill Hill had increased greatly and they have, ever since, been far too high to be anything other than destructive. The very same animals which were once critical in maintaining the species-rich turf have become its enemy".

The yellow flowers of Horseshoe Vetch at Denbies Hillside in Surrey

STATUS

This species has suffered over the long term due to a loss of suitable calcareous grassland, either because it was converted to arable farmland or because grazing ceased when it became uneconomic. Other sites were severely impacted due to the decline of the rabbit population from the 1950s to 1980s as a result of myxomatosis, where the lack of grazing resulted in sites becoming unsuitably rank and overgrown. The fragmentation of remaining sites and the sedentary nature of females has also limited the ability for the species to recolonise areas that have since become suitable again. On a more positive note, almost all Chalk Hill Blue sites now have a level of protection, while others have benefited from agri-environment schemes that have incentivised landowners to manage their land sympathetically for this species.

The long-term trend shows a significant decline in distribution and a moderate increase in abundance, with the more recent trend over the last decade showing a stable distribution and a significant increase in abundance. Using the IUCN criteria, the Chalk Hill Blue is categorised as Near Threatened (NT) in the UK.

LIFE CYCLE

Unlike its double-brooded cousin, the Adonis Blue, the Chalk Hill Blue is single brooded with the first adults emerging in the last week of June and first week of July in most years, peaking in early

August, and with a few adults still flying until late September. Adults sometimes turn up later in the year, as was the case in the middle of October 2018, with mating pairs seen and many assuming that these were individuals of a second brood rather than stragglers from the first brood. At least one subspecies of Chalk Hill Blue is regularly double-brooded on the continent.

Imago

This sun-loving butterfly will use a variety of nectar sources, including Common Bird's-foot-trefoil, Field Scabious, knapweeds, Selfheal, thistles and Wild Thyme. Males will also visit, often in some numbers, moist earth or animal droppings to gather water, salts and minerals. Both male and female also spend a good amount of time basking. At good sites, this species can also be found roosting communally on the lower slopes of a hillside, usually on grass stems, where they can be found facing head-down and with several individuals occasionally using the same stem. This can subsequently make for an incredible spectacle the following morning as the adults open their wings to catch the first rays of the sun.

♂s obtaining minerals from a cow pat | A group of ♂s roosting communally [GN]

Males are particularly active and spend most of their time feeding or searching for a mate, when they can be found patrolling low over the ground in search of their quarry. Any female that flies up is quickly met with, for the male has a strong and rapid flight, and mating occurs without any courtship, with the pair remaining coupled for around two hours. At the height of the flight season it is not unusual to come across mating pairs across a site where they can be found sitting quite openly atop a flowerhead and, when disturbed, will fly a few feet before landing nearby.

When egg-laying, the female butterfly will fly over the foodplant, land, and then crawl among the vegetation whilst searching for a suitable plant. She then lays a single egg either directly on a stem of the foodplant or in nearby vegetation before flying off. Plants rich in nitrogen are preferred since this allows the larva to develop more quickly and may also result in higher quality secretions from both larva and pupa that are an important part of the relationship that this species has with ants. A study by Stephen Henning in the 1980s showed that females of the Roodepoort Copper *Aloeidas dentatis*, from South Africa, were stimulated to lay by a pheromone left on ant trails, and this phenomenon may also apply in the case of the Chalk Hill Blue.

A mating pair with ♂ on the left [IL]

In the article 'Chalkhill Blue on Therfield Heath, Hertfordshire', published in *British Wildlife* in 2003, J.P. Duff and V. Thompson share some interesting observations. Their hypothesis is that mated females have two strategies for avoiding the unwanted attention of amorous males—mated females feed away from the core concentrations of adults in the morning when males are most active, and reserve their egg-laying activities for later in the day when they are less likely to be pestered.

The egg has a flattened top

Ovum

The egg is 0.5 mm in diameter and 0.3 mm high, and has the shape of a compressed sphere, with a coarse network pattern and raised points over its sides, although its top is relatively flat and lacks any significant protuberances. J.W. Tutt described the egg's appearance as resembling *"a flattened cheese with rounded edges"* when viewed from the side. The egg is white when first laid but turns slightly duller over the winter, when the larva is fully developed. The egg is not strongly attached to the surface on which it is laid and will fall to the ground, where it is impossible to spot among chalk or limestone chippings.

Larva

After seven or eight months in the egg, the 1-mm long and pale yellow **first instar** larva emerges by eating a hole in the crown, but it does not eat the remaining eggshell. It then searches for fresh growth of Horseshoe Vetch before it positions itself on a leaflet, usually on the underside. Here the larva grazes on the leaflet surface while leaving the epidermis of the opposite side intact, feeding primarily at night and resting at the base of the stems during the

1st instar: a larva emerging from its egg

1st instar: a larva grazing on the surface of a leaflet

day. The larva is of the typical Lycaenid shape, with a shallow furrow down the middle of its back, and with each of the segments clearly delineated. Two rows of long backward-pointing and pale hairs run along the back, and each hair emanates from a dark base. Shorter pale hairs also run along the base of each side. The larva has black spiracles and two rows of black spots above them on each side of the body. When fully grown, the first instar is 1.3 mm long.

After nine days or so, the larva moults into the **second instar**. It is similar in appearance to the previous instar but now has several more rows of hairs over its body. Also, as the larva matures, it is possible to make out two pale yellow lines that run along the back on each crest, either side of the central furrow, along with another pale yellow line that runs along the base of each side, and this colouring provides a hint of things to come. The fully grown second instar is still tiny and only 2.5 mm long. In *The Moths and Butterflies of Great Britain and Ireland* of 1990, A. Maitland Emmet writes: *"After the first instar, larvae are invariably attended by ants and are sometimes buried under the soil by them during the day"* although, unsurprisingly, the relationship with ants is easier to observe in more mature and larger larvae.

After another nine days, the larva moults into the **third instar** and is now covered in numerous hairs that each emanate from a black base, with the longest hairs found on the back. The larva is

2nd instar: the mature larva has pale yellow lines on its back and the base of each side

2nd instar: the larva has more hairs than the previous instar

3rd instar: the pale yellow markings are much more prominent

yellowish immediately after moulting, but soon develops a bluish-green ground colour with two rows of yellow markings along the back and along the base of each side, that are much more prominent than the similar markings found in the previous instar. The fully grown third instar is 5 mm long.

After around 12 days, the larva moults into the **fourth instar** and is similar in appearance to the previous instar, although the colours are richer, the pale yellow markings are more clearly defined, and the black spiracles are especially prominent. When fully grown, the fourth instar is just under 9 mm long.

After two weeks, the larva moults for the last time into the **fifth instar** and, while similar to the previous instar, now has much shorter hairs all over its body and a pale patch on the first segment. The shiny black spiracles are also most prominent in this instar and the yellow markings now contrast against a clear green ground colour. Larvae will now eat all parts of the plant, including flowers and even stems, and continue to feed mainly at night, remaining tucked away at the base of the plant during the day where ants are thought to occasionally bury them in earth. It is possible to find the nocturnal larvae on mild nights at good sites, when it is easier to look for attendant ants, such as the Yellow Meadow Ant *Lasius flavus*, before searching for a larva.

The ant-attracting organs of pore cupolas (all over the body), a honey gland (on the 10th segment) and tentacle organs (on the 11th segment) are most developed in this instar. In their paper 'Functional analysis of the myrmecophilous relationships between ants (Hymenoptera: Formicidae) and lycaenids (Lepidoptera: Lycaenidae)', published in *Oecologia* in 1988, Konrad Fiedler and Ulrich Maschwitz determined that the amount of secretion provided by Chalk Hill Blue larvae and consumed by their host ants (in exchange for the protection provided by the ants) was so great that the relationship was truly symbiotic with the ants depending on the sugary secretions for survival: *"Using data from the literature on ant metabolism, it is known that these carbohydrate secretions may contribute significantly to the nutrition of attending ants. The myrmecophilous relationship between the larvae of P. coridon and ants should therefore be regarded as a mutualistic symbiosis"*.

After two weeks in this instar, the 16-mm long larva is fully grown and, when ready to pupate, will retire to the base of the foodplant where it remains hidden among the roots and stems, or in a crevice in the ground.

Pupa

Like the larva, the 12-mm long pupa produces secretions that ants find attractive and it is thought that ants will carry pupae away and bury them in earth. The pupa can also produce a sound by rubbing two abdominal segments together, which also attracts the

attention of ants. The pupa lacks any cremastral hooks and, when first formed, is a translucent brownish-yellow, eventually turning a darker brown as the butterfly develops, with the adult becoming clearly visible through the pupal case when fully formed. After around four weeks, this characteristic butterfly of chalk and limestone grassland emerges, when it can often be found in the company of others as it dries its wings.

Bibliography

The following books are referenced in the species descriptions

Agassiz, D.J.L., Beavan, S.D. & Heckford, R.J. (2013) *Checklist of the Lepidoptera of the British Isles*. Royal Entomological Society, St. Albans.

Albin, E. (1720) *A Natural History of English Insects*. Innys, London.

Asher, J., Warren, M., Fox, R., Harding, P., Jeffcoate, G. & Jeffcoate, S. (Eds) (2001) *The Millennium Atlas of Butterflies in Britain and Ireland*. Oxford University Press, Oxford.

Barrett, C. (1893) *The Lepidoptera of the British Islands*. Reeve, London.

Blencowe, M. & Hulme, N. (2017) *The Butterflies of Sussex*. Pisces Publications, Newbury.

Buckler, W. (1886-1901) *The Larvae of the British Butterflies and Moths*. Ray Society, London.

Coleman, W.S. (1860) *British Butterflies*. Routledge, Warne & Routledge, London.

Curtis, J. (1823–1840) *British Entomology*. London.

Donovan, E. (1807) *The Natural History of British Insects*. Rivington, London.

Dowdeswell, W.H. (1981) *The Life of the Meadow Brown*. Heinemann Educational Books, London.

Dutfield, J. (1748) *New and Complete Natural History of English Moths and Butterflies*. M. Payne, London.

Feltwell, J. (1982) *Large White Butterfly – the biology, biochemistry and physiology of* Pieris brassicae *(Linnaeus)*. Dr W. Junk, The Hague.

Ford, E.B. (1945) *Butterflies*. Collins, London.

Frohawk, F.W. (1924) *Natural History of British Butterflies*. Hutchinson, London.

Frohawk, F.W. (1934) *The Complete Book of British Butterflies*. Ward Lock, London.

Furneaux, W. (1894) *Butterflies and Moths*. Longmans, London.

Harding, J. (2009) *Discovering Irish Butterflies and their Habitats*. Jesmond Harding.

Harris, M. (1775) *The English Lepidoptera*. Robson, London.

Harris, M. (1766) *The Aurelian*. London.

Haworth, A. (1803–1828) *Lepidoptera Britannica*. John Murray, London.

Henriksen, H.J. (1982) *The Butterflies of Scandinavia in Nature*. Skandinavisk Bogforlag, Odense.

Heslop, I.R.P., Hyde, G.E. & Stockley, R.E. (1964) *Notes and Views of the Purple Emperor*. Southern Publishing, Brighton.

Howarth, T.G. (1973) *South's British Butterflies*. Warne, London.

Humphreys, H.N. & Westwood, J.O. (1841) *British Butterflies and their Transformations*. William Smith, London.

Huxley, C.R. & Cutler, D.F. (1991) *Ant-Plant Interactions*. Oxford University Press, Oxford.

Lewin, W. (1795) *The Papilios of Great Britain*. J. Johnson, London.

Linnaeus, C. (1759) *Systema Naturae*, 10th edition. Laurentii Salvii, Stockholm.

Maitland Emmet, A. & Heath, J. (Eds) (1990) *The Moths and Butterflies of Great Britain and Ireland*. Harley Books, Colchester.

Morris, Revd F.O. (1853) *A History of British Butterflies*. Groombridge, London.

Newman, E. (1871) *An Illustrated Natural History of British Butterflies*. Tweedie, London.

Petiver, J. (1717) *Papilionum Britanniae*. London.

Petiver, J. (1695–1703) *Musei Petiveriani* ['Petiver's Museum']. London.

Petiver, J. (1702–1706) *Gazophylacii naturae et artis* ['Treasure Chest of Nature and Art']. London.

Pollard, E., Hooper, M.D. & Moore, N.W. (1974) *Hedges*. Collins, London.

Ray, J. (1710) *Historia Insectorum*. Royal Society, London.

Rennie, J. (1832) *A Conspectus of the Butterflies and Moths found in Britain*. Orr, London.

Samouelle, E. (1819) *The Entomologist's Useful Compendium*. Thomas Boys, London.

South, R. (1906) *The Butterflies of the British Isles*. Warne, London.

Stace, C. (2010) *New Flora of the British Isles*, 3rd edition. Cambridge University Press, Cambridge.

Thomas, J.A. & Lewington, R. (2014) *The Butterflies of Britain and Ireland*, 3rd edition. British Wildlife Publishing, Oxford.

Tutt, J.W. (1906) *A Natural History of the British Lepidoptera*. Swan Sonnenschein, London.

Verity, R. (1908) *Rhopalocera Palaearctica*. Florence.

Vismes Kane, W.F. de (1901) *A catalogue of the Lepidoptera of Ireland*. West Newman, London.

Wilkes, B. (1742) *Twelve New Designs of English Butterflies*. London.

Wilkes, B. (1749) *The English Moths and Butterflies*. London.

Wilkes, B. (1749) *One Hundred and Twenty Copper-Plates of English Moths and Butterflies*. London.

Williams, C.B. (1958) *Insect Migration*. Collins, London.

Willmott, K., Bridge, M., Clarke, H.E. & Kelly, F. (Eds) (2013) *The Butterflies of Surrey Revisited*. Surrey Wildlife Trust, Woking.

WEBSITES

butterflylifecycles.com
A website dedicated to this book.

butterfly-conservation.org
The website of Butterfly Conservation, the UK's leading charity for butterflies, moths and the environment.

ukbutterflies.co.uk
A website dedicated to 'Building a community of responsible butterfly enthusiasts'.

dispar.org
The publishing arm of the UK Butterflies website.